ION BEAM SURFACE
LAYER ANALYSIS

Volume 1

ION BEAM SURFACE LAYER ANALYSIS

Volume 1

Edited by

O. Meyer, G. Linker, and F. Käppeler

Nuclear Research Center
Karlsruhe, Germany

Springer Science+Business Media, LLC

Library of Congress Cataloging in Publication Data

International Conference on Ion Beam Surface Layer Analysis, Karlsruhe, 1975.
  Ion beam surface layer analysis.

Proceedings of the International Conference on Ion Beam Surface Layer Analysis
held at Karlsruhe, Ger., Sept. 15-19, 1975.
  Includes bibliographical references and indexes.
  1. Thin films, Effect of radiation on—Congresses. 2. Ion bombardment—Congress-
es. 3. Straggling (Nuclear physics)—Congresses. 4. Backscattering—Congresses. 5. Sur-
faces (Technology)—Congresses. I. Meyer, Otto. II. Linker, Gerhard. III. Käppeler,
Franz. IV. Title.
QC176.84.R3I57 1975   530.4′1   76-2606
ISBN 978-1-4615-8878-8   ISBN 978-1-4615-8876-4(eBook)
DOI 10.1007/978-1-4615-8878-8

Proceedings of the first half of the International Conference on Ion Beam Surface
Layer Analysis held at Karlsruhe, Germany, September 15-19,1975

©1976 Springer Science+Business Media New York
Original published by Plenum Press, New York in 1976
Software reprint of the hardcover 1st edition 1976

Preface

The II. International Conference on Ion Beam Surface Layer Analysis was held on September 15-19, 1975 at the Nuclear Research Center, Karlsruhe, Germany. The date fell between two related conferences: "Application of Ion-Beams to Materials" at Warwick, England and "Atomic Collisions in Solids" at Amsterdam, the Netherlands.

The first conference on Ion Beam Surface Layer Analysis was held at Yorktown Heights, New York, 1973. The major topic of that and the present conference was the material analysis with ion beams including backscattering and channeling, nuclear reactions and ion induced X-rays with emphasis on technical problems and novel applications. The increasing interest in this field was documented by 7 invited papers and 85 contributions which were presented at the meeting in Karlsruhe to about 150 participants from 21 countries. The oral presentations were followed by parallel sessions on "Fundamental Aspects", "Analytical Problems" and "Applications" encouraging detailed discussions on the topics of most current interest. Summaries of these sessions were presented by the discussion leaders to the whole conference. All invited and contributed papers are included in these proceedings; summaries of the discussion sessions will appear in a separate booklet and are availble from the editors.

The application of ion beams to material analysis is now well established. The results of the conference in Karlsruhe have shown the progress in analytical problems, such as depth resolution and depth profiling and the necessity of using combined techniques for a complete material analysis. The backscattering technique has been extended to new materials such as biological samples, to more complicated layered structures and to interfacial reactions. Progress has been achieved in the application of backscattering and channeling to investigate bubble and blister formation in metals as well as in lattice location determination of light atoms in heavy mass host materials by nuclear reactions. The advantages of ion-induced X-ray analysis have been pointed out and this will certainly lead to specific applications and to the combination with other techniques.

The sponsorship and the financial support of the Gesellschaft für Kernforschung mbH was vital to the success of the Conference. We particularly wish to acknowledge help and advice from the International Advisory Committee consisting of: G. Amsel (Paris), R. Behrisch (Garching), L.C. Feldman (Bell Labs), J.W. Mayer (Caltech), S.T. Picraux (Sandia), E. Wolicki (Naval Research) and J.F. Ziegler (IBM).

The operation and success of the discussion sessions is due to G. Amsel (Paris), J. Biersack (HMI, Berlin), J. Cairns (Harwell), W.K. Chu (IBM), L.C. Feldman (Bell Labs), F. Folkmann (Darmstadt), S. Kalbitzer (MPI Heidelberg), J. Poate (Bell Labs) and B. Scherzer (Garching).
The Committee responsible for organizing this conference consisted of W. Gläser, F. Käppeler, G. Linker, O. Meyer and G. Schatz. Special thanks are due to Mrs. E. Maaß and Mr. P. Emmerich for their help and excellent skill in organization.

Karlsruhe, Germany   O. Meyer
September 1975
                                                      G. Linker

                                                      F. Käppeler

Contents of Volume 1

I. ENERGY LOSS AND STRAGGLING

III.   APPLICATIONS OF BACKSCATTERING
       AND COMBINED TECHNIQUES

IV. EQUIPMENT

Contents of Volume 2

VI.   RELATED TECHNIQUES

Energy Loss and Straggling

THE TREATMENT OF ENERGY-LOSS FLUCTUATIONS IN SURFACE-LAYER ANALYSIS

BY ION BEAMS

J. W. Butler

U. S. Naval Research Laboratory

Washington, D. C. 20375

ABSTRACT

The present paper proposes a modification of current energy-loss theory to improve the accuracy of predictions of the distributions of energy losses of low-energy protons in medium- and high-atomic-number materials. This paper also presents experimental measurements of the distributions of energy losses for 1-MeV protons after penetrating thin targets of nitrogen and xenon. The agreement between predicted and experimental distributions is good.

INTRODUCTION

When surface layers are probed by charged particles, the particle energy scale is ordinarily used to obtain the depth scale. However, the energy-loss process is statistical in nature, resulting in fluctuations in energy losses and a corresponding degradation of depth resolution. Therefore, one needs an accurate theory of energy-loss processes to transform from an energy to a depth scale. But for protons used in nuclear-reaction analyses or Rutherford backscattering analyses on targets with atomic number (Z) higher than about 14 (silicon), existing methods of treating the problem of fluctuations in energy losses among the different particles in an analyzing beam are inadequate. For example, conventional theoretical distributions tend to be too broad. Furthermore, although the distribution of energy losses for an incident monoenergetic proton beam in a thin surface layer of a solid target is quite asymmetric, practitioners either ignore fluctuations altogether or else they usually make the simplifying assumption that the distribution has a symmetric gaussian shape. Present methods of calculating the correct asymmetric

3

distribution are valid for medium- and high-Z materials only at high energies (>10^2 MeV). The present paper (a) proposes a modification of current theory so that the energy-loss distributions for low-energy protons on medium- and high-Z targets can be predicted with reasonable accuracy, (b) describes experimental measurements of the distributions of energy losses for 1-MeV protons on a low-Z target (nitrogen) and a medium-high-Z target (xenon), and (c) compares theory with experiment.

CONVENTIONAL THEORY

Bethe's equation [1] for the average energy loss ΔE_a of a heavy charged particle with speed v = βc and charge Ze in a thin layer of matter of atomic mass A and thickness t may be written in the following unconventional form:

$$\Delta E_a = 2\xi(\ln \frac{E'_m}{I} - \beta^2),\tag{1}$$

where

$$\xi \equiv \frac{2\pi z^2 e^4}{m_e v^2} \frac{NZ}{A} t\tag{2}$$

and

$$E'_m \equiv \frac{2m_e v^2}{1-\beta^2} .\tag{3}$$

In (1), m_e is the mass of the electron, N is Avogadro's number, I is the average ionization potential of the target material, and E'_m is the maximum energy that can be transferred to a stationary free electron in a collision with the incident particle.

Bethe's equation was derived under the assumption that the incident particle speed is much greater than the Bohr-orbit speed v_K of the K-shell electrons in the target material. However, much of the application of ion beams to surface-layer analysis involves cases in which the target K-shell electrons have speeds comparable to or greater than the incident particle speed. For example, the K electrons in silicon have a Bohr-orbit speed about equal to that of a 3-MeV proton; and the K electrons in tantalum have a speed about equal to that of a 120-MeV proton. Futhermore, it is obvious that (1) is not applicable to situations in which $I \geq E'_m$ because in such cases the logarithm term is either zero or negative. The assumptions under which (1) was derived require that $E'_m \gg I$.

For situations in which (1) is applicable, the distributions of energy losses for incident particles have been calculated by Landau [2] for very thin targets and by Symon [3] and Vavilov [4] for moderately thin targets. However, no satisfactory theory exists for cases in which (1) is not applicable (i.e., for cases in which the incident-particle velocity is comparable to or less than v_K.)

PROPOSED MODIFICATIONS OF PRESENT THEORY

The quantity I of Bethe's equation for the average energy loss
and the corresponding quantity in the theories of Landau, Symon, and
Vavilov are the logarithmic mean of the minimum excitation potentials
of the various electrons in the target atom, weighted according to
their relative transition probabilities. Bloch [5], using the Thomas-
Fermi model of the atom, showed that I is approximately proportional
to Z and that I is independent of the incident ion energy. But
Bloch's analysis was also based on the assumption that the incident
particles are much faster than the target K-shell electrons. Because
of the impracticality of calculating the relative transition probabilities from basic principles, the value of I has been obtained
empirically; i.e., measured values of the energy-loss rate have been
used with (1) to calculate the value of I. Hence all current energy-
loss theories are at best semi-empirical.

The present proposal is to modify the traditional definitions
of I and Z in (1) so as to include situations in which some atomic
electrons are faster than the incident particles. In such cases,
these faster (and hence tightly bound) electrons do not effectively
participate in the energy-transfer mechanism. In effect, their
transition probabilities are essentially zero; hence they should not
be counted in the definitions of I and Z. Equation (1) thus becomes

$$\Delta E_a = 2\xi_{eff}\left(\ln \frac{E_m'}{I_{eff}} - \beta^2\right), \qquad (4)$$

$$\text{where} \qquad \xi_{eff} \equiv \frac{2\pi z^2 e^4}{m_e v^2} \frac{NZ_{eff}}{A} t. \qquad (5)$$

CALCULATIONS CONCERNING NONPARTICIPATING ELECTRONS

The justification for neglecting certain tightly bound electrons
in the energy-transfer process may be obtained in part from calculations of the relative fractions of energy received by the various
electrons in the target atom.

Figure 1 shows the fraction of the total energy transfer that
is associated with K-shell vacancy production for selected target
materials bombarded with protons with energies E between 0.1 and 100
MeV. Figure 1 is based in part on (i) a K-shell vacancy cross-section
calculation by Garcia et al. [6] and (ii) a total stopping power
tabulation of Williamson et al. [7]. Figure 1 shows, for example,
that the two K electrons in vanadium absorb on the average only about
$10^{-2}\%$ of the total energy transferred to the entire atom by a 1-MeV
proton. Hence the K-shell electrons of vanadium are essentially
nonparticipants in the energy-transfer process for incident protons
near or below about 1 MeV.

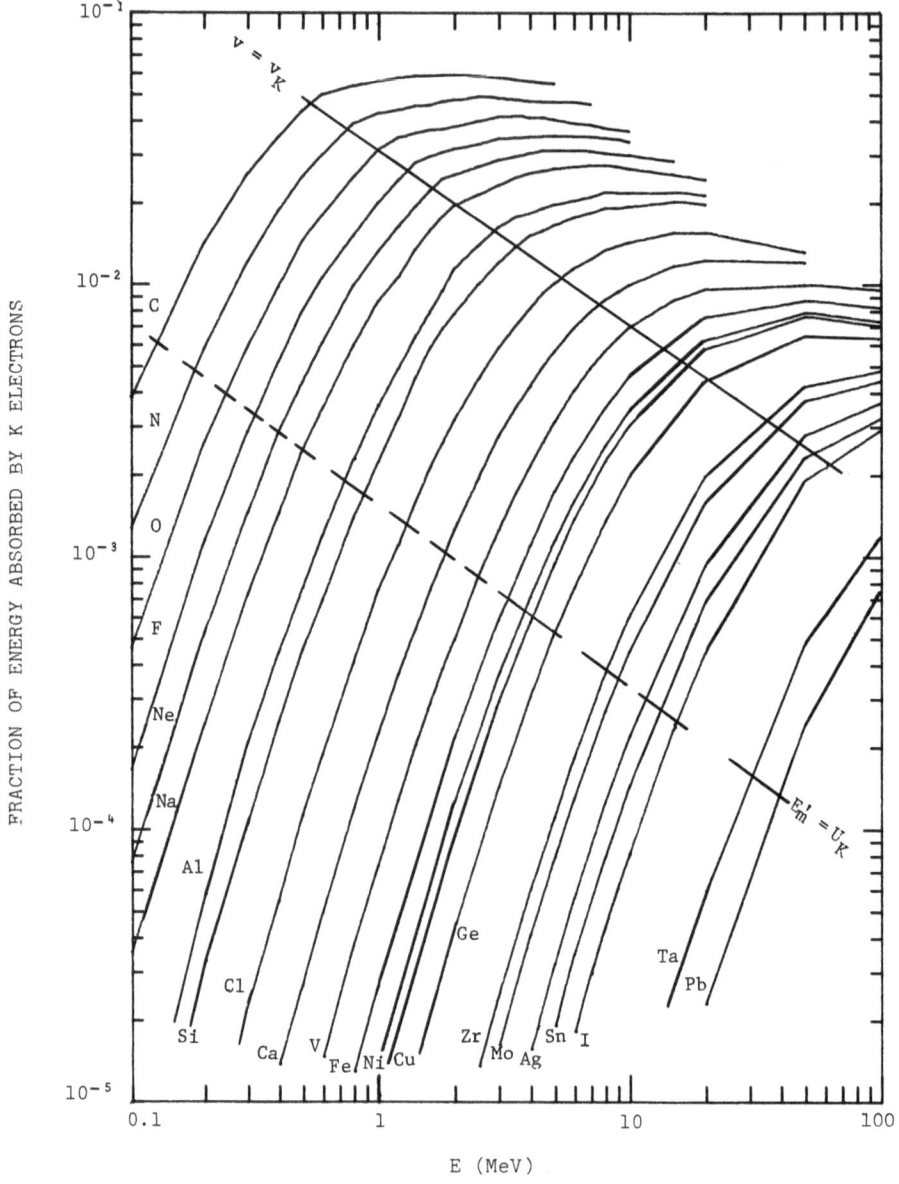

Figure 1. The fraction of the total energy transfer that produces K-shell vacancies, for selected elements, as a function of incident proton energy. The dashed line across the curves indicates the E,Z combinations for which $E_m' = U_K$. The solid line across the curves indicates the E,Z combinations for which $v = v_K$.

The family of curves in Figure 1 defines a surface in E,F,Z space. The dashed line lies on this surface and represents the locus of points for which E_m' is equal to the binding energy U_K of the K electrons. Above and to the right of this line $E_m' > U_K$; and below and to the left of this line $E_m' < U_K$. Note that the average contribution of the K electrons to the energy-transfer process is roughly 0.1% when $E_m' = U_K$.

The solid line across the curves represents the locus of points for which $v = v_K$. Note that the average contribution of the K electrons to the energy-transfer process is roughly 1% when $v = v_K$. But since K electrons participate only in relatively large energy-transfer collisions, the K electrons are involved, on the average, in much fewer than 1% of the collisions. Furthermore, for the same reason, for very thin targets the K electrons influence only the high-energy-loss end of the distribution--they have almost no effect on the shape of the distribution near the peak or near the most probable energy loss.

Figure 1 and the associated foregoing discussion concern the K-shell electrons, but analogous figures and considerations apply to the L, M, and N shells.

For all of the reasons mentioned above (including the assumption that $v \gg v_K$ in the derivation of the equations of Bethe and Bloch) it appears reasonable to define Z in (2) so as to exclude nonparticipating electrons. For present purposes we assume that electrons for which $v_K > v$ do not effectively participate in the energy-transfer process.

Figure 2 shows for selected elements (indicated by the atomic number on the curves) the effective number of participating electrons as a function of incident proton energy. The binding energies of the various electrons in the atom as given in the table of Bearden and Burr [8] were used in the calculations of the curves in Figure 2.

Equations (4) and (5) have been used along with the curves in Figure 2 and the tabulation of Williamson et al. [7] to calculate the effective value of the ionization potential I_{eff} as a function of E as shown in Figure 3.

These effective values of Z and I may be substituted for the conventional values in the equations of Landau, Symon, or Vavilov to give the necessary parameters (most probable energy loss, width of the distribution, and the asymmetry of the distribution, in addition to average energy loss) for the distribution of energy losses. [It is perhaps worth noting here that in Bethe's equation for the average energy loss, unrealistically high values of both Z and I would tend to compensate each other (because the former is in the numerator and because the latter is in the denominator), thus

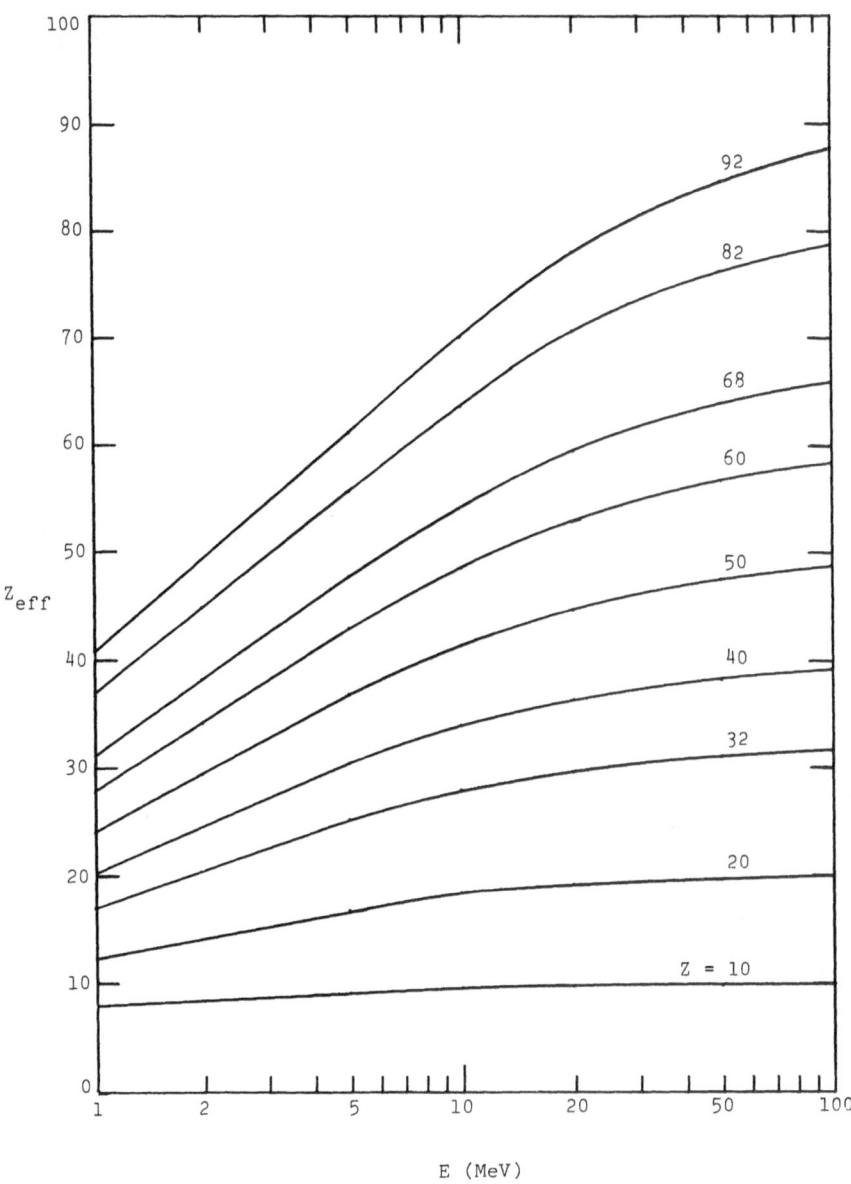

Figure 2. The effective number of participating electrons in an atom as a function of incident proton energy, for selected elements (identified by their atomic numbers). Subshell effects have been smoothed out.

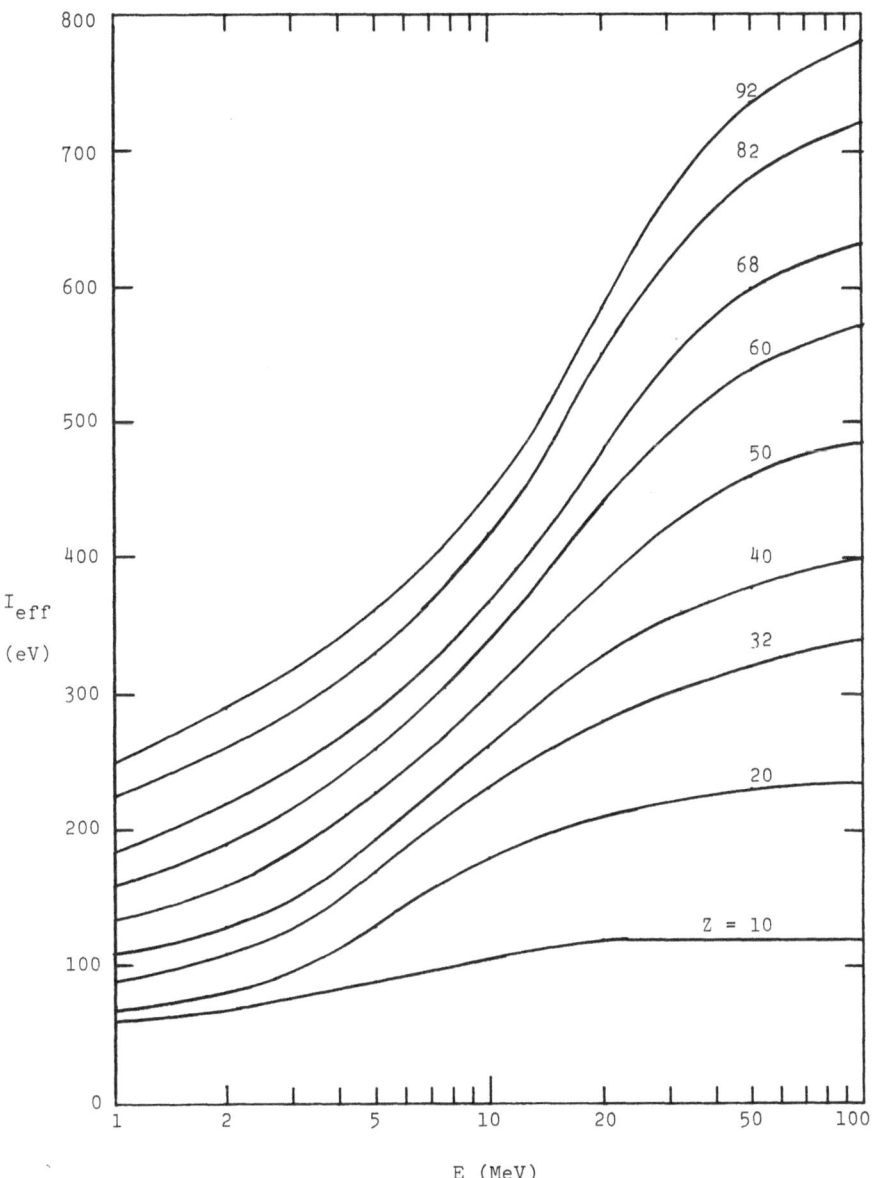

Figure 3. The effective ionization potential for selected elements, as a function of incident proton energy.

leading in some cases to approximately correct values of the average
energy loss in spite of much too large values of both Z and I. However, this fortuitous compensation does not occur in the equations
for the most probable energy loss or the width of the distribution.
For example, in the Landau regime of layer thickness the width of a
distribution is proportional to Z and is independent of I; and in the
Bohr regime, the width is proportional to $Z^{\frac{1}{2}}$ and is likewise independent of I.]

EXPERIMENTAL MEASUREMENTS

The ultimate justification for any theoretical concept, including that of Z_{eff} and I_{eff}, can come only from a comparison of theory
and experiment. Therefore, as a test of this concept a number of
comparisons have been made between theory and experiment, with
generally good agreement. The present paper describes two such
comparisons: the energy-loss distribution of 1-MeV protons incident
on a target with Z = 7 (nitrogen) and another with Z = 54 (xenon).

The experimental procedure involved passing a 1-MeV beam of
protons from the NRL 5-MV Van de Graaff accelerator through (i) a
2-meter electrostatic analyzer set to give a beam-energy resolution
of 0.02% and (ii) then through a differentially pumped gas cell
(effective length: 20 cm) containing the desired material for slowing the protons. The energy distribution of the transmitted beam was
then measured by means of a narrow p,γ resonance as an energy filter.
The resonance chosen is the 992-keV resonance in the $^{27}Al(p,\gamma)^{28}Si$
reaction. The target was an evaporated layer of aluminum on a smooth
layer of copper which in turn was deposited onto a disk made from
microscope-slide glass. The average energy loss of the beam in the
target was about 150 eV. The full width at half maximum of the
experimental resonance curve was less than 300 eV when no gas was in
the cell, as shown in Figure 4. Since the energy filter on the transmitted beam was fixed in value, the distribution of output energies
was measured, in effect, by a sweep of the input energy through the
desired range.

The broken curve of Figure 5 shows the experimental distribution
of energy losses when the cell contained 160 μm pressure of nitrogen,
corresponding to an average energy loss of 1125 eV. Note that the
most probable energy loss is significantly less than the average
energy loss. Since $v = 1.4 v_K$ for 1-MeV protons on nitrogen, it is
reasonable to expect conventional theory to describe with adequate
precision the distribution of energy losses in this experiment.
The solid curve is the Landau distribution with the conventionally
recommended value of I = 100 eV for nitrogen and with the full value
of Z = 7. The finite resolution of the apparatus, represented by the
curve in Figure 4, has not been folded into the solid curve of Figure
5. Folding in the apparatus resolution causes the two curves to agree

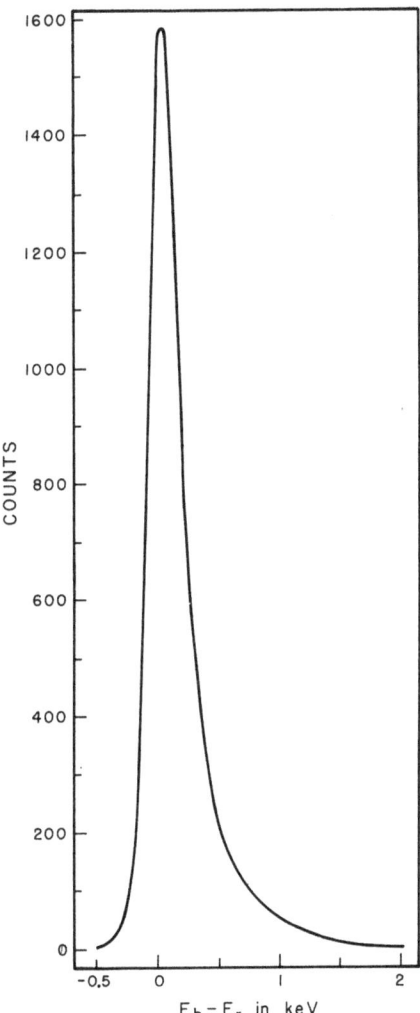

Figure 4. Proton/gamma-ray yield curve
near resonance with no gas in the cell
in front of the target. The abscissa
is stated in terms of the difference
between bombarding energy E_b and
resonance energy E_r. This curve gives
the intrinsic resolution of the
apparatus and technique (about 300 eV).

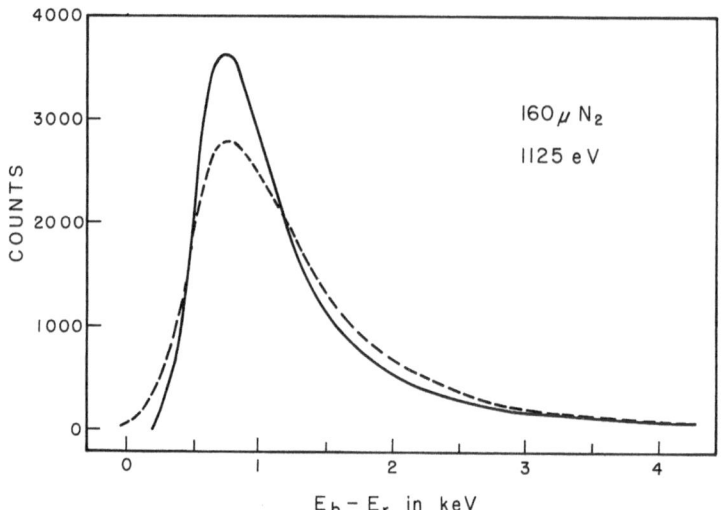

Figure 5. Resonance yield with 160 μm pressure of
nitrogen in the gas cell, corresponding to an average
energy loss of 1125 eV. The broken curve is the
experimental yield, and the solid curve is the
theoretical yield (Landau theory) except the effects
of the finite resolution of the apparatus have not
been folded in. These curves are, in effect, the
distribution of proton energy losses in the gas cell.

within experimental uncertainty. It thus appears that conventional
theory is adequate to predict the energy-loss distribution for 1-MeV
protons on a thin target of nitrogen. (The proposed effective values
Z_{eff} and I_{eff} for 1-MeV protons on nitrogen are the same as the
conventional values Z and I.)

However, since xenon (Z = 54, recommended I value of 555 eV)
has K-electron binding energies of about 35 keV, L-electron binding
energies of about 5 keV, and M-electron binding energies of about
1 keV, it is reasonable to expect conventional theory to fail to
predict correctly the energy-loss distribution of 1-MeV protons
($E'_m \simeq 2$ keV) striking thin targets of xenon; and indeed we observe
such to be the case. The broken curve of Figure 6 shows the measured
energy-loss distribution of 1-MeV protons transmitted through the gas
cell containing enough xenon to provide an average energy loss of
2680 eV. The solid curve labeled "I = 555 eV" shows the distribution
predicted by conventional theory. Note that the distribution is so
broad that it even predicts some energy gains.

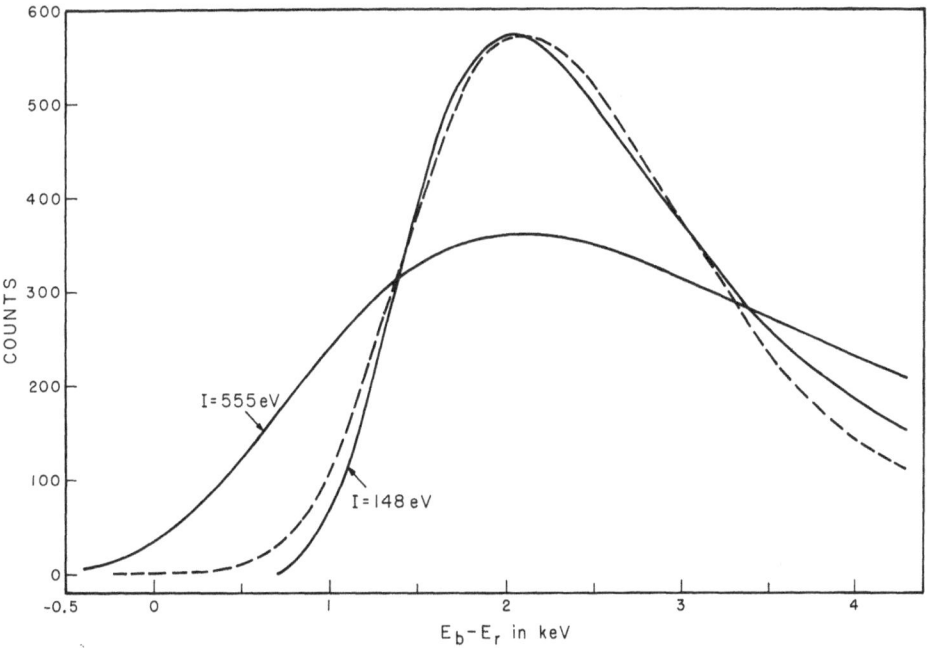

Figure 6. Resonance yield curve with enough xenon in the gas cell
to provide an average energy loss of 2680 eV. The broken curve is
the experimental yield. The solid curve labeled "I = 555 eV" shows
the yield curve or energy-loss distribution predicted by conventional
theory. The solid curve labeled "I = 148 eV" shows the energy-loss
distribution predicted by the theory proposed herein.

From Figure 2 we see that Z_{eff} for Z = 54 and for E = 1 MeV is 26; and from Figure 3 we see that I_{eff} for Z = 54 and for E = 1 MeV is 148 eV. When we use these values in Symon's equations, we obtain the solid curve labeled "I = 148 eV" in Figure 6, in good agreement with the experimental curve.

CONCLUSIONS

Conventional energy-loss theory was derived under assumptions consistent with high-energy applications; and the theory has been successful in that regime. However, in the application of ion beams to surface-layer analysis, relatively low-energy particles are used to probe medium- and high-Z materials; and conventional theory fails in this regime. If the theory is modified so as to take into account the nonparticipation of some target electrons, together with a concomitant redefinition of the average ionization potential, then the modified theory can indeed predict the energy-loss distributions for low-energy protons striking high-Z materials.

REFERENCES

[1] H. A. Bethe, Z. Physik, 76, 293 (1932).
[2] L. Landau, J. Phys. (USSR), 8, 201 (1944).
[3] K. R. Symon, "Fluctuations in Energy Lost by High Energy Charged Particles in Passing Through Matter," Ph. D. Thesis, Harvard University, 1948.
[4] L. Vavilov, Zh. Eksp. Teor. Fiz., 32, 320 (1957); English translation in JETP, 5, 749 (1957).
[5] F. Bloch, Ann. Physik, 16, 285 (1933); Z. Physik. 81, 363 (1933).
[6] J. D. Garcia, R. J. Fortner, and T. M. Kavanagh, Rev. Mod. Phys. 45, 111 (1973).
[7] C. F. Williamson, J. P. Boujot, and J. Picard, "Tables of Range and Stopping Power of Chemical Elements for Charged Particles of Energy 0.5 to 500 MeV," CEA-R 3042 (1966).
[8] J. A. Bearden and A. F. Burr, Rev. Mod. Phys. 39, 125 (1967).

EVIDENCE OF SOLID STATE EFFECTS IN THE ENERGY LOSS
OF ^{4}He IONS IN MATTER

J. F. Ziegler, W. K. Chu* and J. S.-Y. Feng

IBM - Research
Yorktown Heights, New York 10598

ABSTRACT

It is shown that ^{4}He ion stopping cross-sections for elemental
gas targets appear consistently higher than those for elemental
solids when the ion velocity is comparable to the electrons' orbital velocities. The solid/gas differences are proposed as an empirical first order correction to obtain solid target values from gas
target stopping cross-sections. A confirming series of experiments
on the energy loss of ^{4}He ions in oxygen and nitrogen in solid
oxides and nitrides are in agreement with the empirical correction
factor proposed.

The recent widespread use of ion beams to analyze solid state
problems[1] has revitalized interest in particle/solid interactions.
Most of these studies involve the use of H$^+$ or He$^+$ ion beams in the
energy range of 0.2 - 3 MeV. The physics of the energy loss of
these ions in this energy region is not well understood where the
ions are at velocities comparable to the classical outer-shell
electron orbital velocities of the target materials. For very high
energy ions (He$^+$ > 3 MeV), theoretical calculations[2] based on an
impulse approximation are accurate in predicting stopping cross-
sections to a few percent. Here the outer electrons of the stopping
medium appear to be motionless to the fast moving ion. For very
low energies, the interaction between the ion and medium can be
approached with an adiabatic approximation which allows electronic
perturbations in the medium[3]. However, recent calculations for He$^+$
in solids in the medium energy region have failed to produce the
shape of experimental stopping cross-section curves[4].

---

* Present Address: SP Division, IBM Co., Hopewell Junction,
New York.

The possible change in stopping cross section from gas to solid phase because of dielectric effects was first noted by Swann[5]. By including dielectric screening corrections, Fermi[6] estimated that in the intermediate energy range solid phases should have lower stopping than gases "in the order of several percent." A recent calculation based on Lindhard and Winther's statistical approach compares gas and solid stopping by using the wave functions appropriate to each[4]. It finds only a few percent difference between calculated gas and solid stopping cross-sections for He^+.

The first reported experimental "physical-state" effect[7] was for H^+ in solid carbon and carbon in hydrocarbon gas form. The differences in stopping cross-sections were large, reaching 20% greater for H^+ (100 keV) in the gases. The same divergence has been found for He^+ in carbon[8], but for carbon in complex molecular gases there are no unique values[9-11]. Bonding effects within the molecular gases have been introduced to explain the observed complex stopping cross sections.

The effects on stopping cross-section of the physical state of the stopping medium (solid or gas) was not originally considered when "Bragg's Rule" was proposed[12]. This rule states that the stopping power of a compound is merely the sum of the stopping effects of each individual element, weighted by their abundance in the compound. This means that for the stopping power of SiO_2, for example, one may add together Si and O contributions without regard for the fact that the stopping power of O has only been measured as a gas.

Because of the practical applications of ion beams for analyzing solids there have been more papers written about the validity of Bragg's Rule in the last two years than in the preceding sixty years[1,13,14]. The results have been contradictory, with some authors finding accurate agreement with Bragg's Rule predictions[14], and others finding serious disagreement[1]. In these cases the authors were mostly concerned with solid compounds containing oxygen and nitrogen, whose stopping powers have been measured only in the gas phase as O_2 and N_2.

A recent review[15] of experimental stopping cross-sections of He^+ in elemental matter tabulates values which allow us to compare experimental stopping cross-sections of He^+ in a large number of gases and solids. This reference also includes theoretical values based on Lindhard-Winther stopping theory, evaluated using Hartree-Fock-Slater wave functions. Division of the experimental stopping cross sections by theoretical values allows most of the systematic differences due to the mass and atomic number of the stopping medium to drop out, and second order effects become more evident. Although the theoretical values are accurate to about 20%, we use them only as <u>normalizing factors</u> so we may compare, for example, ^4He energy loss values in oxygen to those in carbon. Shown in Figure 1 and 2

are the results of arithmetically averaging He$^+$ stopping cross-
section ratios (experiment ÷ theory) for 10 elemental gases
(Figure 1) and also for 32 elemental solids (Figure 2).

We show the many curves of the 32 normalized stopping cross-
sections so the reader can be aware that there is a significant
spread of values, and the two <u>sets</u> of curves do overlap in part.

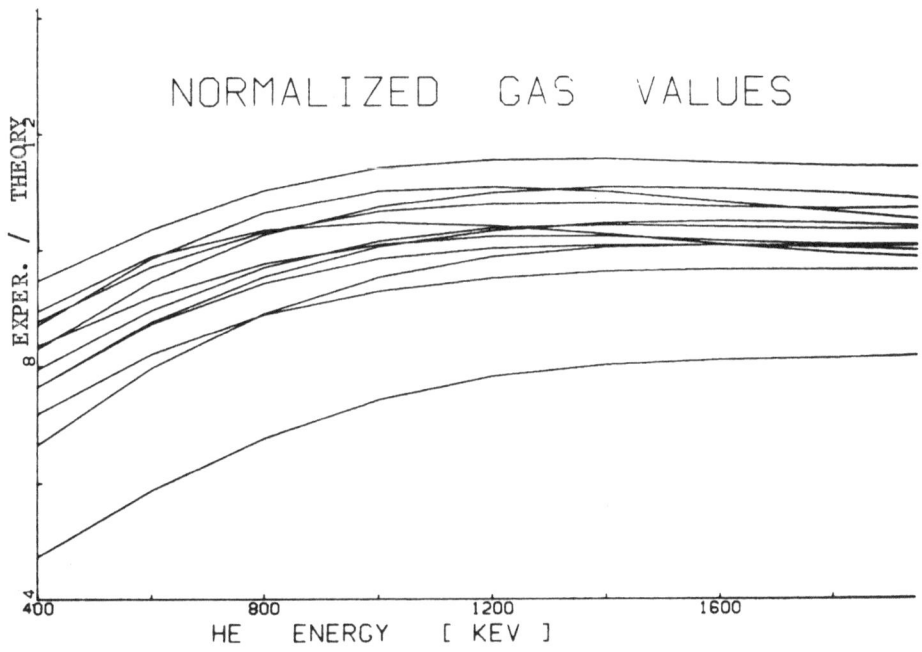

Figure 1. Experimental stopping cross-sections for ^{4}He ions in 10
gases are shown for energies of 400 - 2000 keV, and normalized by
dividing by calculations of stopping cross-sections based on
Lindhard-Winther formalism (see Ref. 15). This normalization
removes primary effects (such as target atomic number and mass) and
allows secondary effects to be more noticeable.

However, we can simplify the comparison of these sets by taking the
average of each set. These mean values are plotted in Figure 3.
For He$^+$ energies above 1.5 MeV (the He$^+$ velocity is ~5X the Bohr
velocity) the ratios for both solids and gases approach the same
constant value. Below 1 MeV, the theory deviates more from experimental values, but significant differences appear between the gas
average and the solid average.

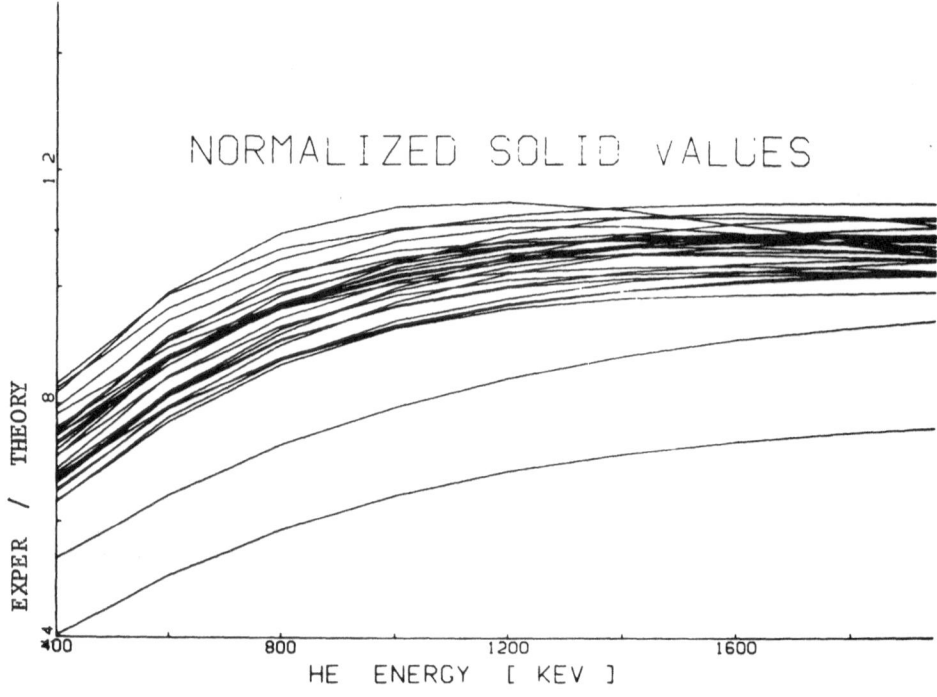

Figure 2. Experimental stopping cross-sections for ^{4}He ions in 32 elemental solids are shown in a manner similar to those for gas targets in Figure 2. The two lowest curves are for ^{4}He in Si and Mg.

In Figure 4 we plot the ratio between the gas average ratio and the solid average ratio. As the ^{4}He ion velocity approaches that of the outer shell electrons the ratios of the gas stopping cross sections become much <u>larger</u> than those of the solids. One can note that the He$^+$ ions may have a higher average charge state in the solids because the high collision frequency can promote electrons through a series of small collisions. Such a higher charge state should increase Coulomb interactions and hence energy loss. Since the normalized energy loss in solids appears to be smaller than losses in gases, one suspects dielectric screening effects could be responsible for this deviation.

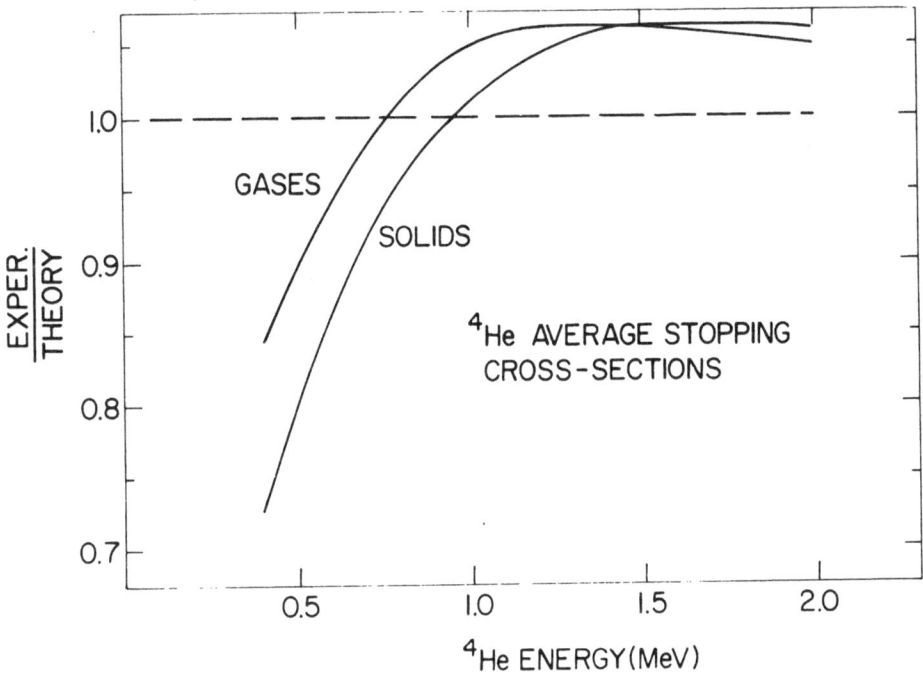

Figure 3. Experimental values divided by theoretical values of
stopping cross-sections of He$^+$ ions in gases and solids. The values
for solids are averaged over 32 elemental solids, and for gases are
averaged over 10 elemental gases. The theoretical values are from
Ref. 15. This normalization attempts to separate out primary
effects due to target atomic number and mass, and allow secondary
differences between solids and gases to become evident. The distribution of individual elements which make up both the solid and gas
curves is wide with some elements falling 15% from the average.
However, there are statistically significant differences between
the gases and solid average ratios.

We propose that this difference between gases and solids as
shown in Figure 4 may be used as an approximation to obtain a
physical-state correction to stopping cross-sections. That is, one
can obtain the stopping cross-sections for He$^+$ in an elemental
solid (e.g., oxygen in solids) by dividing the equivalent gas
values by a gas/solid correction similar to that plotted in
Figure 4.

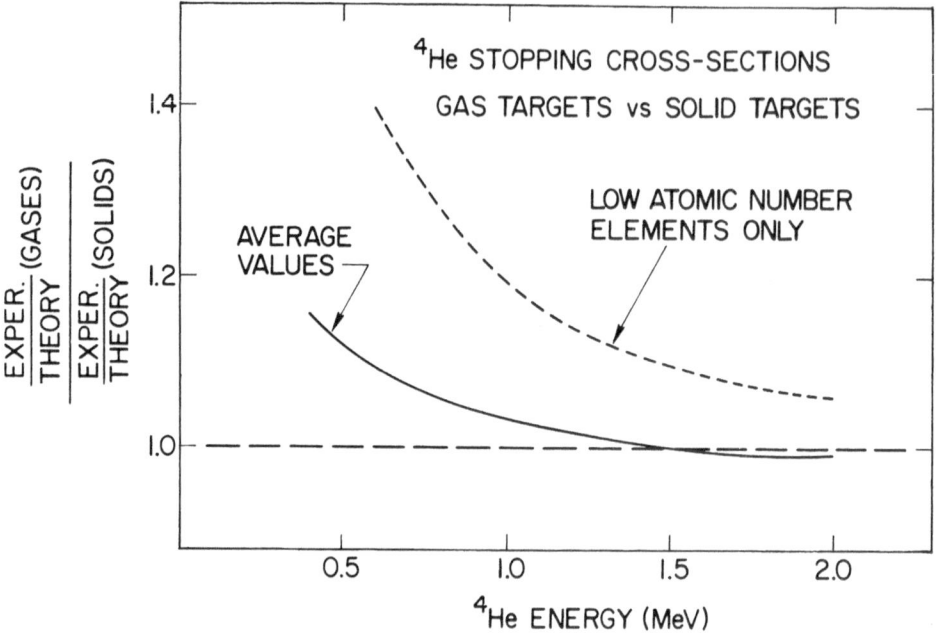

Figure 4. The ratios plotted in Figure 3 are directly compared by
dividing the ratios for gases by those for solids. This difference
is proposed in the text as a physical-state correction factor for
stopping cross-sections of He⁺. If the gas to solid ratio is
calculated using only elements whose atomic number is below 15, a
much larger difference is found. This is shown above as a dashed
curve.

     We also find that by plotting each individual gas relative to
similar atomic number solids (in a manner similar to that done for
all gases and solids in Figure 4) a second feature appears. The
gas/solid difference is larger for low Z elements than for high Z
elements. The low Z gases (such as N and O) deviate from low Z
solids (such as C and Mg) much more than high Z gases from solids
(such as I to Sb and Te). This may be because for high atomic
number atoms the outer electrons which participate in binding constitute only a small part of the total number of electrons. It is
believed that outer shell electrons absorb proportionally more of
the energy. For purposes of calculating solid state corrections for
O and N in the experiments to be discussed below, we have used as
gases: N, O, F, and Ne; and for solids: Be, C, Mg, Al, and Si.
This low Z correction is plotted as a dashed line in Figure 4.

To evaluate the accuracy of this proposed correction we have measured the stopping cross-sections of He$^+$ passing through oxygen in solids using the self-consistent method discussed by others[7,8]. This method consists of taking a series of compounds (Fe_2O_3, Fe_3O_4, SiO_2, MgO, Al_2O_3), finding the differential energy loss for He in these compounds using standard techniques[13,16], and then using established values for the metals (Fe, Si, Mg, and Al) and Bragg's Rule to extract a self consistent stopping cross section for oxygen in solids. We find that the results agree with the gas values for oxygen divided by the correction term shown in Figure 2.

One sensitive way to show this agreement is to evaluate the difference in energy loss between a pure metal solid, and its oxide[17]. This is shown in Figure 5 for an oxide of SiO_2 grown on pure Si. By backscattering from such a target we simultaneously

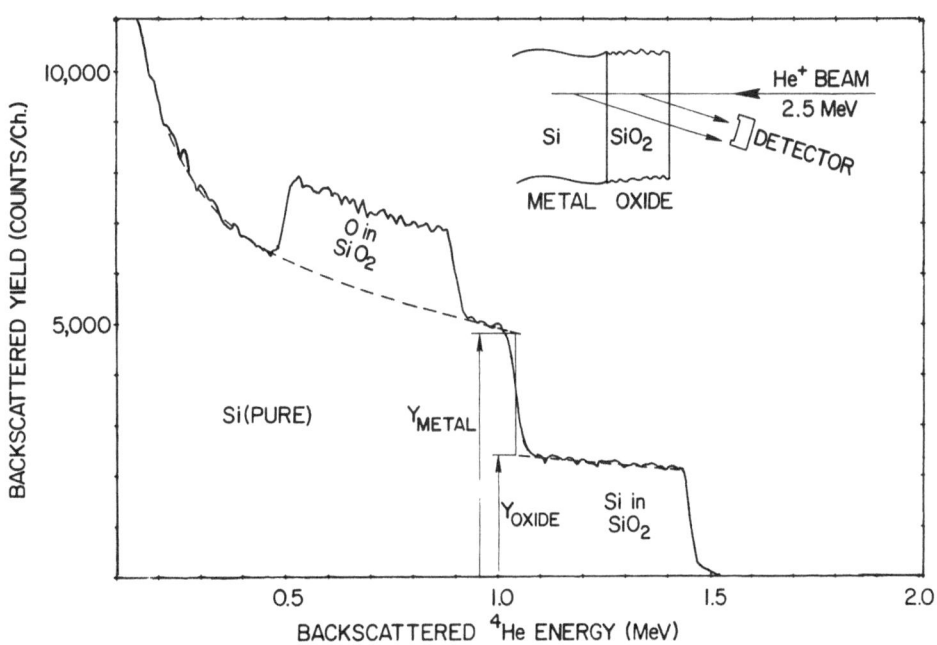

Figure 5. The experimental arrangement for the backscattering of He$^+$ from a typical target is shown schematically. Also shown is the spectrum of backscattered He$^+$ from a target of SiO_2 or Si. The common energy backscattering yield ratios tabulated in Table 1 are defined above as "Y_{METAL}" and "Y_{OXIDE}". The drop in yield from Si (pure) to Si in SiO_2 can be related to the energy loss contribution of the oxygen. This technique is discussed in detail in Ref. 13.

measure the stopping factors for pure Si and for SiO_2, and we can
cleanly extract the <u>difference</u> which the oxygen in the oxide contributes to the energy loss (assuming the metal contribution scales
with its concentration). Most absolute experimental parameters are
eliminated using this approach and we can concentrate on the relative energy loss changes. Shown in Table I are the experimental
values for the drop in yield (Y_{METAL}/Y_{OXIDE}), the calculated values
using the "corrected" oxygen gas values, and the values using
stopping cross-sections from oxygen gas measurements. The results
show disagreement between experiment and calculations using oxygen
gas values, and agreement with gas values modified with the gas to
solid correction.

Table I

Yield Ratio

Target	Experimental	Calculated with Corrected Oxygen	Calculated with Oxygen Gas Values
Fe_2O_3	1.68	1.69	1.77
Fe_3O_4	1.62	1.61	1.68
MgO	1.70	1.72	1.83
Al_2O_3	2.08	2.13	2.31
SiO_2	2.18	2.29	2.45

Target	Experimental	Calculated with Corrected Nitrogen	Calculated with Nitrogen Gas Values
Si_3N_4	1.81	1.80	1.92

Tabulated are experimental and calculated ratios of the He^+
(2MeV) backscattering yield (Y_{METAL}/Y_{OXIDE}) as defined in Figure 3.
The calculated values are based on the formalism of Ref. [13], using
the stopping cross-sections tabulated in Ref. [15], with column 3
above also incorporating the gas-solid correction proposed in this
paper.

We have also measured one nitrogen solid compound, Si_3N_4, and
again find that the change in silicon yield disagrees with calculations using nitrogen gas values, but agrees with values corrected
for the gas/solid phase change. We are limited to a single nitride
as it appears impossible to obtain other nitride films which can be
shown to be physically stoichiometric and which do not contain up to
a few percent of nitrogen in solution, in gas bubbles, etc.[18]

We caution that the agreement in Table I may not be as good for other compounds if band gap effects are important. The materials we studied were all wide gap insulators, and the correction factor may be quite different for small or zero gap compounds.

As a minor note, we have considered for oxygen gas values one half the stopping cross-section of molecular O_2 gas[8]. This allows the possibility that it may be O_2 which violates Bragg's Rule by twice our correction factor, and that the solids of Table I may be in accord with oxygen stopping cross-sections if it could be measured in the atomic state. This appears improbable because of the generalized gas/solid evaluation shown in Figure 4.

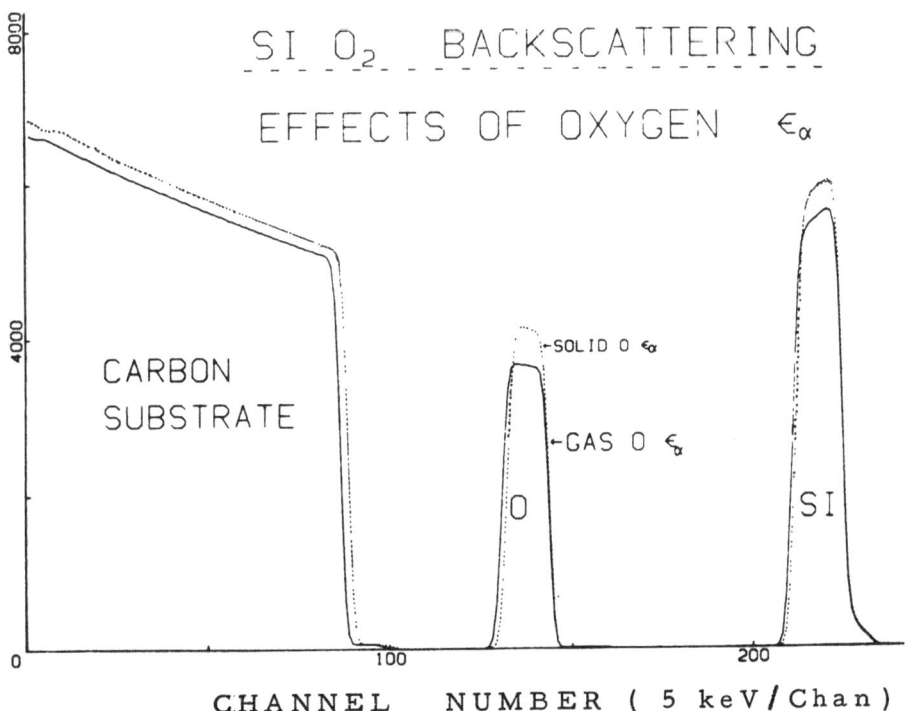

Figure 6. Theoretical backscattering spectra have been calculated for a target of SiO_2 (0.1µm) on a thick carbon substrate. The solid curves shows spectra calculated using stopping cross-sections of ^{4}He in gaseous oxygen, and the dotted curves are generated using the proposed values. Since the proposed cross-sections are lower in value, the peaks are narrower by 10-15%. The Si peak top is sloped because of isotope effects. The calculation is described in detail elsewhere[20]. The parameters used for the calculation are: E_o=2MeV, Q=10µC, θ=170°, Ω=4.11msr, det. resol.=15keV.

We may show the effects of the proposed stopping cross-sections for ^{4}He in oxygen compounds by the theoretical backscattering spectrum shown in Fig. 6. The details of the calculation can be found in the paper in this book entitled "The Uses of Computer Analysis in Nuclear Backscattering" (by Ziegler, Lever, and Hirvonen). The target was 0.1μm of SiO_2 on a carbon substrate. The solid curve shows the calculated spectrum using stopping cross-sections from gaseous oxygen, and the dotted curve is for calculations using the proposed values for oxygen in solids. Since the proposed values are smaller, the peaks are narrower for these latter values. The magnitude of the effect on the peak heights is 10-15%.

In summary, we have found that the energy loss of He^+ in gases appears systematically higher than in solids when the He^+ velocity approaches electron orbital velocities. We have also estimated the energy loss of He^+ in oxygen in solid compounds by backscattering yield measurements, and find these oxygen values differ from oxygen gas values in agreement with the general solid/gas differences. We find that for one solid compound, nitrogen energy loss values differ from nitrogen gas values also as predicted. Numerical values of the solid state stopping cross-sections for N and O will be published elsewhere[19].

REFERENCES

1)   Over 200 recent references to material analysis using nuclear reactions, nuclear backscattering, ion induced X-rays, etc. are found in "Ion Beam Surface Layer Analysis", ed. by J. W. Mayer and J. F. Ziegler, Elsevier Co., Laussane (1974).

2)   See, for example, W. Whaling, "Handbuch der Physik", 34, 193, Springer Co., Berlin (1958).

3)   See, for example, J. Lindhard and M. Scharff, Phys. Rev. 124, 128 (1961), or A. Teplova, V. S. Nikolaev, I. S. Dmitriev, and L. N. Fateeva, Sov. Phys. JETP, 15, 31 (1962).

4)   W. K. Chu, V. L. Moruzzi, and J. F. Ziegler, J. Appl. Phys. 46, 2817 (1975).

5)   W. F. G. Swann, J. Franklin Inst., 226, 598 (1938).

6)   E. Fermi, Phys. Rev. 57, 485 (1940).

7)   C. A. Sautter and E. J. Zimmerman, Phys. Rev. 140, A490 (1965).

8)   P. D. Bourland, W. K. Chu, and D. Powers, Phys. Rev. B3, 3625 (1971), and P. D. Bourland and D. Powers, ibid., 3635 (1971).

9)   D. Powers, W. K. Chu, R. J. Robinson, and A. S. Lodhi, Phys. Rev. 6A, 1425 (1972).

10) D. Powers, A. S. Lodhi, W. K. Lin, and H. L. Cox jr., Thin Solid Films, 19, 205 (1973).

11) A. S. Lodhi and D. Powers, Phys. Rev. A10, 2131 (1974).

12) W. H. Bragg and R. Kleeman, Phil. Mag. 10, 318 (1905).

13) J. S.-Y. Feng, W. K. Chu, and M-A. Nicolet, Phys. Rev. B9, 3881 (1974).

14) J. E. E. Baglin and J. F. Ziegler, J. Appl. Phys. 45, 1413 (1974).

15) J. F. Ziegler and W. K. Chu, Atomic Data and Nuclear Data Tables 13, 463 (1974).

16) O. Meyer, G. Linker, and B. Kraeft, Thin Solid Films 19, 217 (1973).

17) J. S.-Y. Feng, W. K. Chu, M-A. Nicolet, and J. W. Mayer, Thin Solid Films, 19, 195 (1973).

18) Electrochemical Soc. Ext. Abstr. 3 (1966). Over twenty papers discuss the difficulties of obtaining stoichiometric thin-film nitrides.

19) J. F. Ziegler and W. K. Chu (submitted to Jour. Appl. Phys Comm.).

20) J. F. Ziegler, R. F. Lever, and J. Hirvonen, "The Uses of Computer Analysis in Nuclear Backscattering", published in this book.

DISCUSSION

Q: (J.M. Poate) Would you explain the shape of the Si and O peaks in your theoretical SiO_2 backscattering spectrum. The O peak is flat-topped whereas the Si falls with decreasing energy.

A: (J.F. Ziegler) This peak is skewed because of the presence of isotopes (^{29}Si, ^{30}Si) which are present to an abundance of 6 %. You can see that on the high energy side of the Si peak there is a similar"foot" caused by the isotopes (Note: The figure referred to is the last figure (Fig. 5) in the paper "Evidence of Solid State Effects in the Energy Loss of [4]He Ions in Matter").

Q: (J. Davies) You have presented experimental evidence suggesting the existence of a large solid/gas effect in stopping powers. On the basis of this, you then find that the metal/oxide stopping power data are consistent with Bragg's law. Can you suggest an explanation as to why Si or Al seem to exhibit the <u>same</u> stopping power in the oxide as they do in do metal (i.e. <u>chemistry</u> does not affect the stopping power), whereas the <u>physical</u> state seems to have a huge effect.

A: (J.F. Ziegler) First I would point out that our proposed correction does not work for SiO_2, while it does work for Al_2O_3. This appears to mean that it is chemical effects in these two compounds which make the difference rather than solid state effects. The critical experiment, which can go a long way in separating these two contributions, is a measure of the same element in gas phase and then in solid phase. This could be done for O and N and we could possibly see the differences if the binding can be assumed to be undisturbed in the transition between phases. If chemical effects are important, there may be no difference in stopping cross-sections between solid and gas phase. If physical effects are important, there will be dramatic difference as we show in Figure 4.

Q: (B.R. Appleton) Comment that previous conference on Ion Beam Surface Layer Analysis showed that the greatest difficulty in verifying Bragg's law was in obtaining materials with known composition. I suspect that materials studied here are not known sufficiently accurately to verify measurements or calculations.

A: (J.F. Ziegler) We attempted to analyse these films with all the auxiliary techniques we could find. That is one reason this paper took four years to complete. Not only were many samples used, but their stoichiometry was checked by ellipsometry, backscattering, dielectric strength, conductivity, and magnetic properties (Fe_2O_3 and Fe_3O_4).

Q: (B. Scherzer) What is the actual difference of stopping cross section of SiO_2 from Bragg's rule assuming Si and gaseous oxygen values you obtain ?

A: (J.F. Ziegler) This correction is not accurate for SiO_2 as it underestimates the compound energy loss by 5 %. However, the "gas" values overestimate the compound energy loss by 5 %. This one compound shows the worst case of the 6 compounds we measured. Since we had bombarded the sample with 500 keV Si atoms we know that there was no channeling at the SiO_2/Si interface.

Q: (D. Simons) Your last slide showed a calculation of SiO_2 on carbon. Has that experiment been done?

A: (J.F. Ziegler) No, it has not been done. It is difficult to have SiO_2 to adhere to carbon.

EMPIRICAL STOPPING CROSS SECTIONS FOR ^{4}He IONS

R A BARAGIOLA AND J C ECKARDT

CENTRO ATOMICO BARILOCHE, COMISION NACIONAL DE ENERGIA

ATOMICA, SAN CARLOS DE BARILOCHE, ARGENTINA

The use of high energy He ions to analyse solid state materials often requires information on the stopping powers, S, (actually stopping forces) of these ions in matter. As S values have been measured only for some elements, it is of interest to find ways of predicting S for the other elements.

Two successful attempts have been made in the past to produce empirical values in the energy range 400-4000keV. Ziegler and Chu[1] have used the theory of Lindhard and Winther[2] for electronic stopping as extended by Bonderup[3]. This theory is based on two rather simplifying assumptions. First, that it is possible to excite collective oscillations of bound electrons and second, that these electrons behave, in stopping, like in a free electron gas. In spite of this, a good agreement with experimental data was found, which allowed the generation of empirical values from a piecewise adjustment of the theoretical results in different regions of Z_2 values - Z_2 being the atomic number of the target.

The empirical approach of Eckardt et al[4] was based on the remarkably strong correlation found between the Z_2 dependence of S and the Z_2 dependence of the atomic potentials, U, at a fixed radial distance r. The formula:-

$$S(Z_2, v) = C(v) \, r(v) \, U(r, Z_2) \qquad \ldots\ldots 1$$

was derived, where v is the velocity of the projectile and C a fitting parameter. The choice of r was given also by the adjustment procedure.

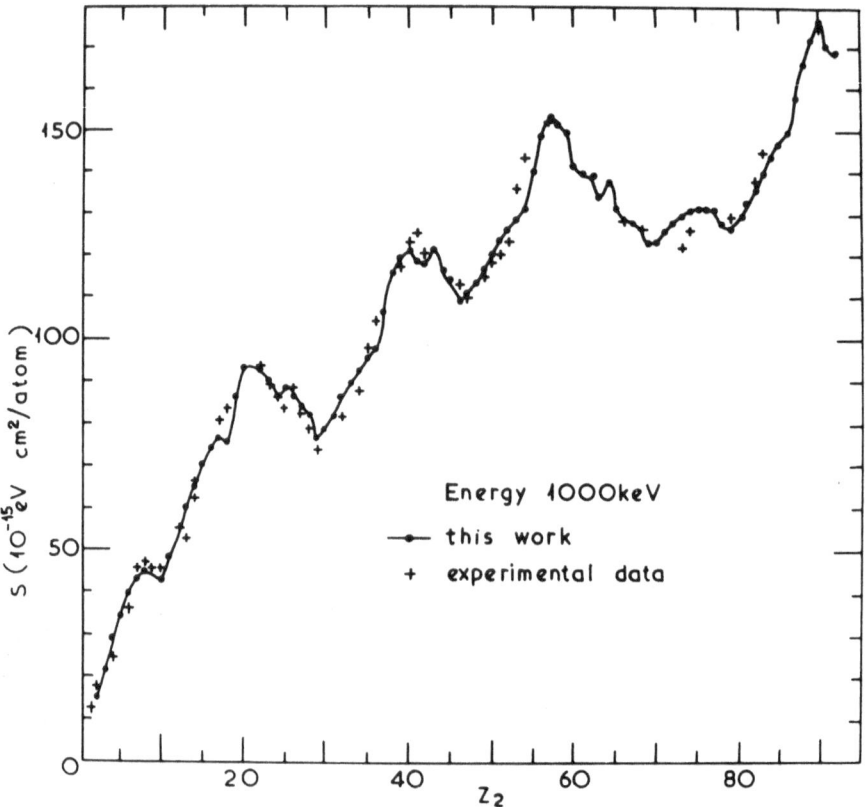

Fig. 1. Comparison between our empirical results and experimental
         data from Ref. 1,6 and 7, for the stopping cross sections
         of 1000 keV ^{4}He ions in targets of atomic number Z_2.

We have found that an even better agreement with experimental
data can be obtained from the relation:-

$$S(Z_2, v) = C(v) [r(U, Z_2)]^{n(v)} \qquad \dots\dots 2$$

where $r(U, Z_2)$ is the radial distance from the nucleus at which
the potential has a fixed value U and where C and n are fitting
parameters which vary with v but not with Z_2. The fits were done
at each velocity as follows. A value of U was chosen and then C
and n adjusted, taking $r(Z_2, v)$ from the Hartree-Fock-Slater tables
by Lu et al[5]. The procedure was then repeated for different values
of the potential until the best fit to the experimental values
was obtained.

The velocity dependence of the stopping power is given mainly by the variation of C with v as n was found to vary only between 2.41 and 2.21 in the velocity range 100-500keV/a.m.u. The inclusion of an additive parameter d(v) in equation 2 was found to give a marginal improvement in fit for the low Z_2 elements.

As no physical basis could be found for the empirical formulae, their value is only given by their ability to accurately predict unknown values of the energy loss. Fig.1 shows the results at 1000keV together with experimental values taken from the compilation by Ziegler and Chu[1] and from more recent publications[6,7]. The agreement with experiment improves at higher energies but is less satisfactory at lower energies. This behaviour also occurs for the other published empirical data[1,4].

REFERENCES

1. J. F. Ziegler and W. K. Chu, Atomic and Nuclear Data Tables 13, 463 (1974)

2. J. Lindhard and A. Winther, Kgl. Danske Videnskab. Selskab, Mat.-Fys. Medd. 34, No. 4 (1964)

3. E. Bonderup, Kgl. Danske Videnskab. Selskab, Mat. Fys. Medd. 35, No. 17 (1967)

4. J. C. Eckardt, W. Meckbach and R. A. Baragiola, Radiation Effects, to be published.

5. C. C. Lu, T. A. Carlson, F. B. Malik, T. C. Tucker and C. W. Nestor Jr., Atomic Data 3, 1, (1971)

6. W. K. Lin, H. G. Olsen and D. Powers, J. Appl. Phys. 44, 363 (1973)

7. J. A. Borders, Radiat. Effects 21, 165 (1974)

DISCUSSION

Q: (W.K. Chu) There is a unique value of potential V at a given radius r so V(r) is well defined, but for a given V there could be several r locations having the same V due to shell structure of Hartree Fock calculation. How do you handle your r(V) function ?

A: (R.A. Baragiola) The r(V) function is single-valued so there are not several r locations for the same V.

DETERMINATION OF STOPPING CROSS SECTIONS BY RUTHERFORD BACKSCATTERING[+]

B.M.U.Scherzer[++], P.Børgesen, M-A.Nicolet, J.W.Mayer

California Institute of Technology, Pasadena

California 91125

ABSTRACT

The stopping cross section, ε, of light ions in solids determines the accuracy of depth distribution measurements by RBS. Errors quoted for individual measurements of stopping cross section are often within 2 -3 % but agreement between different authors is mostly no better than $\pm$ 10 %. Stopping cross sections have been measured by many authors by transmission of ions through thin films or by backscattering from thick targets.

Two basic methods exist to determine stopping cross sections from Rutherford backscattering spectra:
1) ε can be determined from the energy width ΔE of a thin surface film (10,20)
2) ε can be calculated from the height H (counts/channel) of the spectrum (8,11,21,22).
Both methods imply approximations. The errors introduced by these approximations are investigated by applying the different methods to artificial spectra generated by a computer program, assuming Rutherford backscattering and a known stopping cross

+ Work supported in part by a grant of the National Science Foundation (T.Mukherjee).

++ On leave of absence from:
   Max-Planck-Institut für Plasmaphysik
   EURATOM-Association
   D-8046 Garching bei München, Germany

section. The results show that all approximations cause errors which are within a few % if $M_1/M_2 \ll 1$. For larger M_1/M_2 ratios, the ΔE-method is generally more accurate. The computer-generated spectra are further compared to measured backscattering spectra. The agreement is generally good in the upper energy range of the spectrum ($E_1 \simeq 1/2 KE_0$). At low primary energies ($E_{He} < 500$ keV) increasing deviations of the measured spectra from the calculations are observed due to a background produced at least in part by plural scattering.

Backscattering measurements were performed with ^{4}He in the energy range 200-2000 keV on films of Au, Pt, SiO_2, and Ta_2O_5 using both methods. The data obtained for Au and Pt agree well with those of other authors. They are 5 - 10% lower than those given in the tables of Ziegler & Chu[6]. ε for SiO_2 is found 20% lower than obtained from (6) applying Bragg's rule.

The validity of ε measurements depends critically on the calibration procedure. In our case calibrated thin film targets of Au and Pt were used. The mass per area has been measured by microbalance at two different institutions with good agreement.

I. INTRODUCTION

Rutherford backscattering of energetic light ions is a well established method for the investigation of nearsurface regions of solids (1-2). Among other features the method provides a tool for the measurement of depth distribution with a high resolution ($\lesssim 100$ Å). To obtain an accurate relation of energy vs depth, the stopping cross section, ε , for a given combination of ion and target, must be known with some precision. Several tables of stopping cross sections have been published, e.g.(3-6). These data are based only partly on measurements and the agreement between data measured by different investigators is generally not better than within $\pm$ 10%. For this reason it is often desirable to measure the stopping cross section directly on the target under investigation, especially in the case of the less common elements and of compounds where no experimental data are available. This raises the question of what is the best way to obtain reliable stopping cross section data.

Two distinct experimental procedures have been applied in the past: one either measures the energy loss of a monoenergetic particle beam after transmission through a thin film, or one determines the energy of particles scattered back from a surface layer of a thick target. In the transmission method, the evaluation of the experimental data to obtain the stopping cross section is straightforward, but considerable experimental difficulties are encountered in the preparation of self-supported films of sufficiently low thickness and contamination. In the backscattering method, the problem of target preparation is simplified but the extraction of $\mathcal{E}$ values from the experimental data is complex.

The present paper addresses itself to the backscattering method since it can often be applied directly to the backscattering spectra of the target under investigation.

II. DESCRIPTION OF METHODS

In a backscattering experiment (Fig.1a) a primary particle beam of energy E_0 penetrates the target surface at an incident angle α to the surface normal. The particle detector is set at an angle β to the surface normal and in the plane defined by the normal and the incident beam. The detected particles have thus been scattered by an angle $\theta = \pi - (\alpha + \beta)$. On their path in and out of the target, the particles lose energy. If the energy immediately before scattering is E, the energy immediately thereafter is reduced to $K \cdot E$, where K is the kinematic factor ($0 < K < 1$). The scattered particle leaves the target with energy $E_1 < KE$ which is a function of the scattering depth x. Particles scattered from the surface atoms have energy KE_0.

Two basic methods are available to measure stopping cross sections from such an experiment:

(a) The target is a pure thin film of thickness Δx on a substrate of low mass material. The corresponding backscattering spectrum is shown by the full drawn line in Fig. 1b. The measured quantity of interest is the energy width ΔE, of the signal. Additionally, the areal atom density of the film, $N \cdot \Delta x$ (number of atoms/cm^2), has to be determined. Then, according to Warters (10) and Chu et al., (7)

$$\Delta E/(N \cdot \Delta x) = K(\sec\alpha)\,\mathcal{E}(\overline{E}_{in}) + (\sec\beta)\,\mathcal{E}(\overline{E}_{out}) \equiv [\,\overline{\mathcal{E}}\,] \quad (1)$$

$\overline{E}_{in}$ and $\overline{E}_{out}$ are intermediate energies along the ingoing and outgoing particle paths, respectively.

(b) The target is a pure piece of bulk material. In this case, one obtains a spectrum extending to lower energies as shown by the broken line in Fig.1b. The measured quantities are the number of counts/channel, $\Delta A(E_1)$, at the emerging energy E_1, the energy width E_1 of each channel, and the product $Q \cdot \Omega$ of the incident number of primary particles Q times the solid angle Ω of the detector. Then (7,8)

$$\Delta A(E_1)\cos\alpha/(Q\Omega\Delta E_1) = (d\sigma(E)/d\Omega)\left\{\mathcal{E}(KE)/\mathcal{E}(E_1)\right\}/[\mathcal{E}(E)] \quad (2a)$$

where

$$[\mathcal{E}(E)] \equiv K(\sec\alpha)\,\mathcal{E}(E) + (\sec\beta)\,\mathcal{E}(KE). \quad (2b)$$

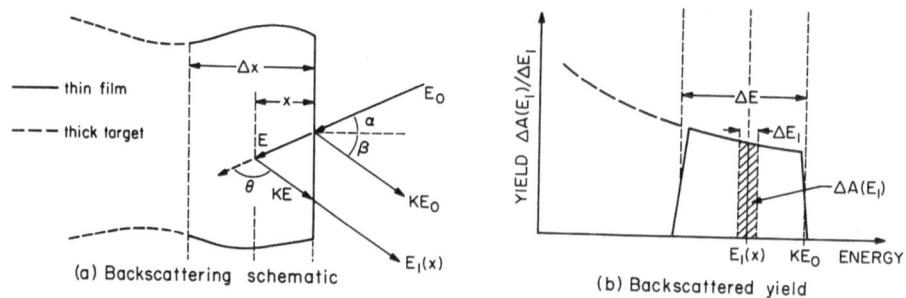

Figure 1.(a) Backscattering experiment schematic in case of a thin film of thickness Δx (full drawn line) and thick target (broken line). (b) Corresponding backscattered energy spectra.

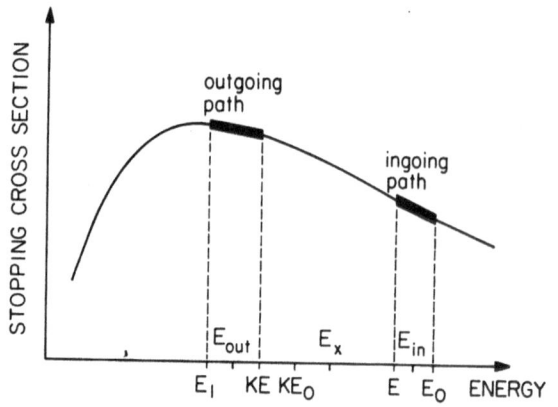

Stopping cross sections in backscattering

Figure 2. Stopping cross section as a function of energy. Energy ranges for ingoing and outgoing particles in a backscattering experiment.

The problem is how to find the stopping cross section $\varepsilon(E)$ from the knowledge of the right hand sides of (1) and (2). Both expressions contain ε at different energies, some of which ($\bar{E}_{in}$, $\bar{E}_{out}$, E) are not directly accessible to measurement. The reason for this is that the energy loss of the particles on the ingoing and outgoing path takes place in two discontinuous energy ranges (Fig.2).

One procedure to find $\varepsilon(E)$ from these expressions is to apply an iterative curve fitting procedure, starting with an assumed energy dependence. This has been done by Lin, et al.(19). The procedure, although potentially the most accurate one, is rather tedious.

In the case (a) of a thin film spectrum, Warters (10) has proposed an analytical solution. He assumes that an intermediate energy E_x between $\bar{E}_{in}$ and $\bar{E}_{out}$ exists at which $\varepsilon(E_x)$ can be found when ε is expanded in a Taylor series about the point E_x. He finds

$$\varepsilon(E_x) = [\bar{\varepsilon}] / \varkappa \qquad (3)$$

and

$$E_x = (KE_0 sec\alpha + E_1 sec\beta)/\varkappa + (\Delta E/2)(\eta sec\beta - K sec\alpha)/(\eta + K), (4)$$

where $\varkappa = Ksec\alpha + sec\beta$ and $\eta \approx \varepsilon(E_1)/\varepsilon(E_0)$. The second term in (4) is a small correction and $\eta = 1$ is a good approximation in most cases.

In the case (b) of a thick target equation (2a) can be solved exactly for $[\varepsilon(E)]$ at the surface, where $E = E_0$ and $E_1 = KE_0$. Then Warters approximation (eq. 3 and 4) can be applied to obtain $\varepsilon(E_x)$. In this case the error is the same as in the ΔE method for very thin films. Behrisch an Scherzer (11) apply an analytical approximation. They assume $\varepsilon \propto E^{\nu}$. The exponent $-1 \leqslant \nu \leqslant (1/2)$ is a constant. With this assumption, Eq.2 can be solved analytically for $\varepsilon(E_1)$.

To test the accuracies of the two latter methods, we have applied them to backscattering spectra for which $\varepsilon(E)$ was known in advance. Spectra were generated by a computer simulation of the Rutherofrd backscattering experiment. The program for the simulation has been described by Feng (12) and has been tested successfully against experimental data. Figure 3a shows the two $\varepsilon(E)$ dependences which have been used for the tests. They are taken from tables of Ziegler and Chu (6) and extended into a $E^{1/2}$ dependence at the lowest energies. Figure 3 b gives the computed spectra for various energies. Note that the primary doses are different at low and high energies. The two methods described above were then applied to these spectra, and the resulting values compared against the original $\varepsilon(E)$ curve. An accurate procedure should reproduce the $\varepsilon(E)$ curve exactly. A good approximate procedure should leave only small discrepancies.

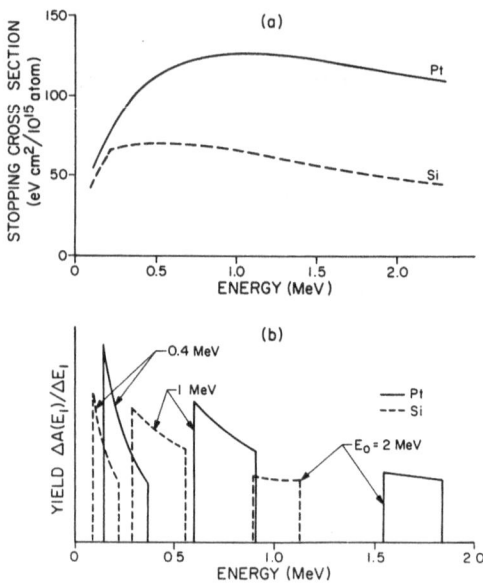

Figure 3. (a) Test dependences of stopping cross sections.
(b) Generated spectra for different primary energies. The primary
dose is different for each spectrum. The 0.4 MeV He on Si-spectrum
is cut off at 100 keV.

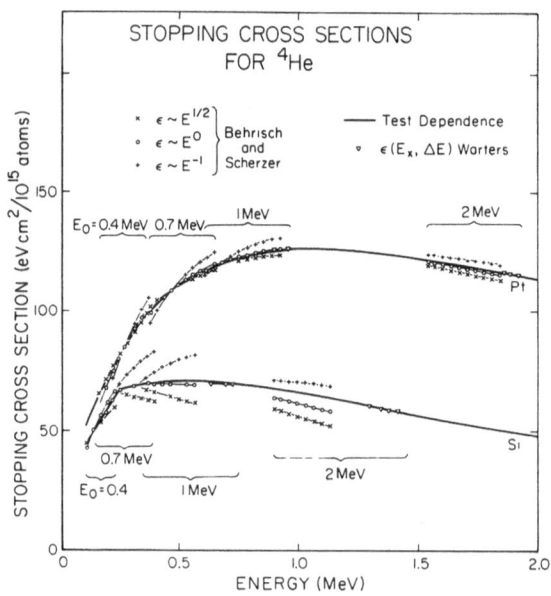

Figure 4. Comparison of calculated stopping cross sections by thin
film and thick target method to the assumed test dependence.

III. RESULTS OF TESTS OF METHODS

The results of the exercise are summarized in Fig.4. In the case (a) of thin films, data were obtained by truncating the spectra at energies corresponding to a few predetermined film thicknesses ranging from 20 to 2000 $\mathring{A}$ for Pt and from 200 to 5000 $\mathring{A}$ for Si. The upper limits give the maximum film thickness used in the simulation. Four different primary energies were chosen.

In all, the results shown in Fig.4 indicate that the thin-film method (a) with the analytical approximation of Warters (10) is most consistently accurate. Typical errors are below 2 %. When K is close to unity the thick-target method (b) with $\nu = 0$, i.e. when ε is set constant to solve Eq.2, yields results which are quite accurate also. The other two assumptions ($\nu = -1, \nu = + 1/2$) are generally less accurate except at very low energies where $\nu = 1/2$ gives very good agreement. This is expected and confirms the validity of our procedure, since the test dependencies of $\varepsilon(E)$ originally assumed actually are proportional to $E^{1/2}$ there.

Note that the thin-film method (a) evaluates ε at higher energies than the thick-target method (b). Also, an actual measurement gives only one $\varepsilon(E_x)$ value for one primary incident energy by method (a), whereas method (b) gives $\varepsilon(E_1)$ over the whole range of emerging energies.

The observed errors in both methods are due to the approximation procedures described above, the numerical errors of the computation being much smaller. The closest agreement between test dependence and $\varepsilon(E_x)$-values (method a) is obtained for the thinnest films (20 $\mathring{A}$), the error being typically less than 0.5 %.

A further method to measure stopping cross sections by backscattering for low mass materials is to deposit the material under investigation in a "sandwich" between two very thin marker layers of heavy mass material (13,14). This method was not specifically investigated here, but the thin-film method (a) is applicable to it as well, and equally accurate results are expected.

IV. MEASUREMENTS

Backscattering measurements were performed with [4]He in the primary energy range from 200 to 2000 keV at the 280 keV He[++] accelerator of Hughes Research Laboratories, Malibu, California, and at the 3 MeV van de Graaff accelerator of the Kellogg Laboratory, at Caltech. Films of evaporated Au, of sputtered Pt, of thermally grown SiO_2 and of electrolytically anodized Ta_2O_5 were used. To assure a proper absolute measurement of the primary doses, secondary electrons were suppressed on both accelerators by maintaining the target at + 300 V against ground and by surrounding the target with a shield kept at a negative

(- 70 V) bias voltage against ground. Careful tests of the
energy calibration of the primary beam and the current integration
system were also performed.

The solid angles of the detectors were determined by measuring the total yield of backscattered particles that hit the
detector from independently calibrated standard targets. Two of
these standards consisted of Pt films sputtered onto SiO_2. Their
areal atom density had been determined by weighing on a microbalance before and after deposition (18). Two Au standards were
produced·by evaporation onto SiO_2. The weighing was performed
on the microbalance of the Geology Department at Caltech. Each
pair of standards claimed an accuracy of $\pm$ 3%. The results obtained from the Au standards were found to be systematically lower
than those obtained with the Pt standards by about 4 %. All
numerical evaluations presented here are based on the Au standard.
The formulae to obtain Ω from the total backscattering yield A
are given in references 7 and 15.

Another method to obtain the surface density of a very thin
standard film consists in comparing the total number of counts,
A_f, originating from backscattering events in the film of the
number of counts $\Delta A(E_1)$ per channel in the signal generated
by backscattering events in the underlying SiO_2, and assuming
a known stopping cross section for SiO_2 (16). In this case (7)

$$(N\Delta x)_f = A_f \, \sigma_{Si} \, \Delta E_1 / (\Delta A(E_1)_{SiO_2} \, \sigma_f [\varepsilon]_{Si}^{SiO_2}$$

(5)

σ_{Si} and σ_f being the Rutherford cross sections of Si and of
the film material, respectively, and

$$[\varepsilon]_{Si}^{SiO_2} = K_{Si} (\sec\alpha) \, \varepsilon(E_0)_{SiO_2} + (\sec\beta) \, \varepsilon(K_{Si}E_0)_{SiO_2}.$$

(6)

Surface densities determined by applying this method are consistently 20 % lower than those obtained by weighing.

Since good agreement was found for two different standards
whose surface density had been determined gravimetrically by
two groups independently, these values of Ω were used in the
following numerical evaluations.

V. RESULTS OF STOPPING CROSS SECTION MEASUREMENTS

Stopping cross section data obtained by methods (a) and (b)
from measured backscattering spectra are shown in Figs.5-7.
For comparison, Ziegler and Chu's (6) semiempirical $\varepsilon(E)$-curves
and the data of other authors (9,14,17,18) are given. For Au
(Fig.5) good agreement is found with $\varepsilon(E)$ of Borders, while Lin,

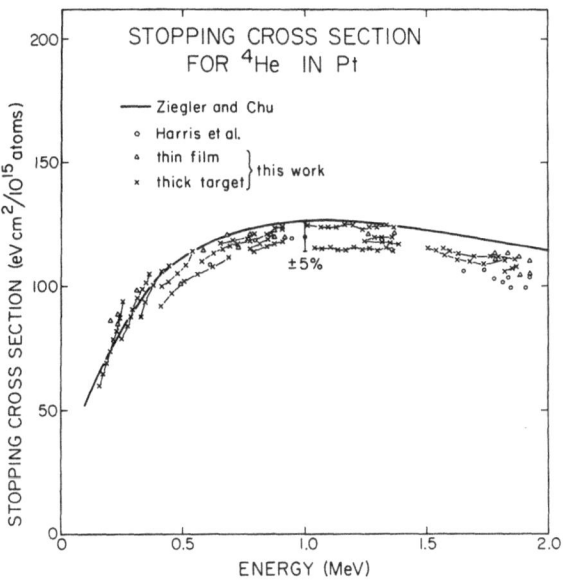

Figure 5. Stopping cross section for ^{4}He in Au.

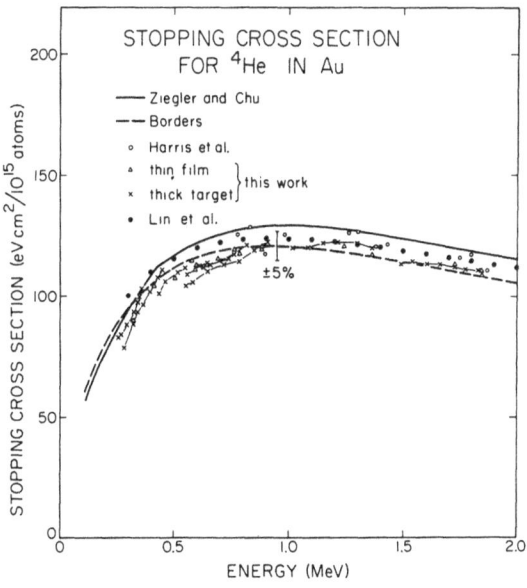

Figure 6. Stopping cross section for ^{4}He in Pt.

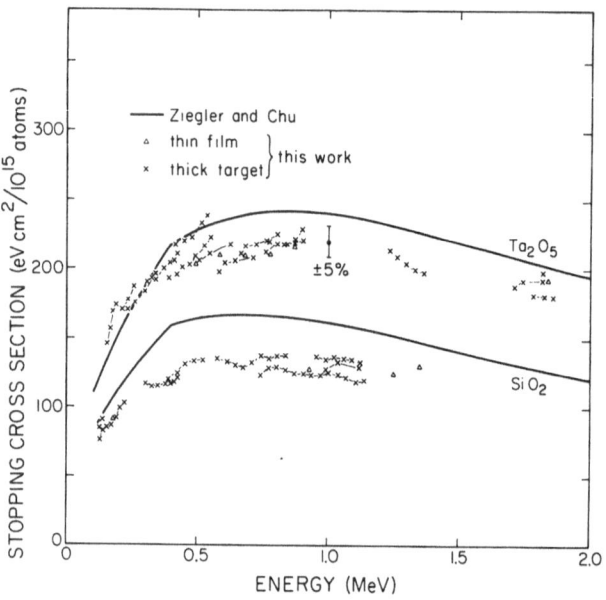

Figure 7. Stopping cross section for ^{4}He in SiO$_2$ and Ta$_2$O$_5$. The cross sections are given per Si and Ta atom, respectively.

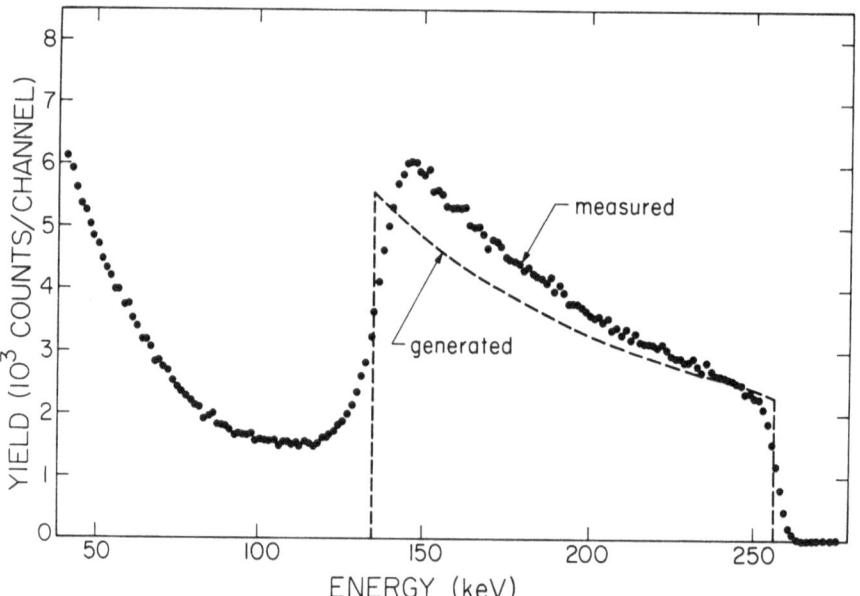

Figure 8. Backscattering spectrum of 280 keV ^{4}He^{++} on 1130 Å of Pt on SiO$_2$ compared to a generated spectrum assuming single collision Rutherford backscattering.

Matteson and Power's (9) data are systematically 3 %, and Harris' (14) about 4% higher. At high energies Ziegler and Chu's $\varepsilon(E)$ curve is about 6% higher, at low energies 3-4 % higher than our data (and typically 1 % higher than Lin, Olson and Power's (19), the latter not shown in the figure). The stopping cross sections of Pt are generally 5% lower than in reference 6. The Pt measurements of Harris, et al.(18) were performed in part on the same target used in this work, but our ε values are typically a few percent higher, as expected from the 4% difference in their and our standard. For SiO_2 the stopping cross sections we find are about 20 % lower than those reported in (16) and those obtained from reference 6 by applying Bragg's rule of additivity. This reflects the discrepancy which was found in the different calibration procedures as reported above. The stopping cross sections for Ta_2O_5 (given in Fig.7 as the stopping cross sections per Ta atom, $\varepsilon^{TaO2.5}$) are 10% lower in the high energy range than those according to (6) applying Bragg's rule. Below 500 keV the agreement is close.

At low energies, the various stopping cross sections derived by the thick target method, $\varepsilon(E_1)$, at a given energy E_1 display a systematic trend; a higher incident energy E_0 leads to a lower value of $\varepsilon(E_1)$. We attribute this effect to the non-negligible background which is present on the low energy side of the signals in the backscattering spectrum (Fig.8). The reason for this "tail" is not completely understood, plural scattering in the film probably playing a major role. This background increases the yield of particles emerging at lower energies. This is seen by comparing the measured spectrum to a generated one which assumes pure single collision Rutherford scattering as shown in Fig.8. The increase of the yield by the tail leads to lower values of ε at large depths if the thick target method is applied. There is as yet no procedure to account for this effect.

VI. CONCLUSIONS

Stopping cross sections can be extracted from backscattering spectra with satisfactory accuracy by rather simple approximate methods. When the material under investigation has a light mass, the thin film method is preferable to the thick target method.

For the determination of absolute stopping cross sections the production of standard targets of high purity and the absolute measurement of areal atom density of these targets is of crucial importance.

Low energy backscattering spectra differ considerably from the single collision Rutherford model because of the background.

ACKNOWLEDGEMENTS

The authors wish to thank Howard Dunlap for his assistance
in running the accelerator at the Hughes Research Labs. Dr.T.Wen
of the Geology Dept., at Caltech, helped in weighing of the
standard Au samples. We also thank Dr.J.P.S.Pringle, Chalk River
Nuclear Laboratories, Chalk River, Ontario, for providing the
anodically oxizided Ta_2O_5 samples.

REFERENCES

1. J.W.Mayer, J.F.Ziegler (Eds.), "Ion Beam Surface Layer
    Analysis", Elsevier Sequoia, Lausanne, 1974.

2. M.ANicolet, J.W.Mayer, I.T.Mitchell, Science 177 841 (1972).

3. S.K.Allison, S.D.Warshaw, Rev.Mod.Phys. 25, 779 (1953).

4. W.Whaling, Handbuch der Physik 34, 193 (1958).

5. L.C.Northcliffe, R.F.Schilling, Nucl.Data Tables A7, 233
    (1970).

6. J.F.Ziegler, W.K.Chu, Nucl.Data Tables 13, 463 (1974).

7. W.K.Chu, J.W.Mayer, M-A.Nicolet, S.U.Campisano and
    E.Rimini, "Backscattering Analysis", paper B of Cantania
    Working Data; A compilation of Tables, Graphs and Formulae
    for Ion Beam Analysis, Cantania, Italy, 1974.

8. W.A.Wenzel, Ph.D.Thesis, California Instiutte of Technology,
    1952. W.A.Wenzel, W.Whaling, Phys.Rev.87, 499 (1952).

9. W.K.Lin, S.Matteson, D.Powers, Phys.Rev. B10, 3746 (1974).

10. W.D.Warters, Ph.D.Thesis, California Instiutte of Technology,
    1953.

11. R.Behrisch, B.M.U.Scherzer, see ref.1, pp.247, and Thin Solid
    Films 19, 247 (1973).

12. J.S.Y.Feng, W.K.Chu, M-A.Nicolet, see ref.1, pp.227 and
    Thin Solid Films 19, 227 (1973).

13. W.White, R.M.Müller, J.Appl.Phys.38, 3660 (1967).

14. J.M.Harris, M-A.Nicolet, Phys.Rev. B11, 1013 (1975).

15. R.Behrisch, B.M.U.Scherzer, P.Staib, see Ref.1, pp.57, and
    Thin Solid Films 19, 57 (1973).

16. D.A.Thompson, W.D.Mackintosh, J.Appl.Phys. 42, 3969 (1971).

17. J.A.Borders, Radiation Effects 16, 253 (1972).

18. J.M.Harris, W.K.Chu, M-A.Nicolet, see ref.1, pp.259, and
    Thin Solid Films 19, 259 (1973).

19. W.K.Lin, H.G.Olson, D.Powers, Phys.Rev. B8, 1881 (1973).

20. J.F.Ziegler, M.H.Brodsky, J.Appl.Phys.44, 188 (1973).

21. W.K.Chu, J.F.Ziegler, I.V.Mitchell, W.D.Mackintosh,
     Appl.Phys.Letters 22, 437 (1973).

22. E.Leminen, A.Fontell, Radiation Effects 22, 39 (1974).

DISCUSSION

Q: (W. Gibson) Could you comment on the effect of energy straggling -
especially the non symmetric straggling expected near the surface
in your calculated results ?

A: (B. Scherzer) No - this has not been taken into account - it is
the next step in the calculation.

Q: (J. Baglin) I would like to comment that the Backscattering Standards Project, using identical samples such as you describe, has
shown disturbing discrepancies in measurements from different labs.
The reasons are still not clear ... possibly solid angles, possibly
charge collection being poorly known.

A: (B. Scherzer) Yes. We also suspect real trouble with solid angles.

Q: (R. Behrisch) What is the reason for the $\pm$ 5 % scatter of your
results at different days?

A: (B. Scherzer) There are a couple of possible reasons: Firstly we
have had some trouble with the solid angle of the detector which may
have changed when other experiments were done between two measurements of ϵ. Secondly inhomogeneities of the target surface in hitting
different spots or some channeling effects due to texture cannot be
excluded.

Q: (J.F. Ziegler) One major point in using thick target yield is
an examination of target texture (defined as the degree of alignment of the metal grains). These texture effects can be quite large.
H.H. Anderson and I have measured peak reductions well over 10 %
for Au evaporated on amorphous substrates such as SiO_2. Such effects would directly change the energy loss calculated from thick
target yields. Such channeling also occurs in Bi and Ag. It is necessary to keep the incident beam away from the channels of such
texture if one expects to obtain accurate stopping cross-sections.

A: (B. Scherzer) Similar to Borders (see his comment) we have
looked for texture effects by tilting the target to different angles

toward the primary beam without finding any within about ± 2-3 %.
Since channeling tends to reduce the backscattering yield the ε-
values obtained from a textured sample may be too high. The same
is true for impurities. Our results on the other hand are generally
at lower limit of data obtained by other authors.

Comment: (J.A. Borders) Our Au stopping cross section measurements
included a detailed study of the Au scattering peak energy width
as the tilt angle of the sample with respect to the incident beam
was changed. These results indicated that sample "texture" effects
were not present to an accuracy of about ± 2 %.

Q: (J. Biersack) You mentioned a steeper slope of S(E) at low energies, than the $\sqrt{E}$ dependence predicted by theory. If multiple or
plural scattering background is subtracted, would you obtain better
agreement with the $\sqrt{E}$ law ?

A:(B. Scherzer) We were not able to find a simple way to account
for the background. There is no obvious relationship between the
enhancement of the yield in the peak, as obtained by comparing
it to the generated spectrum, and the background at lower energies.

Q: (G. Dearnaley) I should like to draw attention to the work of
Schmid and Ryssel (Nucl. Instr. & Meth., July 1974) and Barragàn
on the effects of surface topography upon backscattered energy
spectra. Some oxides grow upon metals with a structure which can
distort spectral shape: submicron asperities are enough.

A: (B. Scherzer) Surface roughness mainly influences the high energy edge of the backscattering spectrum. In our case the slope of
this edge corresponded very well to the detector resolution showing that any roughness effects are below the limit of detection by
this method. Also surface roughness does not influence the backscattering yield at larger depth (500 - 1000 Å) at which we calculate ε by the thick target method.

DEPTH PROFILING OF IMPLANTED [3]HE IN SOLIDS BY NUCLEAR REACTION AND RUTHERFORD BACKSCATTERING

J.Roth, R.Behrisch, W.Eckstein, B.M.U.Scherzer

Max-Planck-Institut für Plasmaphysik, EURATOM-Associa-

tion, D-8046 Garching b.München, Germany

ABSTRACT

Depth profiles of [3]He-ions implanted into a niobium single crystal have been measured by the ^{3}He(d, α)H nuclear reaction and by Rutherford backscattering (RBS) of deuterons assuming Bragg's rule of additivity of stopping cross sections for niobium and helium. The two profiles are obtained simultaneously using a primary beam of 500 keV deuterons and measuring the energy distributions of emitted α-particles and backscattered deuterons. The ^{3}He profiles obtained by both methods agree in their general shape, that obtained by the nuclear reaction method being broader and lower than the one obtained by RBS. Some possible reasons for the disagreement are discussed.

INTRODUCTION

Depth profiling of light ions (hydrogen, helium, lithium etc) implanted into solids at energies between several hundred eV and several MeV is of importance for the investigation of problems arising in the first wall of plasma devices and of a future fusion reactor (1-4).

For an understanding of the processes involved e.g. blistering, gas trapping and reemission depth distributions of the implanted ions have to be known with some accuracy (5,6). A number of different nondestructive methods has been applied recently (5,7-11) to different types of ion-target combination. With two of these, i.e. the Rutherford backscattering method (RBS) using Bragg's rule (11) and the nuclear reaction ^{3}He(d,α)H

method (5), depth distributions of implanted ^{3}He-atoms can be
obtained in one and the same measurement if the target is bombarded by deuterons and the energy distributions of backscattered
deuterons and of emitted α-particles are measured simultaneously.

This paper presents results obtained from a comparison of
depth distributions of 15 keV ^{3}He ions in niobium measured by
both methods.

DESCRIPTION OF METHODS

The nuclear reaction method is shown schematically in Fig.1a.
A niobium target implanted with ^{3}He is bombarded by a 500 keV deuteron beam. In collisions with implanted ^{3}He particles the
reaction ^{3}He(d, α)H takes place. The reaction products, an α-particle and a proton, have original energies E_α and E_H. The α-particles starting in a direction with an angle β toward the surface
normal are detected and energy analyzed in a surface barrier
detector. According to the depth **Δt** below the surface at which
the reaction takes place, the α-particles lose energy

$$\Delta E = \int_0^{t/\cos\beta} \frac{dE}{dx}\ dx,\ \text{where dE/dx is the differential energy loss}$$

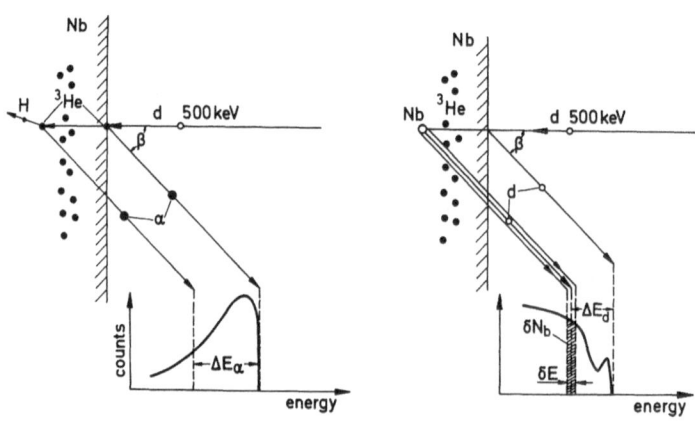

Nuclear reaction: ^{3}He(d,α)H   Rutherford backscattering

a   b

Figure 1. Depth profiling methods schematic.
    a) ^{3}He(d, α)H nuclear reaction
    b) Rutherford backscattering (RBS) of
       deuterons

of the α's in niobium. If dE/dx is known the energy scale of the
spectrum can be transformed into a depth scale. The intensity
of detected α's is directly proportional to the density of
^{3}He-atoms in the target. Thus, using the cross section for the
nuclear reaction, a depth profile of the ^{3}He-density in the
target is obtained. A more detailed description of this method
can be found in reference 6, 12,13,and 20.

The Rutherford backscattering method for depth profiling
(11,14) is shown schematically in Fig.1b. 500 keV deuterons
penetrate the surface at normal incidence. Those deuterons
scattered by the niobium atoms at an angle of $\theta = \pi - \beta$ are
energy analyzed. The number of deuterons δNb scattered into
an energy interval δE corresponding to the energy width of one
channel of the multichannel analyzer is used to calculate the
electronic stopping cross section as a function of emerging
energy (11,15), $\varepsilon(E)$. From the energy difference ΔE_d between
deuterons scattered at the target surface and those scattered
inside the target, the depth Δt at which the scattering occured
can be calculated. From this the stopping cross section as a
function of depth is obtained. We assume validity of Bragg's
rule of additivity of the stopping cross sections of ^{3}He and Nb:

$$\varepsilon_{Nb\ He_x} = \varepsilon_{Nb} + x\,\varepsilon_{He} \qquad (1)$$

where $\varepsilon_{Nb\ He_x}$ is the stopping cross section of the helium
niobium mixture, ε_{Nb} and ε_{He} are the stopping cross sections of
pure niobium and helium for deuterium respectively, and x is the
number of helium atoms per niobium atom. The stopping cross
section ε_{Nb} of pure niobium is obtained from the backscattering
spectrum of the unimplanted sample (15) and ε_{He} is taken from
the tables of Northcliffe & Schilling (13). Then the number
of ^{3}He-atom /Nb atom can be obtained from eq.(1) as a function
of depth.

MEASUREMENTS

The measurements were performed at the combined low and high
energy acceleration set up described in (6).The single crystalline
niobium target was implanted at near normal incidence (12^0off the
surface normal to avoid channeling)with 15 keV ^{3}He ions.

For analysis the target is bombarded with 1 MeV D_2^+ corresponding to 500 keV D^+ ions. The spectra of the α-particles and the
backscattered deuterium were obtained simultaneously at an angle
$\beta = 75^0$ to the surface normal. Again an angle of incidence of 12^0
to the surface normal is chosen to avoid channeling.

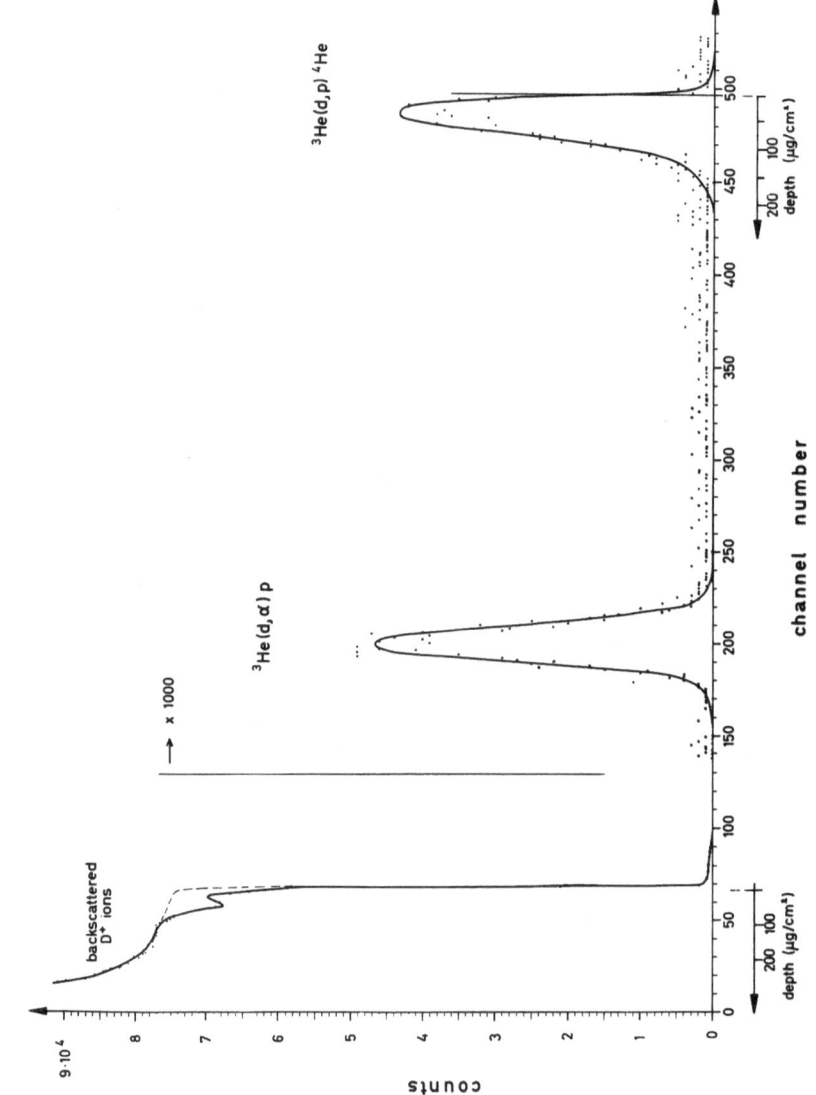

Figure 2. Typical energy distributions of α-particles from the ^{3}He (d, α)H nuclear reaction and backscattered deuterons obtained by bombarding a niobium target implanted with ^{3}He with 500 keV deuterons.

A typical spectrum is shown in Fig.2. The α-particle spectrum forms a peak at high energies, the backscattered deuteron distribution is found at the low energy side. In between a proton peak is obtained due to ∿14 MeV protons generated in the nuclear reaction. This peak occurs at this low energy because the protons lose only a small fraction of their energy in the depletion layer of the detector. Depth scales in μg/cm^2 are given for the deuteron and the α-spectrum.

RESULTS AND DISCUSSIONS

Depth distributions of 15 keV ^{3}He-ions in niobium measured with either method are compared in Fig.3. The upper part corresponds to a total helium dose of 10^{18} cm^{-2} the lower part to 7×10^{18} cm^{-2}. The lower dose is somewhat higher than the critical dose for blistering whereas at the higher dose a surface layer of a constant saturation density has formed (5). The agreement of the general shapes of the distributions is surprisingly good taking into account the different approximations made. These are:

1) For the nuclear reaction method:

    a) The depth scale is determined taking stopping cross sections for α-particles in pure niobium from (18). The additional stopping due to the implanted ^{3}He has been neglected. Evaluation of this effect would decrease the width of the distribution by roughly 5-10%.

    b) The cross section of the nuclear reaction is known only within $\pm$ 10% (22).

2) For the Rutherford backscattering method:

    a) The energy loss of 500 keV deuterons in pure niobium was obtained by RBS to 18 eV/Å. This is about 10-15 % higher than values from (15) and (17) for protons at the same velocity. This may be due to impurities in the target, to crystalline effects (channeling) which are difficult to avoid, and to an incorrect value of the detector solid angle (15).

    b) The formulae for the evaluation of stopping cross sections from backscattering spectra contain an assumption on the energy dependence of the stopping cross section. Due to this the calculated cross sections contain an error which depends on the degree of agreement between assumed and actual energy dependence (19) and is difficult to assess in the present case.

Thus, the more serious errors probably arise in the evaluation
of the backscattering spectra. Nevertheless it cannot be excluded
that the discrepancy between the two distributions is at least
partly due to a deviation from Bragg's rule of additivity. This
would not be surprising since the density of helium atoms in
the implanted layer is extremely high exceeding locally solid
state densities (21).

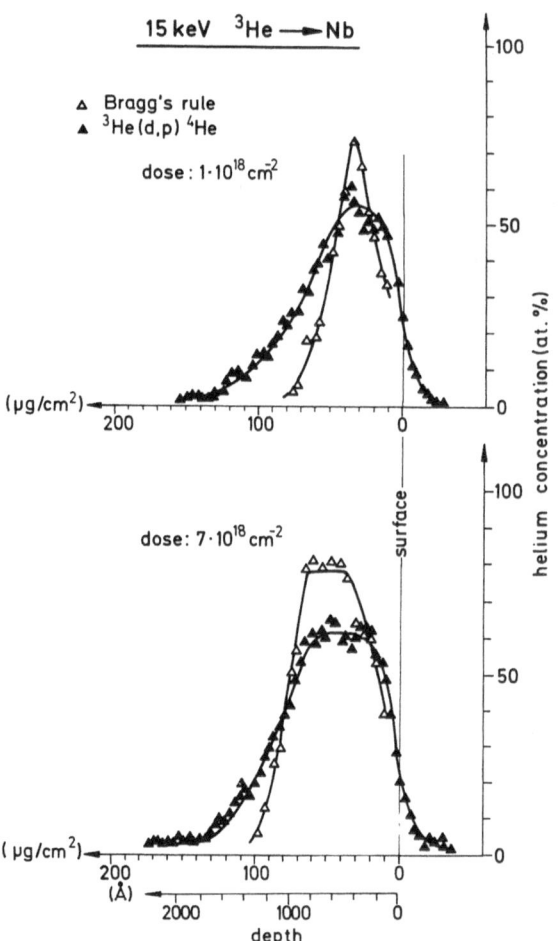

Figure 3. Depth profiles of two different doses of 15 keV ^{3}He-ions
implanted in niobium single crystal. ($\triangle$) Rutherford backscattering method ($\blacktriangle$) ^{3}He(d,α)H nuclear reaction method.

ACKNOWLEDGEMENTS

The authors wish to thank Dr.H.Vernickel for his constant interest in this work. The technical assistance of R.Heilmeier, H.Schmidl, S.Schrapel, and H.Wacker in planning and performing these measurements is gratefully acknowledged.

LITERATURE

(1)   H.Vernickel, Proc.I.Topical Meeting Techn.Controlled Nucl.Fusion, San Diego, 1974

(2)   R.Behrisch, B.B.Kadomtsev, Proc.IAEA Conf.Plasma Phys. and Contr.Nucl.Fusion Res., Tokyo, 1974

(3)   M.Kaminsky, see ref.2

(4)   B.M.U.Scherzer, J.Vac.Sci.Tech., submitted for publ.

(5)   R.Behrisch, J.Bøttiger, W.Eckstein, U.Littmark, J.Roth, B.M.U.Scherzer, Appl.Phys.Letters 27, 199 (1975)

(6)   R.Behrisch, J.Bøttiger, W.Eckstein, J.Roth, B.M.U.Scherzer, J.Nucl.Mat. 56, 365 (1975)

(7)   R.S.Blewer, a) Applications of Ion Beams to Metals, pp. 557 (ed. S.T.Picraux, E.P.Ernisse, F.L.Vook),Plenum Press, New York 1974
        b) J.Nucl.Mat.53, 268 (1974)

(8)   B.Terreault et al., Sympl.on Rad.Effects on Sol.Surf. August 25-26, 1975, Chicago/Ill.

(9)   J.C.Davies, J.D.Anderson, J.Vac.Sci.Tech.

(10)   J.Biersack, D.Fink, a) see reference 7, pp.307
        b) Atomic Collisions in Solids II, pp 737, (ed.S.Datz, B.R.Appleton, C.D.Moak), Plenum Press, New York 1975

(11)   J.Roth, R.Behrisch, B.M.U.Scherzer, Applied Phys.Letters 25 643 (1974)

(12)   A.Turos, L.Wielunski, A.Barcz, Nucl.Instr. Meth.113, 605 (1973)

(13)   R.A.Langley, S.T.Picraux , F.L.Vook, J.Nucl.Mat. 53, 257 (1974)

(14)   W.K.Chu et al.,Proc.Int.Conf. Ion Beam Surf.Layer Analysis, pp.421, Elsevier Sequoia S.A.Lausanne (1974)

(15)   R.Behrisch, B.M.U.Scherzer, Thin Sol.Films 19, 247 (1973)

(16)   W.H.Bragg, R.Kleemann, Phil.Mag.10, 318 (1905)

(17)   L.C.Northcliffe, R.F.Schilling, Nucl.Data Tables A7,233, (1970)

(18)   J.Ziegler, W.K.Chu, Nucl.Data Tables 13, 463 (1974)

(19)   B.M.U.Scherzer, P.Børgesen, M.-A.Nicolet, J.W.Mayer, this conference

(20)   W.Eckstein, R.Behrisch, and J.Roth, this conference

(21)   J.Roth, International Conference Appl.of Ion Beams to Metals, Warwick, Sept.1975

(22)   J.L.Yarnell, R.H.Looling, W.R.Stratton, Phys.Rev.90,292 (1973)

DISCUSSION

Q: (A. Turos) 1.) What is the sensitivity of ^{3}He + d reaction?
2.) What depth resolution could be achieved?

A: (J. Roth) As the sensitivity increases and the depth resolution decreases with the solid angle of the detector, a compromise has always to be found depending on the requirements. In the present case a depth resolution of about 50 Å has been achieved at a sensitivity of about 3 at.% for ^{3}He in Niobium. For the limits of the depth resolution and sensitivity of this method please see the paper by W. Eckstein et al. in this conference proceedings.

Q: (D. Simons) When you analyze your data do you unfold to go from a yield distribution to a profile distribution ?

A: (J. Roth) No. That could be a source of error.

ENERGY LOSS STRAGGLING OF PROTONS IN THICK ABSORBERS

B. H. Armitage and P. N. Trehan*

U.K.A.E.A., Harwell

Didcot, Oxon OX11 ORA, United Kingdom

1.   INTRODUCTION

The energy loss of charged particles transmitted through an absorber is subject to fluctuations due to the statistical nature of the excitation and ionization processes involved. For an initially monoenergetic beam of charged particles incident on an absorber of uniform thickness there will be an energy distribution function associated with the emergent particles. Consider first the case where the fractional energy loss in the absorber $\Delta E/E < 0.1$ but where ΔE is very much greater than the energy lost in a single collision. Here for charged particles of initial energy E incident on an absorber of thickness x, the distribution of energy loss after passing through the absorber depends on the energy loss spectrum for a single collision. In passing through the absorber the energy does not vary appreciably and the single collision spectrum will not alter much. In this case, originally treated by Bohr[1], the energy loss distribution is Gaussian. However, if we now consider a thicker absorber the energy can no longer be regarded as constant over the whole thickness and the single collision spectrum will change. This consideration has been taken into account by Tschalar[2] who has calculated energy straggling distributions for protons in passing through absorbers for which $\Delta E/E$ is up to 0.8.

In the present work we are concerned with the energy straggling of protons in passing through a variety of thick absorbers where the calculations due to Tschalar[2] are expected to be applicable. The

*On leave from Panjab University, Chandigarh 14, India.

measurements have been made at incident proton energies from 6 to
12 MeV with seven different absorbing materials from Al to Au with
thicknesses varying from 20 to 250 mg cm^{-2}. The only measurements
reported in this energy range are with thin Al absorbers[2] and
with a single thickness of Si[4]. Other measurements have been
made with protons of higher energy[5,6].

2.   EXPERIMENTAL ARRANGEMENT

The proton beam from the Harwell tandem accelerator was
incident on a 0.012 mm Au foil after passing through a 1.0 mm
diameter collimator (fig. 1). For a proton beam of initial energy
12.36 MeV, 0.4 MeV was lost in the foil, and due to multiple
scattering only a small proportion of the incident $\sim$50 nA of
protons reached the rectangular defining slit. The latter was set
at either 0.04 x 0.5 mm^2 or 0.3 x 0.5 mm^2, and the arrangement
enabled a convenient counting rate ($\sim$10^3 sec^{-1}) to be obtained
at the Si semiconductor detector.

The technique consisted of transmitting protons through
absorbers of uniform thickness and measuring the energy distribution

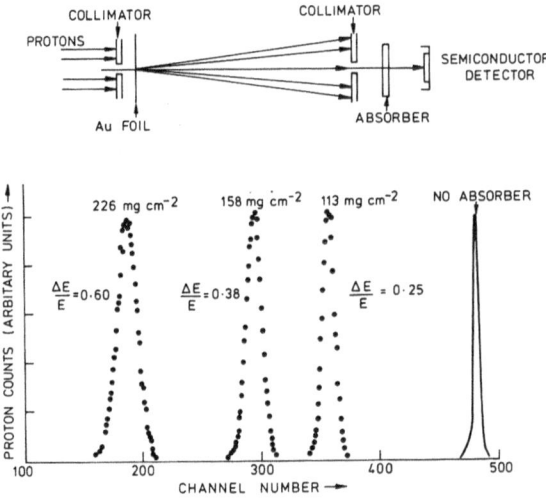

Fig. 1   Experimental arrangement (upper) illustrating the method of
          obtaining a low intensity proton beam at the absorber by
          means of a Au scattering foil. Energy distribution (lower)
          obtained with an 11.96 MeV proton beam after passage through
          Ni absorbers of thickness 226, 188 and 113 mg cm^{-2}. The
          energy distribution in the absence of an absorber is also
          shown.

of the emergent protons with the Si detector. Interchange of
absorbers was facilitated by the use of a frame on which up to
20 absorbers could be mounted. The frame could be moved in two
directions in the plane perpendicular to the beam by remotely
controlled stepping motors. The stepping motor controller also
provided a display of the frame position coordinates.

3.   EXPERIMENTAL PROCEDURE

Energy straggling distributions were obtained for Al, V, Ni,
Mo, Ag, Ta and Au. Measurements were made at incident energies of
11.96, 8.94 and 5.89 MeV for all seven absorbing materials, and a
total of 37 absorbers of varying thicknesses were employed. Some
of the absorbers could not be used at the lower incident energies
owing to complete stoppage of the incident protons. The basic
method was to observe the proton energy distributions with and
without an absorber in place (fig. 1). The straggling distribution
could then be obtained in principle by subtraction of the latter
from the former. Contributions to the absorber-out proton energy
distribution arise from straggling in the Au foil as well as
detector resolution, initial beam energy spread and slit scattering.
Owing to instability in the proton beam from the accelerator, and
consequent variations in the slit scattering the proton energy
distribution in the absence of an absorber was subject to some
variation. Accordingly, this function was measured frequently
during the course of the measurements.

Energy calibration of the detector was made with the proton
beam at a number of known energies, and also with a precision
pulser. During the course of each measurement of the energy
straggling distribution, energy loss in the absorber (ΔE) was
determined from the centroids of the absorber-in and absorber-out
energy distributions using an on-line Honeywell DDP-516 computer.
The full width at half maximum (Ω) of the proton straggling
distribution was calculated by subtraction in quadrature of the
fwhm of the absorber-in and absorber-out energy distributions.

The largest single source of error arises from the previously
mentioned effect of beam instability on the relative shape of the
absorber-in and absorber-out energy distributions. It is estimated
that such errors range from $\pm$ 1% at higher values of $\Delta E/E$ to as
much as $\pm$ 20% at the lowest measured fractional energy losses.
For 90% of the measurements, however, this error is less than $\pm$ 5%.

Some attempt to measure non-uniformity in the absorbers was
made by making measurements at a number of positions on each
absorber. In general it was found that the relative centroid shift
corresponded to thickness variations of < $\pm$ 1% except in the case
of the thinnest foils where the corresponding figure was $\pm$ 3%. This
procedure is of course inadequate in that it does not take into

account non-uniformities and inhomogeneities on a scale less than
0.04 x 0.5 mm² which was the minimum size of collimator used.
Although the measured non-uniformity of the Ta absorbers was no
worse than that of the other elements studied it was found that the
proton distributions were exceptional in that a high energy 'tail'
was exhibited.   This result indicates strongly that the Ta absorbers
suffer from small scale non-uniformities or inhomogeneities and as
a consequence it would appear advisable to treat the Ta data with
some caution.   No account has been taken of the possible effects of
porosity[7] on the widths of the measured proton distributions.
However, except in the case of Ta there is no direct evidence to
suggest that such effects are likely to be significant in the
present work.

3.   COMPARISON WITH THEORETICAL CALCULATIONS

Proton distributions obtained with 11.96 MeV protons after
passage through Ni absorbers are compared with Gaussian line shapes
of the same FWHM in fig. 2.   The 226 mg cm⁻² absorber ($\Delta E/E = 0.6$)
is clearly Gaussian in shape.   Since the absorber-out measurement
exhibits a significant non-Gaussian tail (due to slit scattering)
it would seem that the corresponding tail observed with the
45 mg cm⁻² absorber ($\Delta E/E = 0.1$) is also an instrumental function

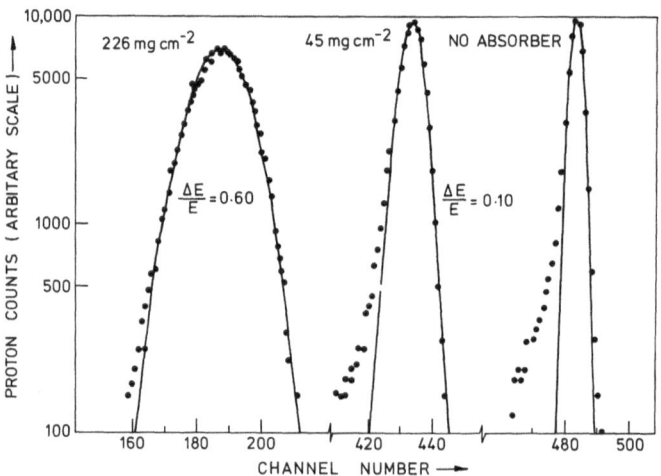

Fig. 2   Energy distributions obtained with Ni absorbers of thickness
45 mg cm⁻² and 226 mg cm⁻² compared with Gaussian curves
(continuous lines).   The energy distribution in the absence
of an absorber is also compared with a Gaussian curve

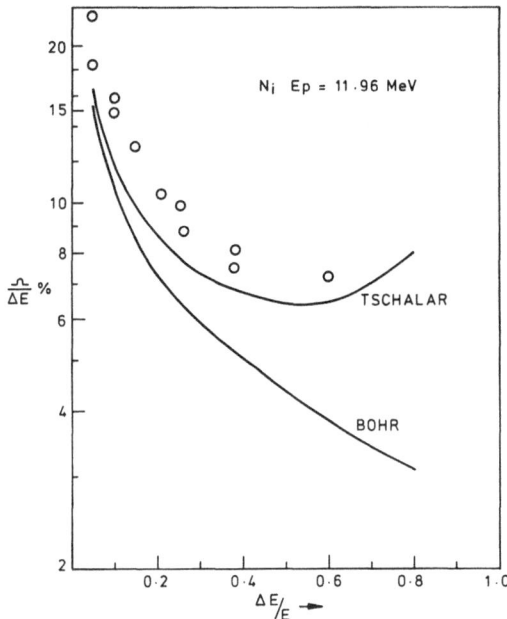

Fig. 3   Energy loss straggling of protons in Ni absorbers plotted
         as a function of the fractional energy loss. The
         experimental points are compared with the theoretical
         predictions of Tschalar and of Bohr

unrelated to straggling in the absorber.

In Tschalar's calculations the standard deviation of the
energy distributions are obtained. However, since the measured
energy distributions have low energy tails it is more appropriate
to extract the FWHM (Ω) than the standard deviation. Consequently
the calculated values of the FWHM from Tschalar's theory are used
for comparison with the experimental data. As far as the present
work is concerned Tschalar's theory gives such low values for the
third moment that no significant error is introduced by assuming
a Gaussian shape for the calculated distributions.

Values for the mean excitation potential, which are required
in Tschalar's theory, were obtained for the materials studied in
this work by interpolation from mean excitation potentials for
various elements given by Fano[9]. The excitation potentials

Element	Al	V	Ni	Mo	Ag	Ta	Au
Mean excitation potential (eV)	163	245	298	420	471	710	761

obtained by this method are given in the table.

The energy loss straggling of 11.96 MeV protons in Ni
absorbers is compared in fig. 3 with calculations due to Bohr and
also Tschalar. Following ref. 8 the energy loss straggling is
expressed as $\Omega/\Delta E$ and plotted against $\Delta E/E$. As expected the
Bohr calculations progressively fail to describe the experimental
situation as $\Delta E/E$ increases. The calculations of Tschalar do not
appear to suffer from this defect and the relatively small
discrepancies between these calculations and the experimental
data for Ni appear to be independent of $\Delta E/E$.

As can be seen in fig. 4 the experimental values for straggling
are nearly always higher than those predicted in the calculations
due to Tschalar. Here again the discrepancies do not appear to
have a $\Delta E/E$ dependence. If the ordinates of the calculated points
(averaged over measurements made at the three incident energies)

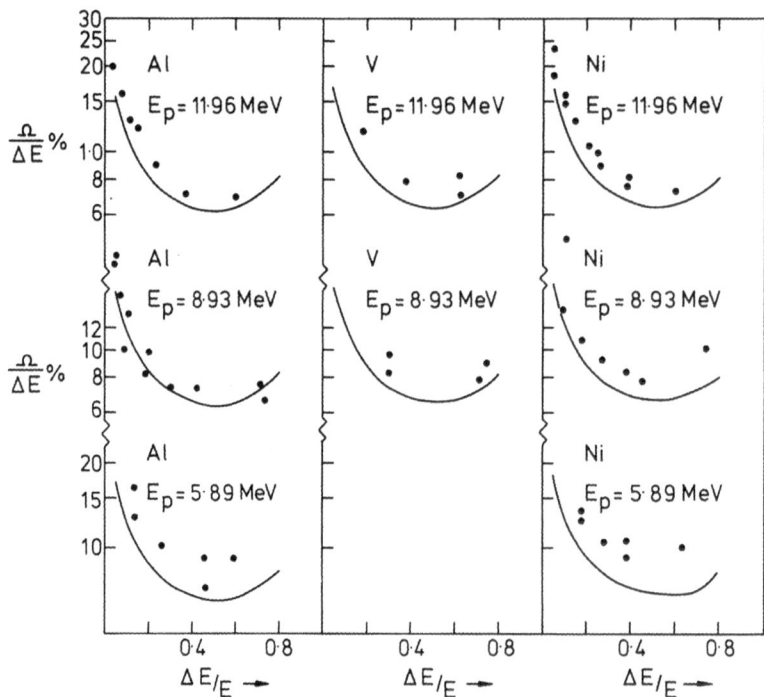

Fig. 4   Energy loss straggling of protons in various absorbers
         plotted as a function of the fractional energy loss.
         The experimental points are compared with the theoretical
         predictions of Tschalar.

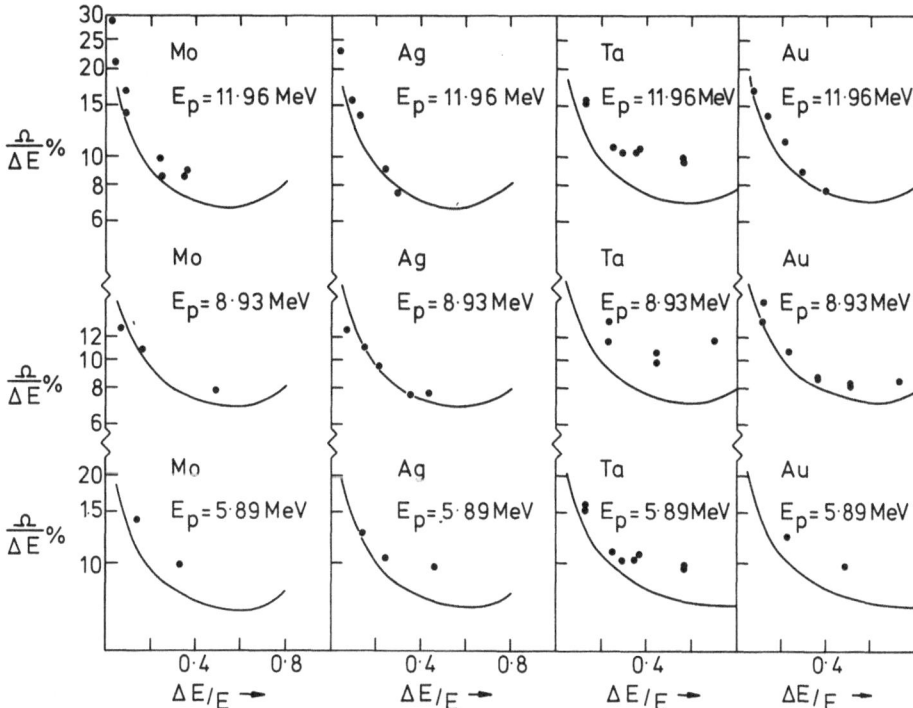

Fig. 4 (continued) Energy loss straggling of protons in various
absorbers plotted as a function of the fractional energy
loss. The experimental points are compared with the
theoretical predictions of Tschalar.

are increased by between 10% for Ag to 25% for Ni, good agreement
can be obtained with the experimental data. However, the errors
on individual measurements together with the observed scatter of
the data indicate that such a procedure is of little significance.
The average figure of 20% required to match the experimental data
is, on the other hand, likely to be more significant. It is
interesting to note that the Ta data, which has been set aside
owing to the skewness of the straggling distributions, is 30%
higher than that indicated by the calculations of Tschalar.

In the present work the objective has been to compare the
experimental data with Tschalar's calculations. No attempt has
been made to take into account the suggestion[10] that the single

collision cross section should be altered to take into account
atomic binding of the interacting electrons. However, Tschalar
and Maccabee[5] have measured straggling on Au with 49 MeV
protons and found that by increasing the calculated values by
9% to take atomic binding into account, good agreement was
obtained with experiment. Finally, in the energy range covered
by the present work Kolata and Amos[4] obtained good agreement
with Tschalar's calculations for 5 to 7 MeV protons on Si without
making any such correction.

References

1.   N. Bohr, Phil. Mag. $\underline{30}$ (1915) 581.

2.   C. Tschalar, Nucl. Inst. and Methods $\underline{61}$ (1968) 141 and
      $\underline{64}$ (1968) 237 and Rutherford Laboratory Reports RHEL/R146,
      1967 and RHEL/R164, 1968.

3.   N. Nam and W. Schafer, Nucl. Inst. and Methods $\underline{100}$ (1972)
      217.

4.   J. J. Kolata and T. M. Amos, Phys. Rev. $\underline{176}$ (1961) 484.

5.   C. Tschalar and H. D. Maccabee, Phys. Rev. B, $\underline{1}$ (1970) 2863.

6.   J. A. Penkrot, B. L. Cohen, G. R. Rao and F. H. Fulmer,
      Nucl. Inst. and Methods $\underline{96}$ (1971) 505.

7.   J. A. Cookson, B. H. Armitage and A. T. G. Ferguson, Non-
      Destructive Testing, August 1972.

8.   V. V. Avdeichikov, E. A. Ganza and O. V. Lozhkin, Nucl.
      Inst. and Methods $\underline{118}$ (1974) 247.

9.   V. Fano, Ann. Rev. Nucl. Sci. $\underline{13}$ (1963) 1.

10.   M. S. Livingston and H. A. Bethe, Rev. Med. Phys. $\underline{9}$ (1937)
      261.

DISCUSSION

Q: (M. Hufschmidt) For what purpose did you mount a foil between
the two collimators (first slide).

A: (B.H. Armitage) To reduce the intensity of the proton beam.

Q: (J. Baglin) In a very recent experiment, we have observed
straggling of MeV protons in very thin Si films. The results are

preliminary. We seem to see distributions which agree nicely with
Vavilov theory if experimental data are given a 20-30 % compressed
ΔE scale, i.e. the same factor for thin Si as you find for high Z,
thick-target cases.

A: (B.H. Armitage) I think more attention needs to be taken in reducing the errors associated with our straggling distributions before we can really see if there is an analogy here, where our measurements are made with very thick absorbers.

Q: (L.C. Feldman) Does the Tschaler theory predict a Gaussian energy loss distribution for large ΔE/E?

A: (B.H. Armitage) This is practically true in that the calculated
values for the third moment of the distribution are very low.

Q: (W. Möller) Did you compare your results with the more simple
formula of Symon?

A: (B.H. Armitage) My understanding is that Symon's calculations
are not presented in a very convenient form for comparison with
experiment.

ENERGY DEPENDENCE OF PROTON STRAGGLING IN CARBON

D.Olmos, F.Aldape, J.Calvillo*, A.Chi, S.Romero

Instituto Nacional de Energía Nuclear, México

and J.Rickards**

Instituto de Física, University of Mexico

ABSTRACT

The straggling of protons in a thick carbon target has been measured at three different energies. The method used consists of observing backscattering spectra of thick targets. When the bombarding energy is slightly above the well known resonances in $^{12}C(p,p)^{12}C$, they appear clearly in the continuum. As the energy is raised, the resonance occurs deeper within the target and therefore appears wider due to straggling. This procedure was carried out in the vicinity of resonances at 0.46, 1.73 and 4.79 MeV. At the highest energy the value of straggling is close to the prediction from Bohr's theory. At the lower energies it becomes higher until it reaches about five times the Bohr prediction, which is energy independent. Different targets follow the same trend with energy, but give slightly different values of straggling; this effect could be due to channeling, surface roughness or porosity. At the higher energy, straggling is linear with the square root of thickness, as expected from the Bohr theory. At the two lower energies there is a marked deviation from linearity.

*   Also at Centro de Estudios Nucleares, University of Mexico.
** Instituto Nacional de Energía Nuclear Consultant.

INTRODUCTION

Although energy straggling of charged particles in general is considered a nuisance because it spoils resolution, it is important to carry out measurements of this effect because of the insight that might be gained concerning the mechanisms of charged particle interaction with matter. This has recently become important because of surface analysis using energetic beams of charged particles. There is a recent review on the status of straggling by Chu and Mayer (1).

We have undertaken measurement of energy straggling of protons in carbon at three different energies using backscattering from thick targets in the vicinity of $^{12}C(p,p)^{12}C$ resonances that serve as markers.

METHOD

The method employed has been described elsewhere (2), so only the principal points will be described here.

From zero to 5 MeV bombarding energy there are three well known resonances in the elastic scattering of protons from carbon, at 0.46, 1.73 and 4.79 MeV. They are shown in figure 1. If a thick carbon target is bombarded at energies slightly above the resonance energy, the backscattered spectrum displays the resonance shape within the continuum, distorted by straggling of the projectile energy as it enters and leaves the thick target. Since the resonance always occurs at the same energy, as the projectile energy is raised, the resonance goes deeper into the target and its shape in the back-scattered spectrum becomes broader. This can be seen clearly in figure 2 for the 4.79 MeV resonance.

This procedure was carried out at some ten energies above each of the resonances mentioned, using a H.V.E.C. AN-700 Van de Graaff for the lower energy (2) and a analysis of the broadening of the resonances was made to extract the value of straggling.

Note that in this type of experiment, the bombarding energy E_1 is well defined by the magnetic analyser and accelerator calibration. E_2 is fixed by the resonance energy, for instance the maximum of the energy distribution. E_3 is also constant for each resonance, and given by kE_2 where k is the kinematic factor for elastic

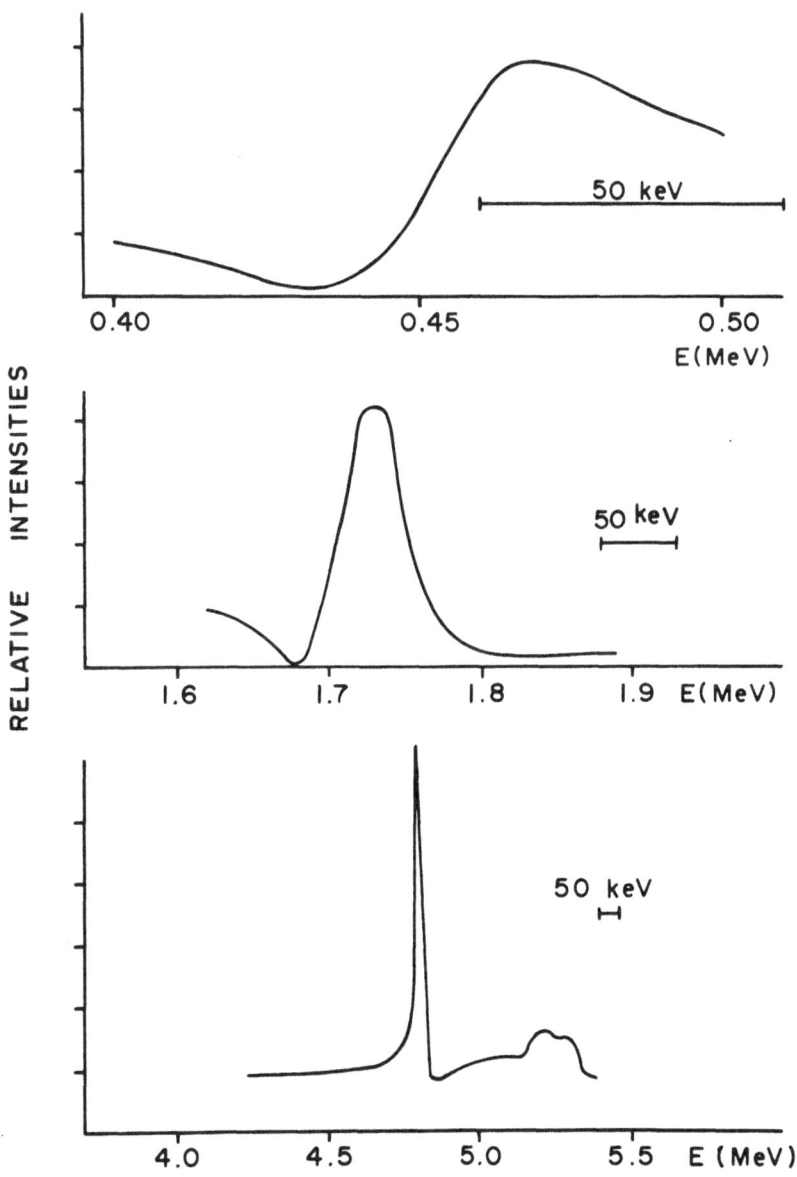

Fig. 1.- The well known resonances in $^{12}C(p,p)^{12}C$ at 0.46, 1.73 and 4.79.

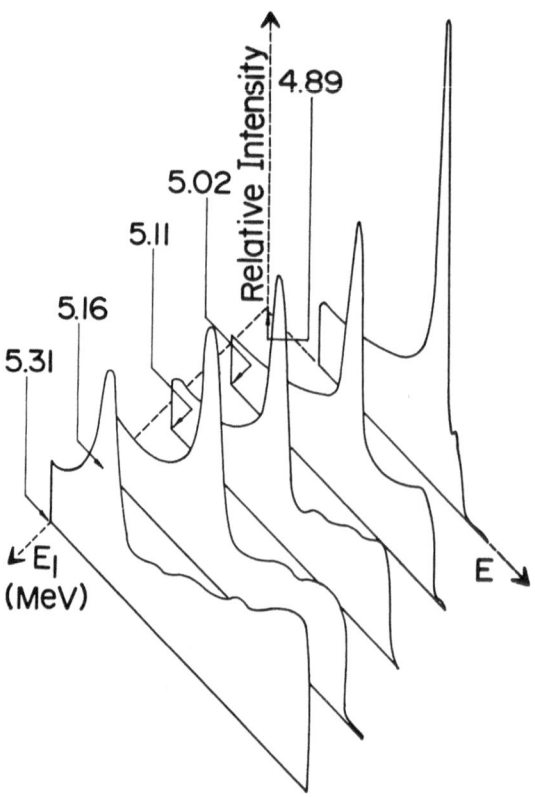

Fig. 2.- Spectra of protons backscattered from a thick carbon target at several bombarding energies, with the resonance at 4.79 MeV dominating the shape of the spectrum.

scattering. The proton energy reaching the detector, E_4, is also well defined by the detector calibration, so all energies involved are well known. Since θ_1, was chosen equal to θ_2 (see figure 3), the distance t_1 from the surface to the point of scattering is equal to t_2, the distance from the point of scattering to the surface on the way out. The distance travelled by the proton during straggling is given automatically by E_1, E_2, E_3 and E_4. If a well polished surface is used, this eliminates uncertainties in t when the more conventional method of passing through thin films is used.

The conversion of energy differences to distance

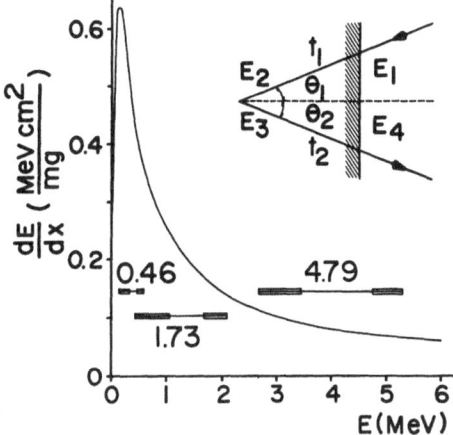

Fig. 3.- Stopping power for protons in carbon, showing the three energy regions studied.

was done using published stopping power tables (3). The curve for proton stopping in carbon is shown in figure 3, and on this are marked the three energy regions studied corresponding to the three resonances.

The broadening of the energy spectrum is shown schematically in figure 4. The standard deviation of the energy distribution is increased with each step of the physical sequence. It is ΔE_0 before the proton enters the material. This is added quadratically to the straggling in the incoming path Ω_{in} to give ΔE_2 just before scattering. Since ΔE_0 is small, it is disregarded, so $\Delta E_2 = \Omega_{in}$. This is multiplied by k and combined with the natural width of the resonance ΔE_{res} to give ΔE_3 right after scattering. Finally Ω_{out}, the straggling on the way out, and ΔE_1, the detector resolution, are included to give ΔE_4, the standard deviation of the distribution actually observed.

It can be shown that if one assumes all these effects to be independent, one disregards ΔE_0, and one assumes the straggling to be independent of energy as a first approximation ($\Omega_{in} = \Omega_{out} = \Omega$), it can be given by

$$\Omega^2 = \frac{(\Delta E_4)^2 - (\Delta E_{res})^2 - (\Delta E_1)^2}{k^2 + 1}$$

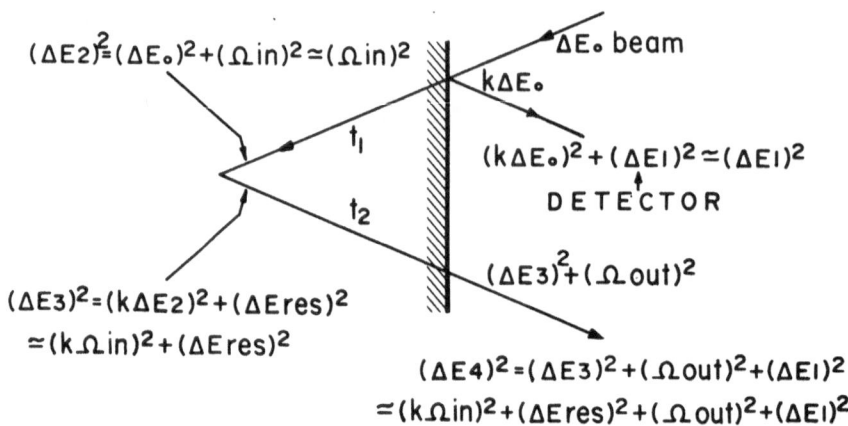

Fig. 4.- Broadening of the energy distribution in each step of the physical process.

RESULTS

Since the energy regions studied are on the downward slope of the stopping power curve it may be assumed that Bohr's theory dominates the process, in which the standard deviation Ω_B of straggling is given by

$$\Omega_B^2 = 4\pi Z_1^2 Z_2 e^4 \, Nt$$

where Z_1 is the atomic number of the projectile, Z_2 the atomic number of the material, e the electronic charge, N the atomic density of the material, and t the thickness traversed (4). It is energy independent.

Therefore plots were made of Ω measured vs $\sqrt{t}$, and straight lines are expected ($t = t_1 = t_2$). Figure 5 shows such a plot for the lowest energy, the different data points corresponding to three targets of different density, 1.49, 1.66 and 1.75 gr/cm^3. A typical error bar is shown, as well as Bohr predictions for the density range used. The straight lines are least square fits.

A target effect, that could be attributed to surface roughness, porosity or channeling (2) is obvious, and

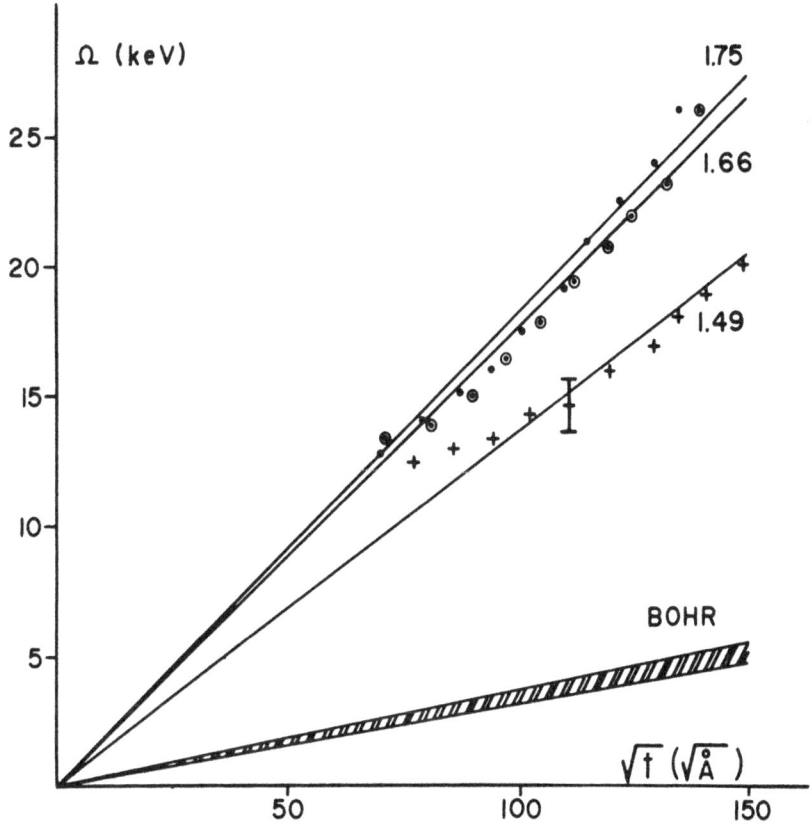

Fig. 5.- Straggling vs $\sqrt{t}$ at the 0.46 MeV resonance, for targets of densities 1.49, 1.66 and 1.75. Note target or density effect, upward curvature, and separation from Bohr prediction.

this effect appeared at the three energies studied. There is always an upward curvature that is well marked at the two lower energies and less so at the higher one. The deviation from Bohr values is also strong for the lower energies and less for the higher one.

Figure 6 is the same type of plot for one target (density 1.49, vitreous carbon), three energies. The closer correspondence to Bohr at 4.79 MeV is clear. The slope of these three fits is plotted against resonance energy in figure 7, to observe the energy dependence and

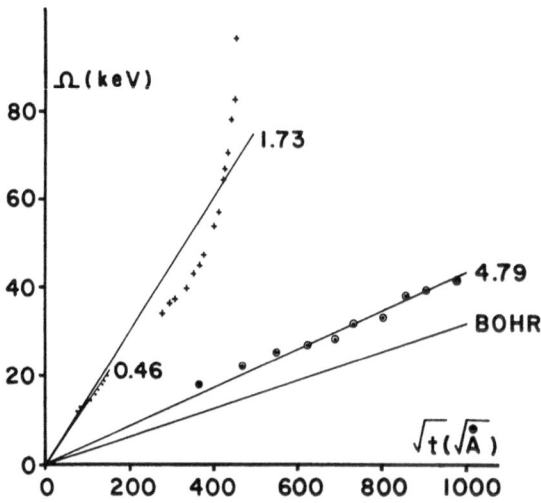

Fig. 6.- Plot of Ω vs $\sqrt{t}$ for a single target (vitreous carbon, density 1.49) at the three energies studied, compared to Bohr values.

compare with Bohr, for the three targets. The vitreous carbon target systematically gives lower values, and at the high energy it is reasonably close to Bohr, which is an indication that this is the best target. Also it is the best polished, least porous and most amorphous target (5). The agreement of this last value with the Bohr prediction confirms the method used.

Although this method implies greater accuracy, it is not enough to eliminate the strong effect target properties have on measuring straggling. The trend of the relative values for the targets is conserved at all energies.

The authors wish to acknowledge the cooperation of the accelerator staffs, in particular W. Domínguez, R. Alba and K. López.

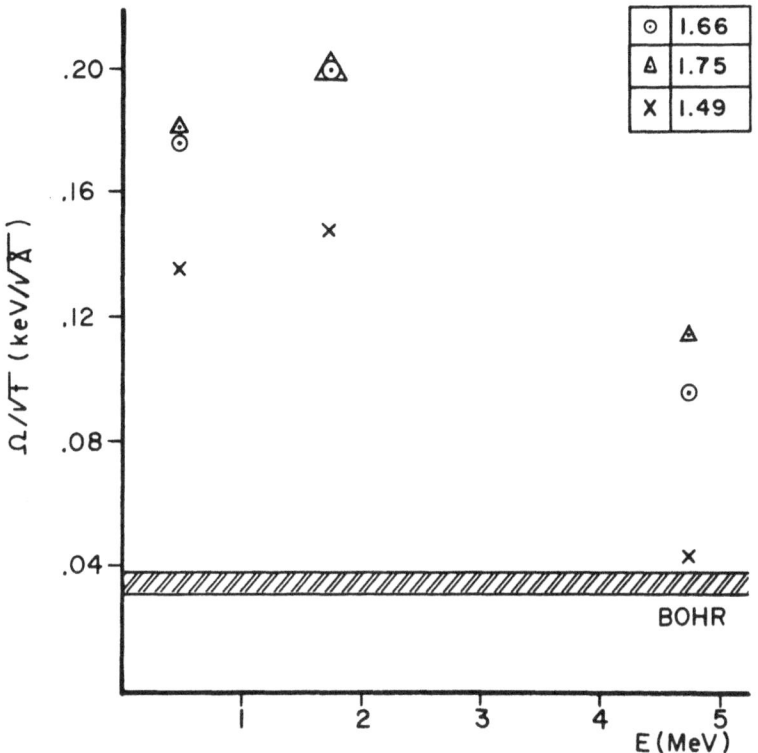

Fig. 7.- All the values measured, shown in a plot of straggling per square root of distance, vs energy.

REFERENCES

1.- W.K. Chu and J.W. Mayer, Catania Working Data, 1974 (unpublished).

2.- J. Rickards, Nucl. Instr. and Meth. (to be published).

3.- L.C. Northcliffe and R.F. Schilling, Nucl. Data Tables A7 (1970) 233.

4.- N. Bohr, Mat. Fys. Medd. Dan. Vid. Selsk. 18, No. 8 (1948).

5.- F.C. Cowlard and J.C. Lewis, J. Matls. Sci. 2 (1967) 507.

DISCUSSION

Q: (W. Pabst) Why don't you use a (p,γ) reaction (instead of a (p,p) reaction) which exist for many materials - straggling would only occur in the ingoing path and so is better defined.

A: (J. Rickards) If I understand the question, you propose carrying out an excitation curve over an appropriate resonance. In this case you would be back to the problem of carefully measuring your foil thickness.

Q: (B.H. Armitage) Did you make measurements with your detector placed at more than one angle to the incident beam?

A: (J. Rickards) We did, but have not analyzed it yet.

ENERGY STRAGGLING OF ^{4}He IONS IN Al AND Cu IN THE BACKSCATTERING

GEOMETRY

M. Luomajärvi, A. Fontell and M. Bister

Department of Physics, University of Helsinki

Helsinki, Finland

The contribution of energy straggling of energetic ^{4}He ions to
the depth analysis by backscattering has been studied. The backscattering spectra were recorded from samples where a thin gold film
was covered with aluminum or copper films of different thicknesses.
The effect of energy straggling was determined from the broadening
of the Au peak. The primary energies of He ions were varied between
0.5 and 2.0 MeV. To determine the thicknesses of the films used in
this work the absolute stopping power curves of Al and Cu for He
ions were also measured.

INTRODUCTION

The use of energetic ion beams for micro-analysis has increased
the interest in stopping power (Refs.1-5) and energy straggling
(Refs.6,7) measurements. E.g. in transforming an energy spectrum
of backscattered particles to a depth distribution of the material
investigated, the energy dependence of stopping power leads to a
nonlinear transformation. The energy straggling limits the mass and
depth resolution and has an effect upon the form of the spectrum obtained from a certain material distribution. The most widely used
bombarding particles in the backscattering analysis are He ions in
the MeV region.

In this work the energy straggling of ^{4}He ions was investigated
in Al and Cu films at primary energies of 500, 1000, 1500 and 2000
keV. Very thick films were also used with the object of observing
the "energy bunching" effect (Ref.8), which means a decrease in energy straggling with growing thickness. A convolution calculation was
performed for a heavy impurity (Au) with a Gaussian depth distribution in Al, in order to get an idea of the effect of straggling on

the energy spectrum observed.

The film thicknesses used in this work were determined with the aid of the backscattering spectra. In these calculations the stopping power values of Al and Cu are needed. As these values for He ions found in the literature have in some cases discrepancies of about 10% (Refs.5,9,10) a stopping power measurement was performed in the energy region of 0.5-2.0 MeV for Al and 0.3-2.0 MeV for Cu.

EXPERIMENTAL PROCEDURE

Bombarding and Detecting Conditions

The backscattering measurements were carried out using the van de Graaff accelerator of our laboratory. The ^{4}He$^+$ ions were deflected through 90° in the energy analysing magnet, which was calibrated through the ^{20}Ne$(\alpha, \gamma)^{24}$Mg resonance reaction at E_α = (1929 ± 5) keV (Ref.11). The beam was collimated to have a diameter of about 0.5 mm. After scattering from the target, helium ions were detected with an annular surface barrier detector mounted coaxially with the beam. The beam intensity was varied between 0.2-6 nA, and the doses needed for any one spectrum varied between 0.2-10 μC. The target-detector distance was between 7-11 cm. The detector was cooled to -30°C, and had an energy resolution of 13 keV (FWHM). The energy calibration of the detecting system was made daily by measuring the backscattering spectra from thick targets of various elements. The channel width of the multichannel analyzer was selected to be about 2.6 keV, while the linearity and stability was observed to be better than one channel. A schematic illustration and further details of the experimental arrangement can be found elsewhere (Ref.4).

Straggling Measurements

The samples used in the straggling measurements were prepared on electrolytically polished, 1 mm thick aluminum plates, which were often also anodically oxidized. On these plates first a thin film of Au (6-20 μg/cm^2) then different thicknesses of Al or Cu was evaporated. Every sample contained the uncovered Au film and three different thicknesses of Al or Cu. During the Al and Cu evaporations, the watercooled sample plates were situated 16 cm above an electron beam evaporator, while a vacuum of 10^{-5} -10^{-6} mmHg was maintained with a mercury diffusion pump. No impurities in the Al and Cu films were observed, when the forms and intensities of backscattering spectra of these foils were compared with the spectra from pure Al and Cu bulk materials. In the intensity measurements a beam chopper and its detector were essential.

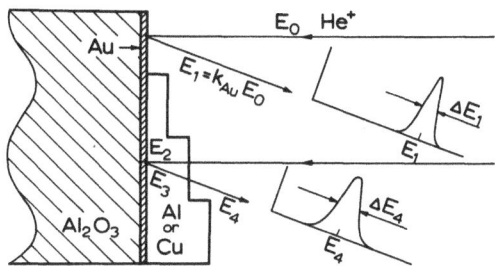

Fig.1. Schematic illustration of the target configuration in a straggling measurement. The FWHM ΔE_1 and ΔE_4 and the energy shift E_1-E_4 are observed.

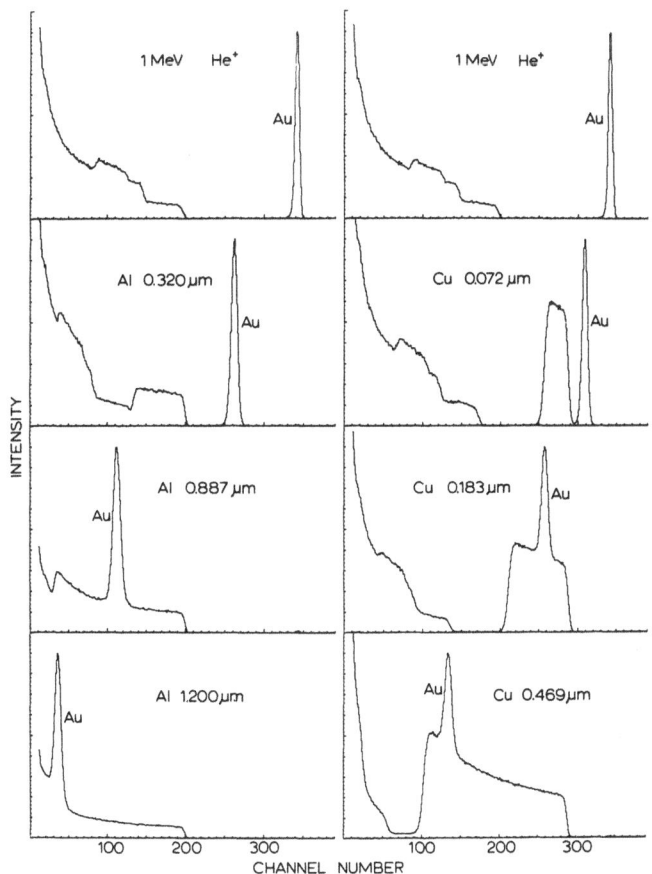

Fig.2. Backscattering spectra measured with 1 MeV ^{4}He$^+$ ions from un-covered and covered Au films on oxidized aluminum. The thickness of the covering is given in the picture and the areal density of the Au film was 11 μg/cm^2 in most spectra. The intensity scale is linear.

The experimental configuration in straggling measurements is illustrated in Fig.1. The mean energy of the incident beam is E_0 and the mean energy after scattering from an uncovered Au film is $E_1 = k_{Au}E_0$ ($\theta = 177°$, $k_{Au} = 0.922$). The mean energies just before and after the scattering from a covered Au film are denoted by E_2 and E_3 and the observed mean energy of the Au peak by E_4. The thicknesses of the films were computed from the observed energy differences E_1-E_4. The areal densities varied between 47-730 $\mu g/cm^2$ in the case of Al and between 65-1315 $\mu g/cm^2$ in the case of Cu.

The spectra of two series of measurements are displayed in Fig. 2. The intensity was measured to about 10^4 counts/channel in the maximum of the Au peak. The effect of energy straggling was obtained by comparing the full width at half-maximum (FWHM) of the Au peak measured through the film (ΔE_4) to that measured by the same uncovered Au film (ΔE_1). In some cases the Au film was so thick that a small correction had to be made due to the change of stopping power of Au when the energy changed from E_0 to E_2. The energy resolution of the detector decreased at low energies to about 11.5 keV, which also caused a small correction to the straggling values.

Stopping Power Measurements

The Al and Cu samples used in the stopping power measurements were prepared on 0.4 mm thick tantalum plates by vacuum evaporation. The material purities were 99.999% according to the manufacturer. The film was collected on a rectangular area of about 5.3 cm^2 at a distance of 28 cm from the evaporation source. Both resistant and electron beam heated sources were used. These evaporations were carried out in an ion pumped system under a vacuum of 10^{-6}- 10^{-8} mmHg (Ultek MX-14). The areal densities, which were determined by weighing on a Mettler M5 microbalance, varied between 46-92 $\mu g/cm^2$ for the four Al films and between 48-129 $\mu g/cm^2$ for the five Cu films used.

The experimental configuration in the stopping power measurements is shown in Fig.3. One measurement consisted of three back-

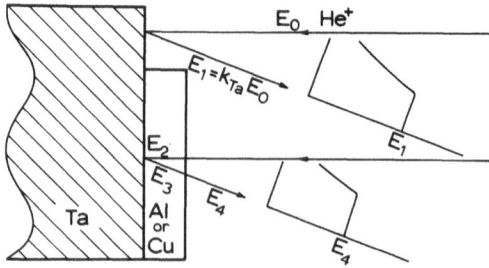

Fig.3. Schematic illustration of the target configuration in a stopping power measurement. The energy shift E_1-E_4 is observed.

scattering spectra. The Ta plate was bombarded through the evaporated film at two different places and also the uncovered area was
bombarded. It was noticed that the areal densities of the films were
constant to within 1%. An intensity of about 4000 counts/channel was
recorded at the tantalum edge, and the stopping power was calculated
from the difference E_1-E_4 determined from the half heights of the
Ta edges.

RESULTS AND DISCUSSION

Straggling Measurements

A series of straggling measurements, like those seen in Fig.2,
give Au peaks with different widths. The peaks measured with the Al
target are displayed in Fig.4, after the substraction of a pure Al
spectrum in the 0.887 and 1.20 µm cases. The peaks do not differ
significantly from the solid line curves, which are Gaussians with
the same area and the same FWHM. The fit is as good also in the case
of Cu films. The system resolution (13 keV) and the effect of the
finite Au film thickness (11 µg/cm^2) add up to $\Delta E_1 = 16.5$ keV in the
measurement presented. Because the shapes are Gaussian, the contribution of energy straggling is obtained from the formula

$$\Delta E_b^2 = \Delta E_4^2 - \Delta E_1^2. \tag{1}$$

In terms of standard deviation this can be written,

$$\Omega_b^2 = \frac{1}{8 \ln 2}(\Delta E_4^2 - \Delta E_1^2). \tag{2}$$

The experimentally obtained value of Ω_b gives the energy straggling

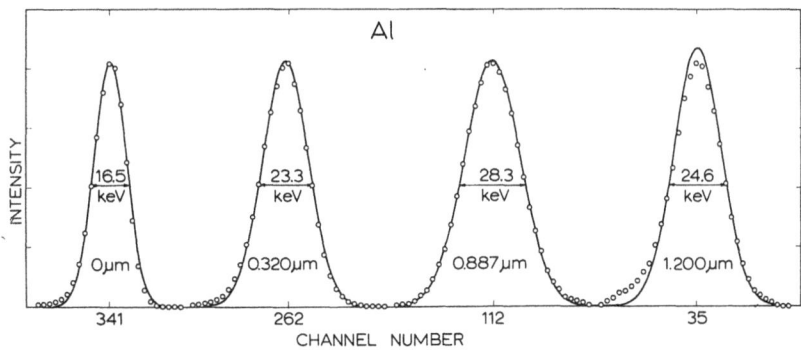

Fig.4. The Au peaks seen on the left hand side of Fig.2 are compared
with Gaussians having the same area and the same FWHM. The "energy
bunching" effect is observed in the thickest aluminum film.

in the backscattering geometry and is expressed by the straggling
generated in the incoming and outgoing paths (Ref.7)

$$\Omega_b^2 = k^2\Omega_{in}^2 + \Omega_{out}^2,\tag{3}$$

where k is the elastic-scattering factor. Applying Bohr's theory
(Refs. 7,12) to the backscattering geometry with perpendicular incidence yields,

$$\Omega_{bB}^2 = \frac{q^4}{4\pi\varepsilon_o^2}Z_1^2 NZ_2(k^2 + \frac{1}{|\cos\theta|})\Delta t,\tag{4}$$

where q is the electronic charge, Z_1 and Z_2 are the atomic numbers
of the incident ion and the target material, respectively, N is the
number of stopping atoms per unit volume, θ is the scattering angle
and Δt is the film thickness.

The results for energy straggling in Al and in Cu are given in
Fig.5. The values of Ω_b are plotted as a function of the square root
of film thickness. A density of 2.70 g/cm^3 for Al and 8.92 g/cm^3 for
Cu were used. The two straight lines are obtained using formula (4).
It is noticed that, at least in the case of Al, the line fits the
experimental points rather nicely up to thicknesses where the "energy bunching" effect (Ref.8) starts to influence the results. This
effect is clearly seen in the case of Al. The correspondingly thick
Cu films cannot be studied by this method, because of the high backscattering intensity from Cu. The Bohr's theory seems to overestimate the value of straggling for copper, even in the thin films.

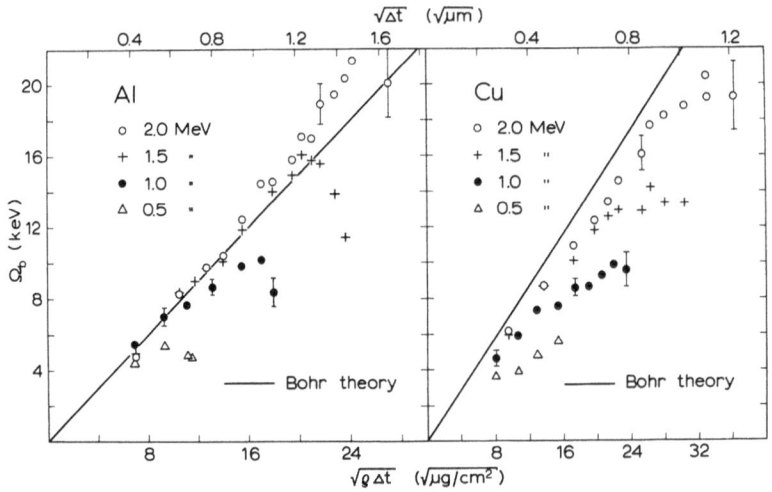

Fig.5. Energy straggling of ^{4}He ions in Al and Cu films at different
incident energies as a function of the square root of the film thickness. Solid lines show the predictions of Bohr's theory.

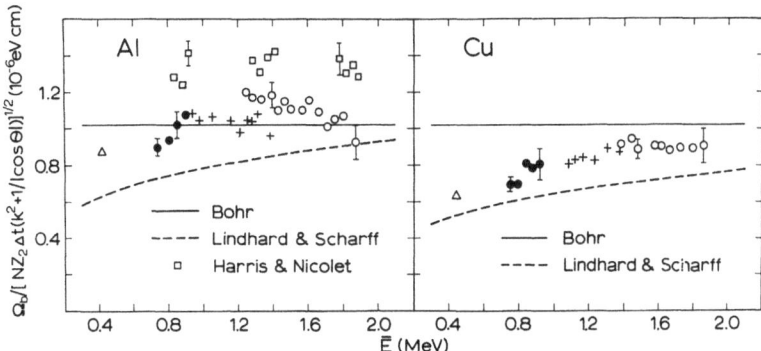

Fig.6. The reduced straggling as a function of the mean energy of
He ions in the Al and Cu films. The following symbols are used for
the results obtained with different primary energies: o 2.0 MeV,
+ 1.5 MeV, ● 1.0 MeV and Δ 0.5 MeV. Comparison with two theoretical
predictions and with another set of experimental results is given.

To get a better picture of the energy dependence of the straggling, the experimental values have been normalized by dividing by
the factor $[NZ_2\Delta t(k_{Au}^2 + 1/|\cos\theta|)]^{1/2}$ and plotted as a function of
mean energy in Fig.6. Only those measurements with no noticeable
"energy bunching" effect are included. The mean energy was calculated with the formula (Ref.13)

$$\bar{E} = (E_1|\cos\theta| + E_4)/(1 + k_{Au}|\cos\theta|). \tag{5}$$

In Fig.6 we also show the theoretical predictions of Bohr's theory
and of Lindhard-Scharff's theory (Ref.14). The latter theory gives
an energy dependent straggling at these rather low projectile velocities and joins smoothly to Bohr's theory at higher energies. In
calculating the Lindhard-Scharff curve the approximation presented
by Eq.(11) in Ref.14 was used. It is noted that our results for Al
lie closer to the theoretical Bohr value than those of Harris and
Nicolet (Ref.7). Any conclusions about the energy dependence can
hardly be drawn in the case of Al. On the other hand, the results
for Cu films show an energy dependence as predicted by Lindhard and
Scharff's theory, although the absolute magnitude is somewhat higher.

A systematic error in the straggling results could be caused by
the diffusion of Au into the sample plate or into the evaporated
layer. The temperature, where diffusion of Au atoms was noticed in
the backscattering spectrum, was determined by bombarding heated
samples. Migration was seen to start between 100-150°C both in Al
and Cu, but it effected the FWHM values very slowly. As the temperature during the measurements was room temperature and the sample
plates were watercooled during the vacuum evaporations, we believe
that diffusion has no effect upon our results.

The accuracy of the straggling measurements depends on the relative thickness of the film. In the case of very thin or very thick films accuracy is poor. It was estimated that the error (standard deviation) in the width of the Au peaks varies between 1.5-2.5%. Unaccuracy in the film thickness was about 3%. These errors lead to an error of ±5% for the medium film thicknesses and of ±10% in the extreme cases.

Stopping Power Measurements

In the backscattering method used in the present work, the energy difference E_1-E_4 was experimentally determined for films of known areal densities. The values of stopping power were calculated from the equation

$$S(\bar{E}_{in}) = \frac{E_1 - E_4}{\rho \Delta t [k_{Ta} + \frac{S(\bar{E}_{out})}{S(\bar{E}_{in})|\cos\theta|}]} , \qquad (6)$$

using first the previously published stopping power values and then an iterative procedure. In the formula the values of stopping power S are taken at the average energies $(E_0+E_2)/2$ and $(E_3+E_4)/2$ in the incoming and in the outgoing paths, respectively, $\rho\Delta t$ is the areal density, $k_{Ta}= 0.915$ and $\theta = 177-178°$.

The stopping power results are presented on the left side of Fig.7. The solid line curves are the fits of the Brice's formula (Ref.15) to our experimental points. The least squares method gave

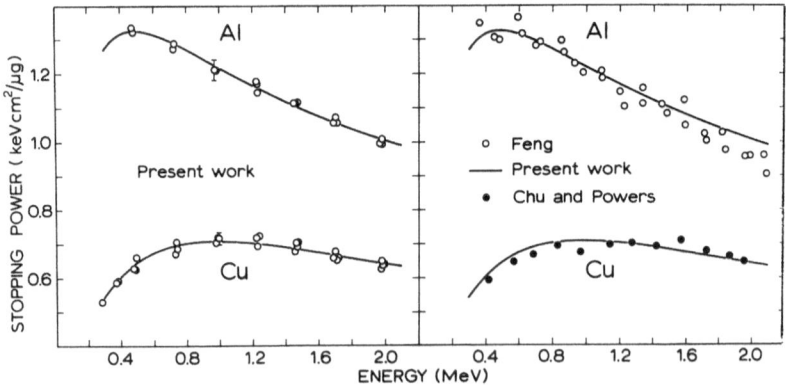

Fig.7. Stopping power of Al and Cu for ^{4}He ions. Our measurements are presented on the left side of the figure. The solid lines are the fits of Brice's formula to our experimental points. On the right side these lines are compared to the results in Refs. 5 and 9.

Table 1. Stopping power and cross sections of Al and Cu for ^{4}He ions.

E (keV)	Al		Cu	
	S (keVcm^2/μg)	ϵ (10^{-15}eVcm^2)	S (keVcm^2/μg)	ϵ (10^{-15}eVcm^2)
300			0.545	57.5
400			0.606	63.9
500	1.327	59.4	0.648	68.3
600	1.316	58.9	0.675	71.2
700	1.296	58.0	0.693	73.1
800	1.272	56.9	0.703	74.2
900	1.246	55.8	0.708	74.7
1000	1.220	54.6	0.709	74.8
1200	1.169	52.3	0.703	74.1
1400	1.122	50.2	0.691	72.9
1600	1.079	48.3	0.676	71.3
1800	1.041	46.6	0.659	69.6
2000	1.007	45.1	0.643	67.8

the following values for the three adjustable parameters in Brice's formula Z = 1.06 (2.29), a = 0.567 (0.345) and n = 2.80 (2.95) for Al (Cu). On the right hand side of the figure these curves are compared with the stopping powers obtained by Feng (Ref.5) and Chu and Powers (Ref.9), who in their publications gave comparisons with earlier experimental results. It can be noticed that the agreement is good, favouring Feng's higher value results for Al at low energies as against those of Chu and Powers (Ref.9). Table 1 lists the values of stopping power and cross section given by the to our experimental points fitted Brice's formula.

The accuracy of the stopping power values is determined mainly by the accuracy in the measurements of the areal densities and of the energy shifts E_1-E_4. The error (standard deviation) in the areal density was estimated to vary between 0.8-1% and that in the energy shift between 1.5-2.5% depending on the film thickness. Thus, including other smaller error sources the accuracy in the stopping power results was estimated to vary from ±2% at the highest energy to ±3% at the lowest energy for the both elements measured.

On the Depth Distribution Analysis

The depth distribution of impurity atoms is obtained in the first approximation straight from the backscattering spectrum. A more accurate analysis, where straggling, energy resolution and the energy dependences of scattering cross section and stopping power are taken into account, demands a deconvolution calculation, which may be a

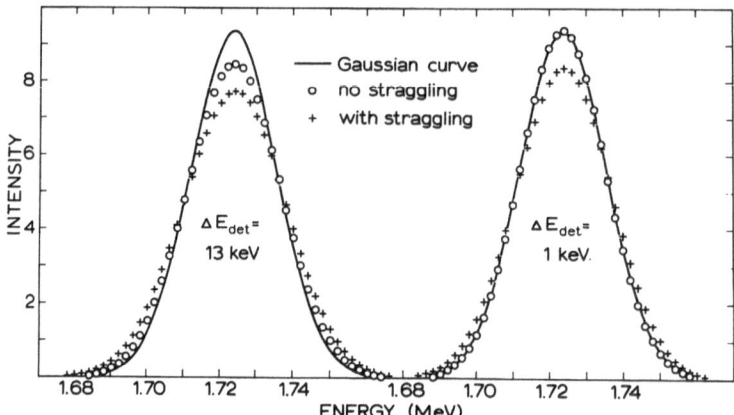

Fig.8. The calculated shapes of the backscattering spectra from a
Au distribution in Al. The Au profile assumed, was a Gaussian distribution with a mean depth of 0.227 μm and a standard deviation of
0.021 μm. The energy spectra are calculated both with and without
energy straggling for two different (13 keV and 1 keV) resolutions
in energy detection. The results are compared to a Gaussian with the
same area and FWHM as the curve obtained with no straggling and
1 keV resolution.

tedious task. To study the importance of these effects the shapes of
2 MeV He backscattering spectra were calculated for an Au impurity
in Al with different assumptions. The impurity had a Gaussian depth
distribution with a mean depth of 0.227 μm and a standard deviation
of 0.021 μm, corresponding to the theoretical projected range distribution of 1 MeV Au ions in Al (Ref.16). The results of calculations
both with and without straggling are shown in Fig.8 for two different energy resolutions (13 and 1 keV) of the detection system. The
1 keV calculation without straggling gives accurately a Gaussian
shape, which indicates that the energy dependence of scattering cross
section and of stopping power have an insignificant effect on the
shape of the spectrum. This example shows that for correspondingly
thick, Gaussian shaped distributions the extraction of the depth
profile from the backscattering spectrum is rather simple. The straggling and energy resolution effects can be taken into account simply
by a quadratical subtraction. When the distributions extend over
large depths the energy dependences of stopping power and scattering
cross section play a more important role.

REFERENCES

1. J.F. Ziegler and W.K. Chu, Atomic Data and Nuclear Data Tables
   13, 463 (1974).

2. W.K. Lin, S. Matteson and D. Powers, Phys.Rev.B 10, 3746 (1974)
3. J.A. Borders, Rad.Effects 21, 165 (1974).
4. E. Leminen and A. Fontell, Rad.Effects 22, 39 (1974).
5. J.S.-Y. Feng, J.Appl.Phys. 46, 444 (1975).
6. J.M. Harris, W.K. Chu and M-A. Nicolet, Thin Solid Films 19, 259 (1973).
7. J.M. Harris and M-A. Nicolet, Phys.Rev.B 11, 1013 (1975).
8. D.A. Sykes and S.J. Harris, Nucl.Instr.Methods 94, 39 (1971).
9. W.K. Chu and D. Powers, Phys.Rev. 187, 478 (1969).
10. D.I. Porat and K. Ramavataram, Proc.Phys.Soc.Lond.78,1135 (1961).
11. P.J.M. Smulders, Physica 31, 973 (1965).
12. N. Bohr, Kgl.Dan.Vidensk.Selsk. Mat.-Fys. Medd. 18, No 8 (1948).
13. R.D. Moorhead, J.Appl.Phys. 36, 391 (1965).
14. J. Lindhard and M. Scharff, Kgl.Dan.Vidensk.Selsk. Mat.-Fys. Medd. 27, No 15 (1953).
15. D.K. Brice, Phys.Rev.A 6, 1791 (1972).
16. W.S. Johnson and J.F. Gibbons, Projected Range Statistics in Semiconductors (1969).

DISCUSSION

Q: (R. Behrisch) Why did you take the high Z material for your backing material, instead of a low Z material used generally ?

A: (A. Fontell) For a heavy material the stopping powers $S(\bar{E}_{in})$ and $S(\bar{E}_{out})$ are close to each other, which makes it easier to calculate $S(\bar{E}_{in})$ from the experimental data.

ENERGY SPREADING CALCULATIONS AND CONSEQUENCES

G. DECONNINCK (*) and Y. FOUILHE (**)

(*) L.A.R.N.,Facultés Universitaires,5000-NAMUR (Belgium)

(**) I.S.N., B.P.257, 38.044-GRENOBLE (France)

Abstract : The spatial resolution in depth profile analysis by nuclear reactions is discussed and practical formula for straggling
calculations are given. Tabulation of Vavilov distributions for
small energy loss and tables of resonance in elastic scattering
cross-sections are given.

1. Energy Spreading

The experimental precision in depth profile analysis using nuclear reactions and backscattering of charged particle is limited
by the energy spreading of the particles travelling in the sample
(fig.1). The total spreading results from incident energy spreading
and from straggling effects, the width of the distribution function
is dependent on the total particle path and on the nature of the
scattering event. These effects have been largely discussed in the
past [1,2] and the energy distribution functions of heavy particles
being slowed down in a homogeneous sample can be calculated. The
aim of this paper is to give practical formulae and tabulations, as
well as confidence limits for depth profile analysis by nuclear reactions. In this technique low energy particles are used (O-3 MeV)
and the non relativistic approximations are valid (β^2=o).

a) Small Proton Energy Loss-Narrow Resonances. For small path
length x (x $\leqslant$ 2 10^{-4} g cm^{-2})Vavilov distribution functions are a
good approximation for straggling [1]. Calculation of these functions
are tedious and time consuming, for these reasons, a complete tabulation is presented and adapted to proton energies E_0 normally used
in depth profile analysis of light elements by nuclear resonant

reactions (p,γ), $(p,\alpha\gamma)$,... The calculation have been made using the Richardson method of iteration, graphs of these functions are represented on fig.2a. The tabulated distributions $\phi(\lambda)$ (Table I) are given as function of the Landau parameter λ which is defined by :

$$\lambda = \frac{\Delta - \Delta_o}{\xi} - 0.423 - \ln \kappa \qquad (1)$$

where Δ is the energy loss of the proton on a pathlength x (g cm^{-2}) and Δ_o is the average energy loss given by : $\Delta_o = xS(E_o)$ where S is the proton stopping power in MeV (g cm^{-2})$^{-1}$, ξ and κ are given by :

$$\xi = \frac{72}{E_o^2} \frac{\bar{Z}}{\bar{A}} \text{ MeV} \quad (2) \quad \text{and} \quad \kappa = \frac{3.3 \ 10^{-4}}{E_o^2} \frac{\bar{Z}}{\bar{A}} x \qquad (3)$$

$\bar{Z}$ and $\bar{A}$ are the average atomic charge and mass number of the atoms present in the sample, for light elements $\frac{Z}{A} \simeq 0.5$. The energy loss distribution function around the mean value Δ_o is $F(W)$ where $W=\Delta-\Delta_o$, it is obtained from the tabulated functions $\phi(\lambda)$ by the transformation $\lambda = W/\xi - 0.423 - \ln \kappa$ after multiplication by the normalizing factor ξ^{-1}. Using Table I and the transformation formula it is possible to obtain the straggling distribution $F(W)$ for different x values and proton energies E_o (fig.2b). In practice, for very small energy losses the FWHM of the distribution is about 30% of the total energy loss Δ_o. An example of application of the Vavilov distributions is given by Dunning et al in ref.3 where depth profile analysis of Al and Na near surface by narrow nuclear resonance are described.

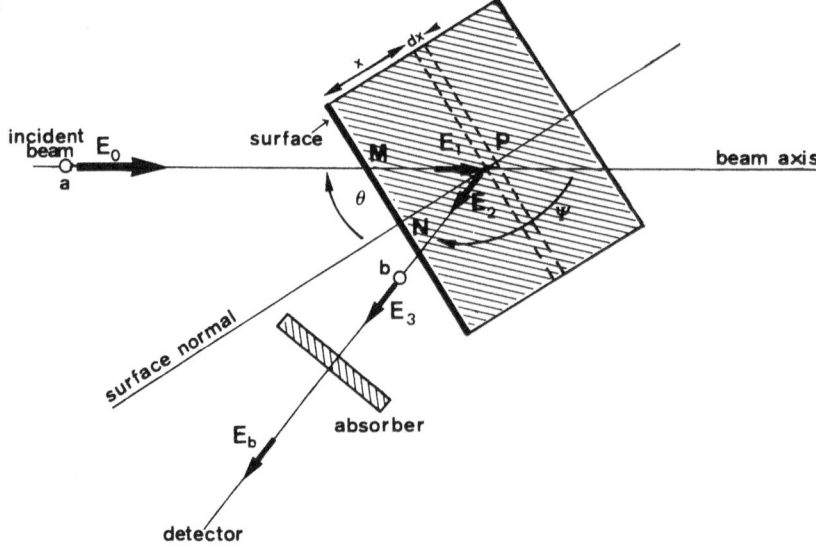

<u>Fig.1</u>. Thick sample excitation yield (geometry).

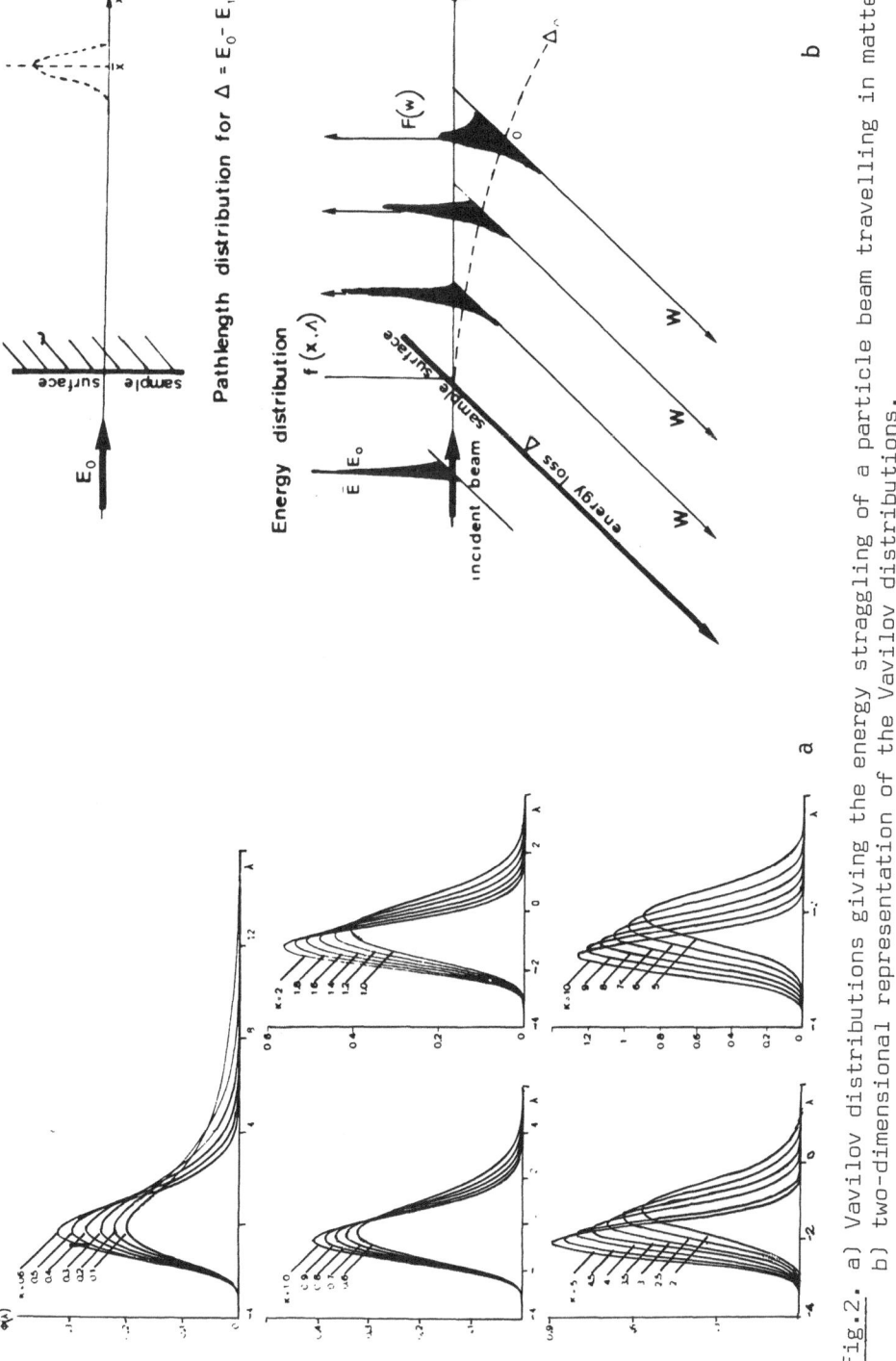

Fig.2. a) Vavilov distributions giving the energy straggling of a particle beam travelling in matter
b) two-dimensional representation of the Vavilov distributions.

TABLE I. Energy straggling distributions $\phi(\lambda)$ calculated from Vavilov's theory for $\beta^2=0$ (low energy particles). The distributions are normalized and tabulated as functions of the Landau parameter λ. The energy loss distributions $f(w)$ are obtained by the substitutions :

$$f(w) = \frac{\phi(\lambda)}{\xi} \qquad \lambda = \frac{w}{\xi} - 0.422 - \ln \kappa \qquad (w = \Delta - \Delta_o)$$

λ	.1	.2	.3	.4	.5	.6	.7	.8	.9	1.0
-3.5	.7439E-03	.8220E-03	.8862E-05	.9824E-05	.1088E-04	.1213E-04	.1350E-04	.1488E-04	.1441E-04	.1821E-04
-3.0	.1065E-01	.1177E-01	.9085E-03	.1004E-02	.1110E-02	.1227E-02	.1354E-02	.1490E-02	.1654E-02	.1630E-02
-2.5	.4801E-01	.5372E-01	.5937E-01	.1349E-01	.1349E-01	.1746E-01	.1941E-01	.2154E-01	.2370E-01	.2624E-01
-2.0	.1111E+00	.1228E+00	.5937E-01	.6502E-01	.7525E-01	.8015E-01	.8857E-01	.9766E-01	.1081E+00	.1193E+00
-1.5	.1473E+00	.1149E+00	.1357E+00	.1500E+00	.1655E+00	.1832E+00	.3020E-01	.2232E-01	.2454E+00	.2453E+00
-1.0	.1960E+00	.2166E+00	.2044E+00	.2259E+00	.2494E+00	.2756E+00	.3046E+00	.3366E+00	.3584E+00	.3455E+00
-0.5	.1977E+00	.2185E+00	.2414E+00	.2664E+00	.2917E+00	.3032E+00	.3473E+00	.3606E+00	.3606E+00	.4012E+00
0.0	.1826E+00	.2018E+00	.2230E+00	.2437E+00	.2407E+00	.3003E+00	.3225E+00	.2320E+00	.3319E+00	.3267E+00
0.5	.1665E+00	.1779E+00	.1966E+00	.2078E+00	.2577E+00	.2643E+00	.2643E+00	.2643E+00	.2393E+00	.2729E+00
1.0	.1373E+00	.1517E+00	.1653E+00	.1659E+00	.2078E+00	.1397E+00	.1966E+00	.1014E+00	.4358E+00	.1245E+00
1.5	.1159E+00	.1281E+00	.1349E+00	.1257E+00	.1552E+00	.1290E+00	.1209E+00	.1014E+00	.8275E-01	.1245E+00
2.0	.9756E-01	.1077E+00	.1069E+00	.1257E+00	.1024E+00	.9057E-01	.1205E+00	.5536E-01	.4135E-01	.6608E-01
2.5	.8266E-01	.9006E-01	.8102E-01	.6349E-01	.6165E-01	.3117E-01	.2075E-01	.55.39E-01	.1808E-01	.1263E-01
3.0	.6932E-01	.7446E-01	.6010E-01	.4277E-01	.2800E-01	.1732E-01	.1018E-01	.1307E-01	.7956E-02	.1766E-02
3.5	.5887E-01	.6043E-01	.6350E-01	.2774E-01	.1642E-01	.9050E-02	.474.3E-02	.5740E-02	.3124E-02	.1664E-02
4.0	.5031E-01	.4401E-01	.3100E-01	.1774E-01	.9295E-02	.4556E-02	.2109E-02	.234.3E-02	.1156E-02	.5377E-03
4.5	.4326E-01	.37.52E-01	.2164E-01	.1090E-01	.5087E-02	.2197E-02	.8957E-03	.9366E-03	.3990E-03	.1651E-03
5.0	.3740E-01	.2680E-01	.1464E-01	.6665E-02	.2707E-02	.1027E-02	.3479E-03	.1499E-02	.1257E-02	
5.5	.3267E-01	.2193E-01	.1000E-01	.3933E-02	.1407E-02	.4463E-03	.1449E-03			
6.0	.2864E-01	.1657E-01	.6646E-02	.2282E-02	.7090E-03	.2062E-01				
6.5	.2527E-01	.1245E-01	.4350E-02	.1299E-02	.3499E-03					
7.0	.2298E-01	.9290E-02	.2813E-02	.7269E-02	.1684E-03					
7.5	.1984E-01	.6483E-02	.1790E-02	.4001E-03						
8.0	.1749E-01	.5057E-02	.113.E-02	.2165E-03						
8.5	.1496E-01	.3682E-02	.7078E-03							
9.5	.1266E-01	.2659E-02	.4368E-03							
10.0	.1054E-01	.1907E-02	.2474E-03							
10.5	.8688E-02	.1359E-02	.1618E-03							
11.0	.5931E-02	.6405E-03								

λ	1.1	1.2	1.3	1.4	1.5	1.6	1.7	1.8	1.9	2.0
-3.5	.2032E-04	.2249E-04	.2528E-04	.2810E-04	.3117E-04	.3451E-04	.3819E-04	.4224E-04	.4671E-04	.5165E-04
-3.0	.2023E-02	.2235E-02	.2471E-02	.2731E-02	.3018E-02	.3335E-02	.3685E-02	.4071E-02	.4496E-02	.4964E-02
-2.5	.2894E-01	.3190E-01	.3532E-01	.3900E-01	.4303E-01	.4744E-01	.5224E-01	.5751E-01	.6322E-01	.6941E-01
-2.0	.1316E+00	.1450E+00	.1594E+00	.1742E+00	.1910E+00	.2090E+00	.2275E+00	.2472E+00	.2678E+00	.2890E+00
-1.5	.2641E+00	.3109E+00	.3463E+00	.3735E+00	.4957E+00	.4275E+00	.5200E+00	.4405E+00	.2478E+00	.5313E+00
-1.0	.4114E+00	.4361E+00	.4589E+00	.4281E+00	.4957E+00	.5104E+00	.5220E+00	.5300E+00	.5380E+00	.5383E+00
-0.5	.3208E+00	.3019E+00	.4238E+00	.4271E+00	.4156E+00	.4076E+00	.2155E+00	.3408E+00	.3661E+00	.3450E+00
0.0	.1008E+00	.3019E+00	.1433E+00	.1638E+00	.1290E+00	.2365E+00	.4066E+00	.1962E+00	.1737E+00	.1541E+00
0.5	.5150E-01	.9117E+00	.1013E+00	.2016E+00	.1233E+00	.1403E+00	.4060E+00	.2277E+00	.0197E-01	.1298E-01
1.0	.2137E-01	.3841E-01	.1017E-01	.6080E-02	.4553E-01	.3816E-01	.1005E+00	.1173E-02	.1720E-02	.5081E-01
1.5	.7977E-02	.5007E-02	.1017E-01	.1857E-02	.1108E-02	.2465E-02	.1008E-02	.1173E-02	.1305E-03	.2660E-02
2.0	.2710E-02	.1527E-02	.3073E-02	.4514E-03	.1108E-02	.6416E-03	.36A3E-03	.2055E-03	.7305E-03	.4491E-03
2.5	.8663E-03	.4738E-03	.2075E-03	.9948E-04	.2307E-03	.1234E-03	.6377E-04	.317A-04	.1165E-03	.6429E-04
3.0	.2661E-03	.1019E-03								

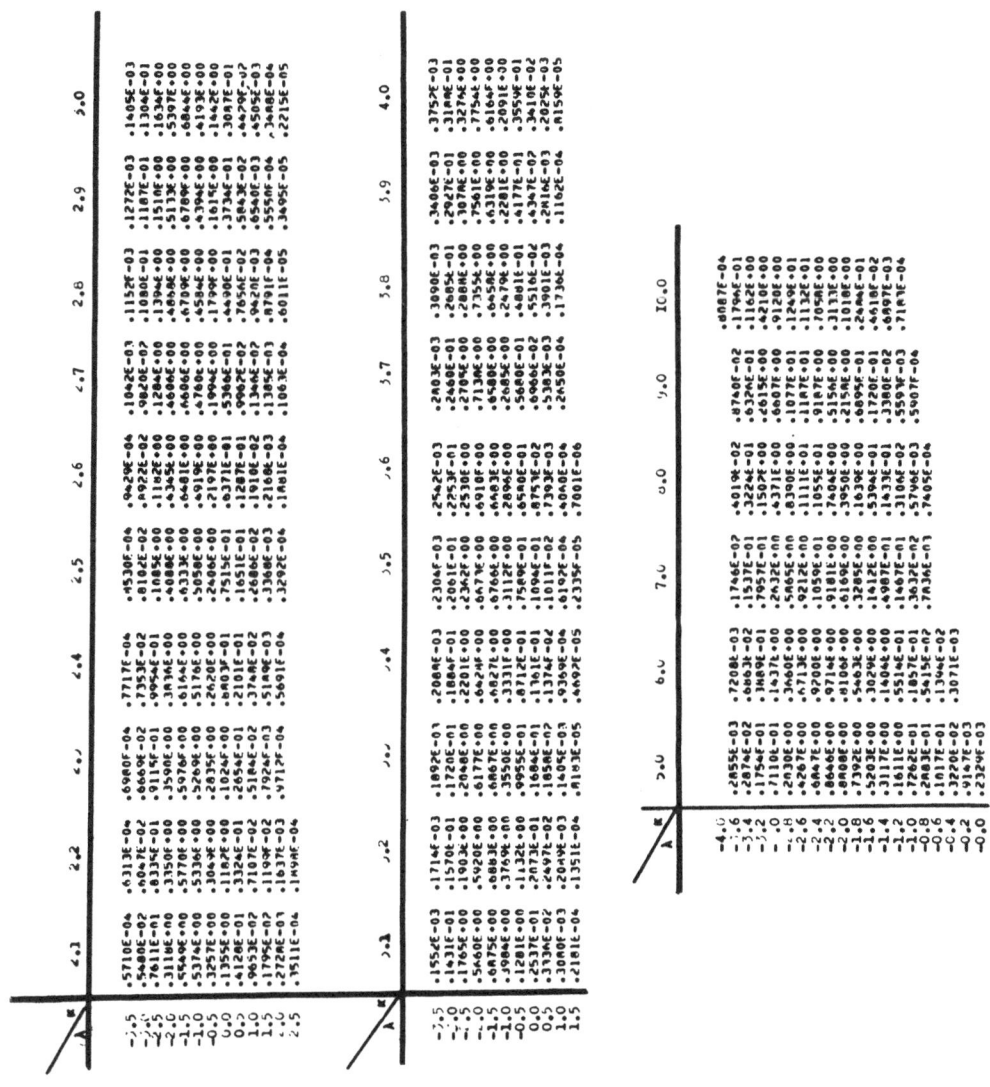

    b) Large energy losses. For large penetration depth
$(x > 2 \ 10^{-4} \ z^{-2} \ g \ cm^{-2})$ gaussian approximation is used with Bohr's
formula for the estimation the FWHM, for a projectile of charge
number z :

$$FWHM = 0,93 \ z \ \sqrt{x \ \frac{\bar{Z}}{A}} \ MeV \qquad\qquad (4)$$

This distribution function will be used in the analysis of particle
spectra from backscattering and nuclear reactions. Formula (4) was
compared with available experimental data, the result is given in
fig.3. Except for heavy elements (Pt) the accuracy of the formula is
found satisfactory in the limits of experimental errors (not shown
on the figure). Of course an oscillatory dependence of the FWHM with
$\frac{Z}{A}$ is probably present but not detectable in now available experimental data, this dependence is observed in stopping power parameter[4].

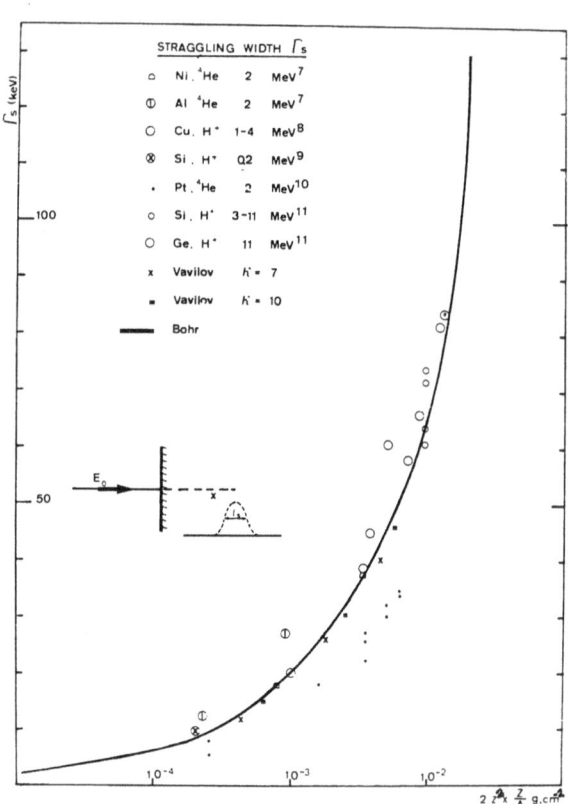

Fig.3. Comparison of Bohr's formula with experimental data.

2. Resonant Scattering of Charged Particles

Resonances are observed in elastic scattering cross sections of protons and ^{4}He ions on light nuclei. Interference with potential scattering is often responsible for deviations from the Breit-Wigner cross section shape, in this case the cross section has rapid varia-tions (anomalies) and cannot be used for analytical purposes. However a few resonances have B.W. behaviour, the resonance energy is then E_R and the FWHM is Γ_R. The more intense resonances and anomalies in elastic scattering of protons and α-particles are reported in Table II. These resonances are apparent in backscattered spectra, their position and width is very sensitive to small variations in stopping power parameters, in particular, the resonance width is considerably enhanced not only by straggling effect but also because of stopping power variations with energy.

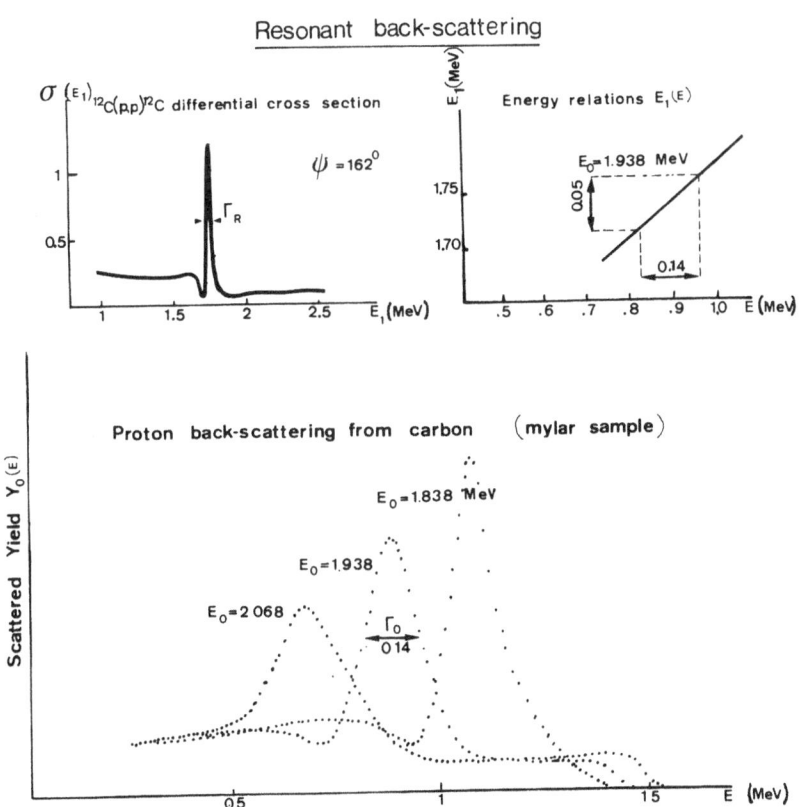

Fig.4. Comparison between measured and calculated resonance width.

TABLE II

RESONANCES AND ANOMALIES IN PROTON ELASTIC SCATTERING (*)

Nucleus	E_R (keV)	Γ_R (keV)	Nucleus	E_R	Γ_R
^{6}Li	1750	500	^{16}O	2660	20
^{7}Li	441	10	^{19}F	699	67,5
	1030	168	^{23}Na	Above 843	
^{9}Be	330	A	^{24}Mg	823	1.3
	980	A		1483	0.3
	998	A		2010	< 10
	1087	3		2410	0.5
	2530	130	^{27}Al	Above 1000	
^{10}B	1150	A	^{28}Si	1652	
	1500	A		2088	18
^{11}B	160	A	^{31}P	1254	1.7
	670	A	^{32}S	1900	8.5
	1400	A	^{40}A	1870	
^{12}C	45.7	A		2470	
	1698	55	^{40}Ca	2390	
^{14}N	1054	6			
	1550	35			
	1744	3.5			
	1805	A			

RESONANCES AND ANOMALIES IN ELASTIC SCATTERING OF ^{4}He IONS

Nucleus	E_R	Γ_R	Nucleus	E_R	Γ_R
^{6}Li	1180	150	^{16}O	2490	
^{12}C	No below 3000			3045	10
^{14}N	1530		^{27}Al	No below 2350	
	2160		^{28}Si	No below 2500	
	2350				

(*) Data are from ref.5-6. The most typical resonance energies
are underlined, resonance curves are often strongly modified
by interferences, strong fluctuation are indicated as
anomalies (A).

The general formula giving the scattered yield is :

$$Y(E_3) = \frac{\rho \ N_0 \ \sigma(\psi,E_1) \ \delta\Omega}{S(E_2) \ [G \ + \ D \ k \ S(E_1)/S(kE_1)]} \qquad (5)$$

where $Y(E_3)$ is the scattered yield (observed particle spectrum) corresponding to a given nucleus A, ρ is the number of target nuclei A
per gram, N_0 the number of projectiles, $\sigma(\psi,E_1)$ is the cross section
for elastic scattering on nucleus A, $\delta\Omega$ is the detector solid angle,
k is the kinematical factor E_2/E_1.

$$G = \frac{-1}{\cos(\theta+\psi)} \quad \text{and} \quad D = \frac{+1}{\cos\theta}$$

Straggling effects are not included in (5), to a good approximation
they are calculated by taking the convolution product of (5) with a
gaussian-function, the half width being given in MeV by :

$$\gamma^2 = (\gamma_0^2 + 0,86 \ z^2 \ Dx \ \bar{Z}/\bar{A})k^2 + 0,86 \ z^2 \ G \ x \ \bar{Z}/\bar{A} + (\delta E)^2 \quad (6)$$

γ_0 is the beam energy spread and δE the detector resolution.

This formula gives the energy straggling of the particles
scattered off the rear edge of a thin layer of thickness x. In the
case of resonant scattering the resonance spreading is not only due
to straggling but also to stopping power effects. A simple calculation shows that the resonance width observed in backscattering is
given approximately by :

$$\Gamma^2 = (\gamma_0^2 + \Gamma_R^2 + 0,86 \ z^2 \ D \ x \ \bar{Z}/\bar{A})(k+ \ \epsilon \frac{G}{D})^2 + 0,86 \ z^2 x \bar{Z}/\bar{A} + (\delta E)^2 \quad (7)$$

where $\epsilon = \dfrac{S(kE_R)}{S(E_R)}$ and Γ_R is the resonance width in the lab.system.

This formula indicates that the resonance width and the straggling
effects are considerably enhanced by the factor $(k+\epsilon \frac{G}{D})^2$. We have
observed this phenomenon in several resonant reactions :
$^{12}C(p,p)^{12}C$, $^{28}Si(p,p)^{28}Si$, $^{16}O(\alpha,\alpha)^{16}O$. In fig.3 the example of
proton backscattering from mylar is given, the following parameters
are used for the calculation : $E_R = 1.740$ MeV, $\Gamma_R = 0.045$ MeV,
application of formula (3) gives $\Gamma=0.139$ MeV for $E_0 = 1.938$ MeV,
the observed value is $\Gamma = 0.140 \pm 0.015$ MeV. This result was also
obtained by solution of equation (5) using numerical methods and
different sets of parameter for E_R, Γ_R, the best fit being obtained
for $\Gamma'_R = 0.045$ MeV.

The results of the calculations are strongly dependant on the
parameter used (E_R, Γ_R, $S(E)_1$) making this technique a sensitive
tool for stopping power parameter determination (test of Bragg

rule in compounds, stoichiometry). The resonance backscattering is also a possible tool for concentration profile determinations by application of formulae (1), the concentration profile is represented by ρ which is obtained from the spectrum by iterative methods.

3. Nuclear Reactions

In the case of nuclear reactions where particles are detected, the total energy spread is calculated exactly in the same way as in the case of elastic scattering.
The reaction yield is given by :

$$Y(E_3) = \frac{\rho \, N_o \, \sigma(\psi, E_1) \, \delta\Omega}{S(E_3)\left[G + D\left(\frac{\partial E_2}{\partial E_1}\right)_E \frac{S_a(E_1)}{S_b(E_2)}\right]} \qquad (8)$$

where a and b are respectively the ingoing and outgoing particles in the reaction A(a,b)B. The straggling distribution function is gaussian with a FWHM given by a formula similar to (6) :

$$\gamma^2 = (\gamma_o^2 + 0.86 \, z^2 Dx \, \bar{Z}/\bar{A})\left(\frac{\partial E_2}{\partial E_1}\right)^2 + 0.86 \, z^2 \, Gx \, \bar{Z}/\bar{A} + (\delta E)^2 \text{ MeV} \qquad (9)$$

Absorbers are often used in order to eliminate backscattered particles and to reduce the counting rate. In this case the particle energy is strongly affected by straggling and by unhomogeneities in the absorber thickness. The straggling introduced by an absorber of thickness y is given by :

$$\Gamma_{abs} = 0.93 \, z \sqrt{y\left(\frac{\bar{Z}'}{\bar{A}'}\right)} \text{ MeV}$$

$\bar{Z}'$ and $\bar{A}'$ are the absorber atomic number and atomic mass numbers, y is in g cm^{-2}. The total straggling will then be given by :

$$\gamma_{tot} = \sqrt{\gamma^2 + \Gamma_{abs}^2}$$

If the absorber thickness is a function distributed around the mean value y with a FWHM δy then Γ_{abs} is given by :

$$\Gamma_{abs}^2 \simeq \left[\frac{S_{abs}(E_3)}{\delta y}\right]^2 + \gamma_{abs}^2$$

S_{abs} being the stopping power parameter for particle b in the absorber material.

References

1. P.V. VAVILOV. Journal of Experimental and Theoretical Physics. 5 (1957) 749.
2. C. TSCHALAR. Nuclear Instrument and Methods. 6 (1968) 141.
3. K.L.DUNNING, G.K.HUBLER, J.COMAS, W.H.LUCKE and H.L.HUGHES. Thin Solids Films. 19 (1973) 145.
4. J.F.ZIEGLER and W.K.CHU, Thin Solid Films 19 (1973) 281.
5. Nuclear Tables Part II. Wunibaldkunz and J.Schintlmeister (1965) Pergamon Press.
6. P.M. ENDT and C. VAN DER LEUN. Nucl. Phys. A 105 (1967) 1.
7. J.M.HARRIS and M.A.NICOLET. Phys.Rev. B 11 (1975) 1013.
8. J.F.CHEMIN, J.ROTURIER and J.Y.PETIT. J.de Phys.33 (1971) 219.
9. J.BOTTIGER and F.H.EISEN. Thin Solid Films 19 (1973) 239.
10. J.M.HARRIS. W.K.CHU and M.A.NICOLET. Thin Solid Films.19(1973)259
11. B.R.APPELTON, C.ERGINSOY and W.M.GILSON.Phys.Rev.16(1967) 330.

ANALYSIS OF NUCLEAR SCATTERING CROSS SECTIONS BY MEANS OF MOLECULAR IONS

Günter Thieme

Inst. f. Angew. Physik

Techn. Universität Hannover

03000 Hannover, Welfengarten 1, Germany

If fractions of molecular ions are detected with surface barrier detectors, the relations between the quantities of signals with different pulse heights provide an analysis of nuclear scattering cross sections. Monte-Carlo-Calculations are compared with experiments, performed with H_2^+-ions.

Molecular ions with energies above 10 kev entering an absorber will be dissociated and the fractions of a molecule will go on nearby tracks, until they are scattered at an atomic nucleus of the absorber. After passing the absorber the fractions of the molecules, (they are mostly atomic ions) will be recorded with a surface barrier detector and a spectrum with some peaks is obtained. Always when one or more fractions of a molecule miss the detector, a smaller portion of the energy is transferred according to the mass ratio of the fractions.

It is most simple to use hydrogen molecule ions, because as with all bimolecular ions of equal atoms, a spectrum with only two peaks is obtained. We shall, therefore, discuss the results obtained with H -ions. The absorbers will be regarded as polycrystalline layers.

The molecular ions are directed normal to the absorber and the deflected atomic ions leave the absorber with angles φ. The absorber is evaporated on a substrate film of formvar. Figure 1) gives a diagram of the scattering

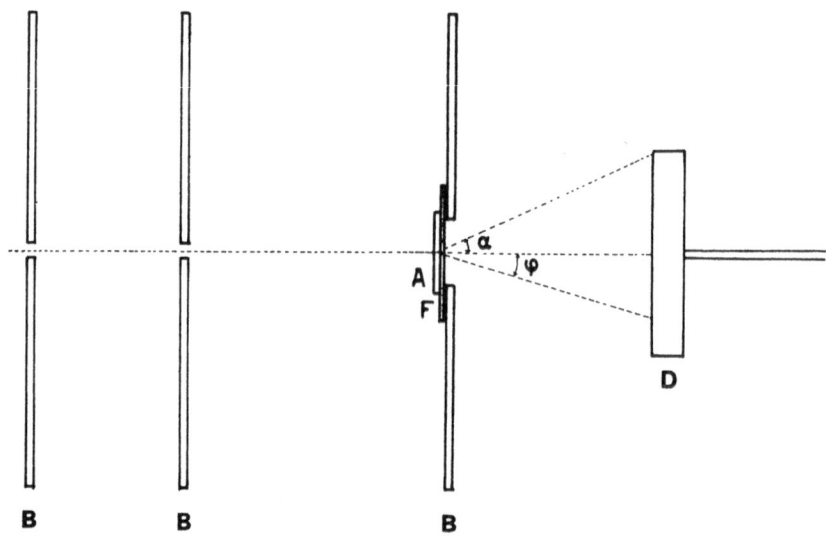

Fig. 1) Scattering geometry, A absorber, F substrate
film, D detector, α aperture angle toward the detector,
φ deflection angle of a particle.

geometry.

Atomic ions of a molecule, striking the detector
simultaneously and transferring their whole energy into
the counter will be called atom pairs, their quantity may
be I_{HH}. I_H is the quantity of molecular ions, striking
the detector with only one atomic ion, I is the quantity
of molecular ions with both ions scattered out of the
aperture angle toward the detector, α. Is w the probability for an atomic ion, to be deflected with an angle φ>α
by passing the absorber, the relative decrease of the atom
pairs in the absorber will be:

$$- \frac{dI_{HH}}{I_{HH}} = 2w(1 - w) + w^2 \simeq 2w$$

$2w(1 - w)$ is the probability that one atomic ion of the
pair is deviated, w^2 is the probability that both ions are
scattered out of the aperture angle.

$$E_H \ (keV) \qquad 35 \qquad 40 \qquad 50$$

$$\sigma_{\varphi,\tilde{\pi}} \cdot 10^{18} \ (cm^2) \qquad 7.0 \qquad 5.7 \qquad 4.6$$

Table 1) Integrated scattering cross sections, obtained by first experiments

Single scattering assumed, the probability w may be represented by the product of the particle density of the absorber n, the absorber thickness dx, and the integrated scattering cross section $\sigma_{\alpha,\tilde{\pi}}$ for scattering of an atomic ion by an angle $\varphi > \alpha$,

$$w = n \ \sigma_{\alpha,\tilde{\pi}} \ dx \ .$$

The quantity of the original molecular ions, detected as atom pairs, is obtained by integration.

$$I_{HH} = (I_{HH})_o \ e^{-2n\sigma_{\alpha,\tilde{\pi}} x}$$

$(I_{HH})_o$ is the quantity of the original molecular ions,

$$(I_{HH})_o = I_{HH} + I_H + I$$

The quantities I_H and I have already been published in a previous paper (1). Low thickness assumed, one is able to neglect I. For gold absorbers I is about 3 % with molecule energies of 70 keV, $\sigma_{\varphi,\tilde{\pi}} = 7 \ 10^{18} \ cm^2$ and a thickness of 5 nm. With a thickness of 25 nm, I is about 40 %.

By neglecting I, the integrated cross section $\sigma_{\alpha,\tilde{\pi}}$ may be derived from the relation I_H / I_{HH} , without knowledge of $(I_{HH})_o$. If the absorber is evaporated on a substrate film, $(I_{HH})_o$ must be substituted by $(I_{HH})_f$, which is I_{HH} after passing the pure substrate film. The scattering cross section is given by the formula:

$$\sigma_{\alpha,\tilde{\pi}} = \frac{\ln((1 + (I_H/I_{HH}))/(1 + (I_H/I_{HH})_f))}{2nx}$$

From experiments with gold absorbers with a thickness between 14.0 and 24.5 nm the values of table 1) are derived. In the paper mentioned (1), the energy of the molecular ions is stated instead of the energy of the atomic ions.

It was tried to simulate the scattering of molecular ions with a Monte-Carlo-Calculation. The energy losses,

caused by electronic interactions, are obtained from the
theory of Lindhard (2). The nuclear interactions are regarded as classical scattering and like Leibfried (3) has
proposed, the interaction potential is approximated by a
cut off Coulomb-Potential $V(r)$, which is adjusted to the
Thomas-Fermi-Potential.

$$V(r) = \begin{cases} E_c \left(\dfrac{a_c}{r} - 1 \right) & \text{for } r \leq a_c \\ 0 & \text{for } r > a_c \end{cases}$$

With the cut off potential one obtains a finit total
cross section $\sigma_{0,\pi}$ and the differential cross section is represented by K_s.

$$\sigma_{0,\pi} = 2\pi a_c^2 = \int_0^\pi K_s \sin(\theta) \, d\theta$$

$$K_s = \frac{\pi}{2} \left(\frac{E_c a_c}{2 E_r} \right)^2 \frac{(1 + E_c/2E_r)^2}{((1+E_c/E_r) \sin^2(\frac{\theta}{2}) + (E_c/2E_r)^2)^2}$$

θ = scattering angle in the centre of mass system
E_r = relative energy

The energy values E_c and the distances a_c are given by
the adjustment to the Thomas-Fermi-Potential, for which
is used the approximation of Sommerfeld, corrected from
March (4). Further details will be published in due course
(5).

The interaction of the two atomic ions of a molecule
is neglected, i. e. the calculation is based on single
atomic ions. In the region of proton energies from 20 to
110 keV the cut off radius a_c varies for gold absorbers
from 0.016 to 0.0095 nm. That means, scattering events
with impact parameters $b > a_c$, respectively scattering
angles $\psi < \psi_c$ are disregarded. For protons with an energy of
40 keV ψ_c is 6.6° in the laboratory system. With this restriction the Monte-Carlo-Calculation gives the result,
that multiple scattering with angles $\psi > \psi_c$ occures with a
low degree of probability only for a low absorber thickness, 5 - 10 nm, as seen in figure 2).

In spite of the disregard of scattering events with
impact parameters $b > a_c$, the calculation uses scattering
angles between 0 and π, according to the cut off poten-

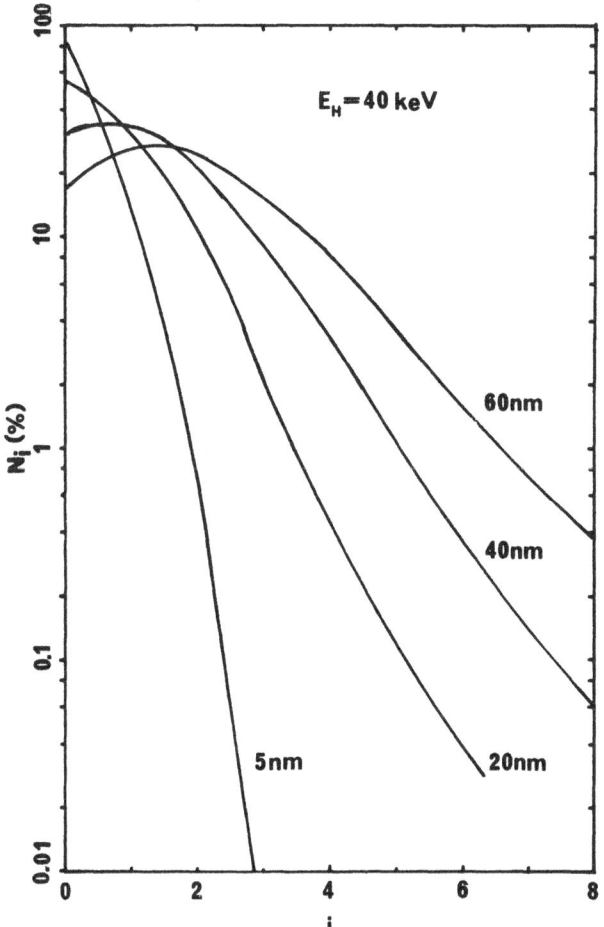

Fig. 2) Quantity of H^+-ions scattered i-times with an angle 6.6° for different Au-layers, expressed in per cent of all H^+-ions, leaving the layer.

tial and gives an angle distribution of the outgoing particles. From this angle distribution the probability u for deflection into an angle $\varphi < \alpha$ can be computed.

$$u = N_{\varphi < \alpha} / N_{all}$$

$N_{\varphi < \alpha}$ = quantity of particles leaving the absorber with an angle

N_{all} = quantity of all particles regarded

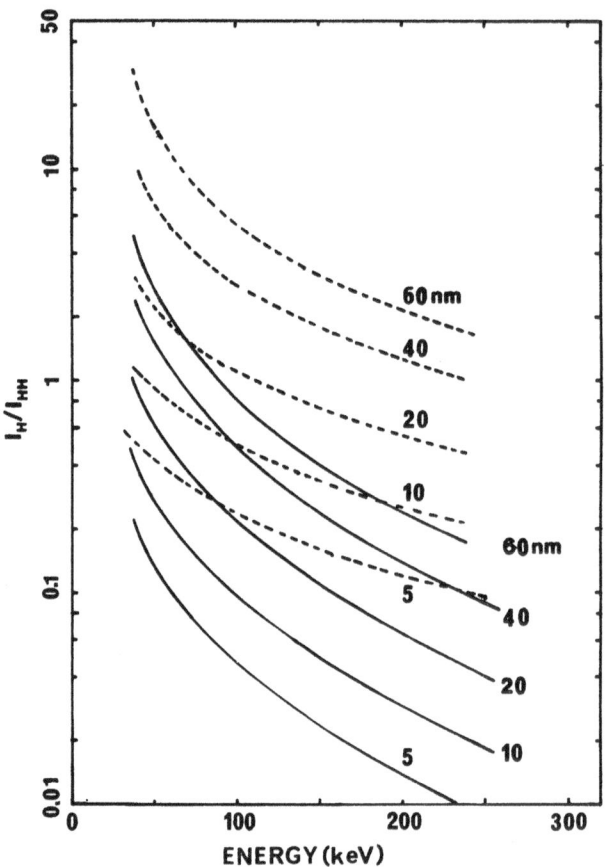

Fig. 3) Relation of molecular ions from which only one
ion is detected I_H to the ions from which both ions hit
the detector I_HH , obtained by Monte-Carlo-Calculation for
an aperture angle of 20°(full drawn line) and of 4°(dotted line), drawn as a function of the energy of the molecular ions.

The relation I_H/I_{HH} can be calculated from u as the relation of the respective probabilities.

$$\frac{I_H}{I_{HH}} = \frac{2u(1-u)}{u^2}$$

With gold absorbers the results are shown in figure 3)
for an aperture angle $\alpha = 20°$ (full drawn line) and also

for $\alpha = 4°$ (dotted line). For $\alpha = 4°$ the used approximation
is no more correct, nevertheless the curves are derived,
to get a comparison with experimental results.

Using substrate films, the probability v for scatte-
ring into an angle $\varphi < \alpha$ has been computed from the relation
$(I_H / I_{HH})_f$ for pure substrate films by a formula equivalent
to the above relation. For the interesting case, absorber
on substrate film, the equation is:

$$(\frac{I_H}{I_{HH}})_{a,f} = \frac{2u(1-u)v + 2v(1-v)u^2}{v^2 u^2} = (\frac{I_H}{I_{HH}})_a \frac{1}{v} + (\frac{I_H}{I_{HH}})_f$$

In figure 4) the influence of the substrate film has been
taken into account. The lowest curve through the measured
points yields $(I_H/I_{HH})_f$ for the pure substrate film. Measurements without any restrictions for absorber thickness
may be compared with computed values, provided that $\psi_c < \alpha$.
Results from preliminary experiments are shown: as expected, they give points above the computed values, the nuclear interaction is not sufficiently accounted for with
the cut off potential method.

The integrated scattering cross section $\sigma_{\alpha,\pi}$ can be
derived from the differential scattering cross section K_s

$$\sigma_{\alpha,\pi} = \int_{\alpha}^{\pi} K_s \sin(\theta) \, d\theta$$

In figure 5) $\sigma_{\alpha,\pi}$ is shown for different values of the relative energy E_r as a function of the scattering angle in
the centre of mass system. The relative energy for protons in gold is 1/2 % lower than the energy of the protons.
An angle of $4°$ in the laboratory system is equivalent to
an angle of $4.02°$ in the centre of mass system. The points
obtained from the experiment by neglecting multiple scattering are approximately 30 % above the values obtained
with the cut off potential method, which gives too low
cross section values for angles $\lesssim \psi_c$.

Scattering measurements made by molecular ions and
surface barrier detectors have the advantage, that only
one system for particle detection is needed, even with
fluctuating particle rate. The method is of special interest in the case of low scattering angles, where the checking of statistical calculations is critical. Single scattering (disregarding scattering angles $< \psi_c$) is a useful
assumption for the determination of integrated scattering

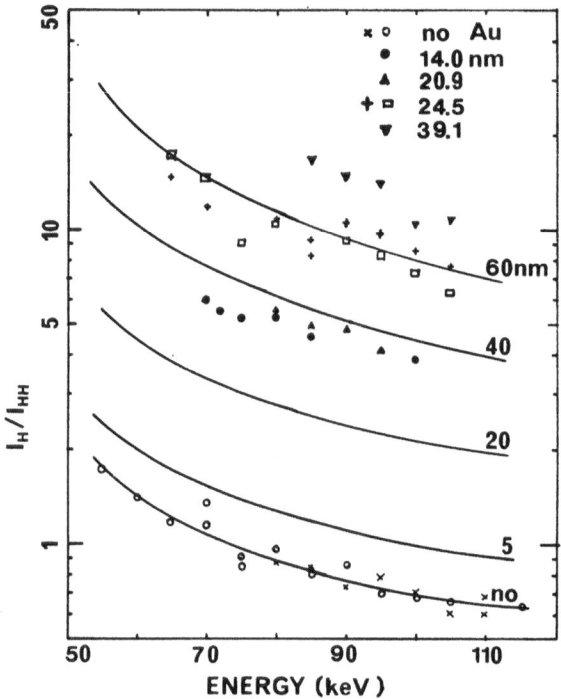

Fig. 4) The relation I_H/I_{HH} ,seen in figure 3) for an
aperture angle of 4°,corrected for the influence of a
substrate film, lowest curve yields for the pure substrate film. Results from preliminary experiments are shown.
For the here used aperture angle the cut off potential
is no more a sufficient approximation.

cross sections by means of molecular ions, provided, the
angle $\psi_c \ll \alpha$, the aperture angle toward the detector and
the thickness of the absorber is small enough, that multi-
scattering has a low probability. As shown, this assumptions are filled for protons with energies $E_H \geq 40$ keV
with $\alpha \gtrsim 20°$ and a gold thickness $x \leq 10$ nm. The determination of this limits for lower aperture angles needs a
better potential approximation as the here used cut off
potential method.

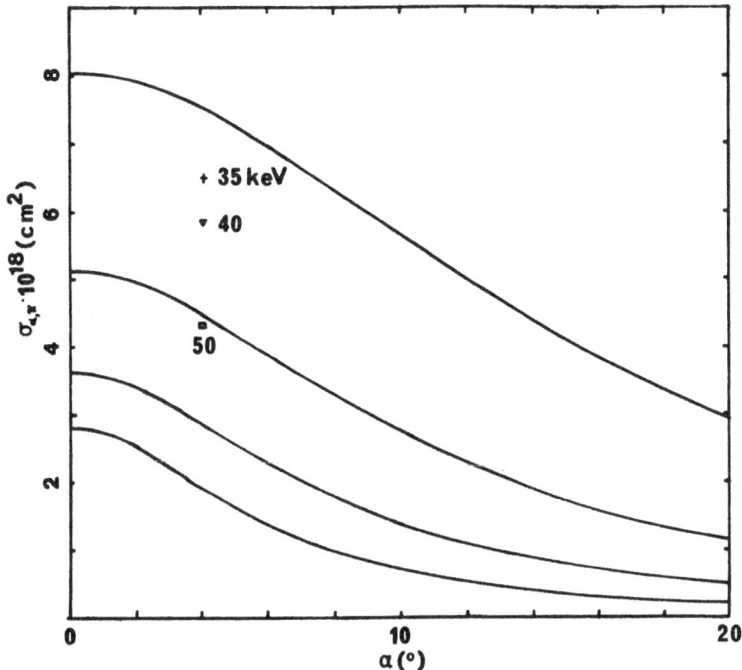

Fig. 5) Integrated scattering cross section, computed
from the differential cross section K_s, compared with
measured values, obtained by neglecting multiple scattering as well as the influence of those particles, scattered out of the aperture angle to the detector.

REFERENCES

(1) G. Thieme, Naturwissenschaften 61, 361, (1974)
(2) J. Lindhard, M. Scharff, H. E. Schiøtt, Mat. Fys.
     Medd. Dan. Vid. Selsk. 33, no. 14, (1963)
(3) G. Leibfried, Bestrahlungseffekte in Festkörpern,
     B. G. Teubner, Stuttgart, (1965)
(4) S. Flügge, Hdb. d. Physik, Bd. XXXVI, p. 128
(5) G. Thieme, Vakuumtechnik, in preparation

DISCUSSION

Q: (W. Brown) Have you taken account on the "proton push" effect between the two outgoing protons? Is it significant in your case?

A: (G. Thieme) The proton push effect has not been taken into account. I think, the effect gives only a low correction for the computed relations I_H/I_{HH}, it may be of more importance for lower angles of deflections.

Q: (W.D. Hofer) Have you never seen any influence from the crystallinity in your thin film targets, namely texture? At these low scattering angles this should show up markedly, as has been observed in multiple scattering investigations by the Aarhus group (H.H. Andersen and J. Bøttiger, e.g. Gausdal Proceedings, 1972).

A: (G. Thieme) The discussed method is very useful for investigation of changes in scattering conditions. The reported experimental results have only preliminary character.

Backscattering Analysis

DETERMINING CONCENTRATION VS. DEPTH PROFILES FROM BACKSCATTERING

SPECTRA WITHOUT USING ENERGY LOSS VALUES

R. F. Lever

IBM System Products Division, East Fishkill

Hopewell Junction, New York 12533

ABSTRACT

In backscattering spectra in which the different elements are well separated, it is possible to obtain a profile of each element vs. the collision energy rather than the energy measured at the detector. This is done by assuming depth-independent values, for each element, for the quantity (energy loss after collision) $\div$ (energy loss before collision). Once these profiles are obtained, a profile of composition vs. collision energy follows directly from the Rutherford scattering cross section, and hence, by integration, a composition vs. depth profile is obtained.

INTRODUCTION

For an exact analysis of backscattering spectra, the energy loss in the solid being analyzed must be accurately known as a function of both energy and composition [1-4]. For 0.5-MeV ^{4}He ions, a convenient tabulation of ε, the stopping cross section, is available [5] for all elements, in units of $eV/(10^{15}$ atoms $cm^{-2})$. Compounds or mixtures of elements are analyzed by assuming Bragg's rule [6], that the stopping cross section per atom of mixture is equal to the sum of the elemental stopping powers weighted by their atom fractions in the mixture. Figure 1 shows some elemental stopping powers taken from reference 1. It is clear that in the experimental region commonly used in ^{4}He backscattering experiments, the elemental stopping powers can vary by approximately one order of magnitude. Hence, in thin

111

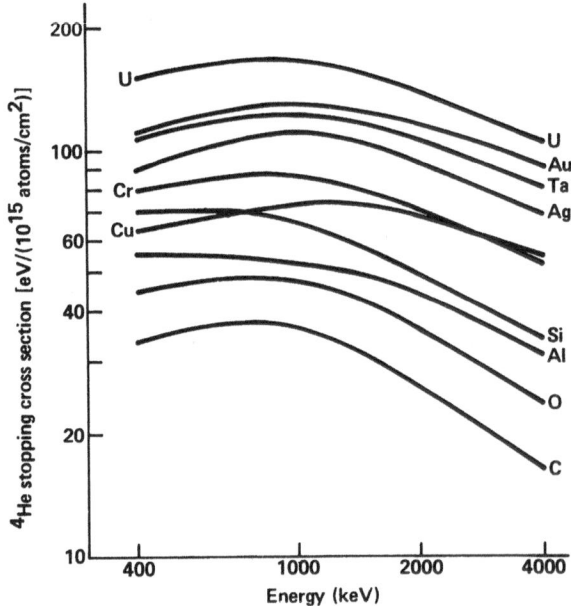

Fig. 1. Stopping cross section for ^{4}He, for various elements as a function of particle energy.

Table 1. Values of K, α, and K + α for 2-MeV ^{4}He scattering from a thin film at normal incidence and a detector angle of 170°

Element	M	K	α	K + α
C	12	0.252	1.35	1.60
O	16	0.362	1.35	1.71
Al	27	0.553	1.16	1.71
Si	28	0.566	1.29	1.86
Cr	52	0.736	1.11	1.85
Cu	63.5	0.78	1.06	1.84
Ag	108	0.86	1.07	1.93
Ta	181	0.92	1.03	1.95
Au	197	0.92	1.03	1.95
U	238	0.94	1.01	1.95

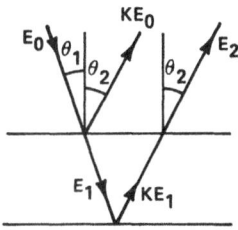

Fig. 2.   Definition of angles and energies
for the backscattering process.

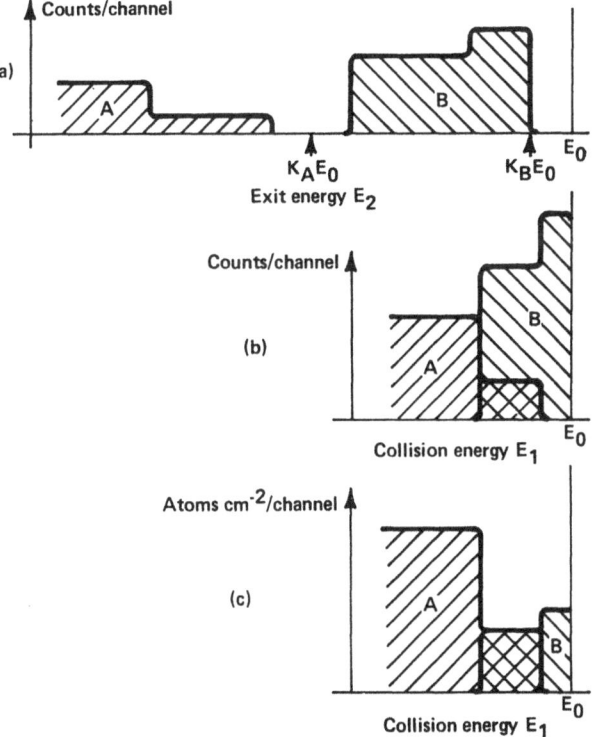

Fig. 3.   Spectrum transformation shceme
(a) Experimental spectrum.   Abscissa is E_2
(b) Abscissa transformed to E_1 by Eq. 3
(c) Ordinate converted to atom cm^{-2}/bin.

films whose composition can vary widely with depth, the relationship between depth and energy can be highly nonlinear. The
purpose of this paper is to point out that a reasonable analysis
of composition versus depth profiles can nevertheless be obtained
without reference to Bragg's rule or to detailed energy loss
curves such as those shown in Fig. 1.

METHOD

Consider particles of energy E_0 incident on a layer of solid
film at an incidence angle θ_1 (Fig. 2). Particles that scatter
from surface atoms will recoil with energy KE_0, where K, the well
known kinematic recoil coefficient, depends only on the mass of
the struck nucleus relative to the incident particle and on the
angle $(\theta_1 + \theta_2)$, where θ_2 is the angle the scattered particle
makes with the film.

Let us now consider a particle which loses an amount of
energy δ before scattering from a target nucleus at some point
beneath the surface. The collision energy E_1 is simply

$$E_1 = E_0 - \delta \tag{1}$$

irrespective of the target atom involved, whereas the backscattered
energy KE_1 will depend on the mass of the target nucleus. The
particle will emerge from the solid and hit the detector with
energy E_2, having lost a further amount of energy Δ, so that

$$E_2 = KE_1 - \Delta \tag{2}$$

The scattering cross section is, of course, known from classical
Coulombic scattering theory in terms of the incident energy E_1,
whereas E_2 is measured by the detector. It is therefore convenient to write E_1 in terms of E_2, as follows:

$$E_1 = \frac{E_2 + \alpha E_0}{K + \alpha} \tag{3}$$

where $\alpha = \Delta/\delta$, the ratio of energy lost after collision to that
lost before collision [7]. The ratio α will depend, in part, on
the ratio of the path lengths in the solid of the incident and
the scattered beam $(\cos \theta_1/\cos \theta_2)$. For normal incidence and a
detector at 170° to the beam direction, where $\theta_1 + \theta_2 = 10^o$ and
$\theta_1 = 0$, this factor alone would make $\alpha = 1.015$. However, reference to Fig. 1 will show that for 2-4 MeV He backscattering, Δ
will be larger than δ, because after recoil the particle energy
is lower and hence ε is higher. For heavy elements where $k \simeq 1$, α
will approach its ideal geometric value $(\cos \theta_1/\cos \theta_2)$; for

recoil from light elements, α will be higher. The magnitude of
this effect for very thin films is shown in Table 1, where α is
equated to the ratio between ε at E_0 = 2 MeV and the value of ε
at an energy of KE_0. It is seen that as K falls with atomic
mass, α rises, so that for most elements $(K + \alpha)$ covers the
rather narrow range 1.7 to 1.95. In thick films $(K + \alpha)$ is not
strictly constant, since the separation between the average
energy before collision and the average energy after collision
will increase and so, in general, will α. It is clear that since
the exiting particle traverses exactly the same range of compositions as before recoil, the value of α will not change widely
with depth even though composition may change so much that the
stopping cross section may vary by several hundred per cent.

SPECTRUM ANALYSIS

This method assumes that the backscattering spectra from
different elements occupy separate energy intervals in the
regions of interest, and that isotope effects may be ignored.
Hence each spectral region requiring analysis will have a unique
and known value of K, the recoil coefficient. The first step is
to convert the abscissa from the exit energy E_2 into the collision energy E_1 by use of equation 3. Equation 3 may also be
written:

$$\frac{KE_0 - E_2}{E_0 - E_1} = K + \alpha \qquad (4)$$

Hence the transformation consists essentially of taking a portion
of spectrum of known K and α, compressing it by dividing its
energy spread by $K + \alpha$, and shifting it so that the point form-
erly at energy KE_0 is now at E_0. This is illustrated in Fig. 3.
A layer of substance B is assumed to overlie a lighter substance
A, with an intervening layer of compound AB. B is assumed to
have double the stopping cross section of A and four times the
scattering cross section. For simplicity, the effects of the
increase in scattering cross section with decrease in energy are
not shown and the compression factor $(K + \alpha)$ is taken as 2. If
the effects of changes in scattering cross section and stopping
cross section were shown, the curves in 3a and 3b would fall
somewhat with energy, whereas those in 3c would rise with energy,
though very slightly. Figure 3a shows a schematic experimental
spectrum of counts per channel as ordinate versus E_2, the exit
energy measured by the detector. Figure 3b shows the result of
transforming spectra A and B according to equation 3 or 4. After
the energy scale is transformed from E_2 to E_1, the collision
energy, the new channel sizes are different for A and for B and
interpolation is required to redraw the transformed spectra with

a common channel width. Figure 3b assumes the same channel width
as in Fig. 3a. It is seen that the number of counts per channel
is doubled because we assumed a value of 2 for $(K + \alpha)$, and that
the spectra overlap in the region of compound AB.

Since in Fig. 3b the collision energy is given, the scattering cross section can be calculated. Hence, given the system
geometry and assuming that the total integrated beam charge has
been measured, one may convert the ordinate from counts/channel
to atoms cm^2/channel. This is shown in Fig. 3c. Since substance
A is assumed to have half the stopping cross section of substance
B, it is clear that the atom density/channel should be double.
Since a compound of composition AB has been assumed, the atom
densities/channel in the region AB are equal.

In order to perform the above transformation, it is necessary to make a reasonable estimate of the values of α, desig-
nated α_A and α_B for recoil from nuclei of types A and B. For
most studies of reaction or diffusion in thin film, an unreacted
film is usually available. Hence, if one α value is known, the
others can be measured by use of equation 4. This is done by
taking equivalent points (viz. the film boundaries) for which E_1
has the same values for substances A and B. Then, from (4) we have

$$\frac{K_A E_0 - E_{2A}}{K_B E_0 - E_{2B}} = \frac{K_A + \alpha_A}{K_B + \alpha_B}$$

where E_{2A} and E_{2B} are the exit energies for recoil at the same
film depth from substances A and B. Since the value of α for the
heavier element is usually readily established (it will usually
lie in the range 1.05-1.1), the other α values are, in effect,
measured from the observed spectrum. In practice, it is frequently more convenient to estimate all the α values, obtain Fig.
3c, and then adjust the α values for the lighter elements until
the spectra overlap correctly. The same α values are then used
for interpreting spectra of the same film after diffusion or
reaction, where equivalent points having the same E_1 may not be
clearly distinguishable. If the α values are made consistent in
this way, the only errors that result from small errors in the
absolute values of α are the small errors in scattering cross
section that result from slightly incorrect values of E_1.

Once a spectrum has been transformed into the form of
Fig. 3c, a simple integration and interpolation yields concentration versus "depth" profiles, where the "depth" is understood to
be the total number of atoms cm^{-2} between the point being measured
and the surface. A conversion to actual depth requires a detailed
knowledge of density as a function of composition for the system
under investigation.

It may be noted that Fig. 3c gives energy loss values as a by-product of the method. If, therefore, accurate energy loss values are known for one of the phases in Fig. 3c, one may check the known value against that obtained from Fig. 3c and adjust the values accordingly.

APPLICATION

The procedure described above has been applied to the analysis of the reaction between chromium and aluminum [8]. Approximately 2000 $\overset{\circ}{A}$ of chromium was evaporated onto about 7000 $\overset{\circ}{A}$ of aluminum film. To protect the chromium against oxidation during annealing, a further evaporation of approximately 1000 $\overset{\circ}{A}$ of Al was deposited on top. Figure 4 shows the spectrum obtained for E_0 = 2820 keV, a detector solid angle of 4.11 msr, and a channel width of 5 keV. The detector angle was at 170° to the beam direction, which was normal to the sample surface, and a total beam dose of 20 μC was measured. Assuming that α = 1.1 for chromium, the α for aluminum was adjusted to 1.28 to give the areal density per channel versus E_1 curve shown in Fig. 5. The larger areal density per channel for aluminum reflects its lower stopping power. Since the channels in Fig. 5 are 5 keV wide, the stopping power can be calculated to be 42 eV/10^{15} atoms cm^{-2} for chromium. Clearly these values do not constitute an accurate measurement, since the α values, though self-consistent, are not known absolutely. At this point the α values could be corrected if the energy loss for aluminum were known exactly. Generally, one would not rely on the chromium film for absolute measurements, because of the difficulty of obtaining a fully dense, pure film, completely impervious to aluminum, during deposition of the protective top layer of aluminum. Figure 6 is obtained by simple integration of Fig. 5 from the surface downward, followed by interpolation so that the result is given as percentage of composition versus film depth in units of 10^{15} atoms cm^{-2}. We see that the surface Al layer contains 7 x 10^{17} atoms cm^{-2} of chromium. The overlap is due to limitations in detector resolution. The film shown in Fig. 6 was then used as input to a computer program [9] designed to generate spectra from experimental data and the Ziegler-Chu energy loss polynomials [5]. This reconstructed spectrum is also shown in Fig. 4, and the agreement is seen to be good. The analysis was first performed on the assumption that E_0 = 2800 keV, and a composition versus depth plot essentially identical with Fig. 6 was obtained. The reconstructed spectrum, however, was significantly displaced from the experimental spectrum for E_0 = 2800 keV; so for consistency the calculation was rerun at E_0 = 2820 keV.

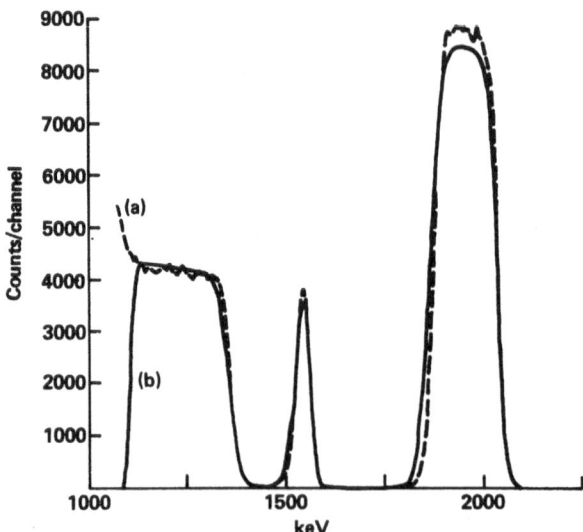

Fig. 4. Backscattering spectra for unreacted Al-Cr system.
(a) Experimental. (b) Theoretical, generated from composition-
depth plot (Fig. 6) by use of tabulated cross sections. From
Reference 1.

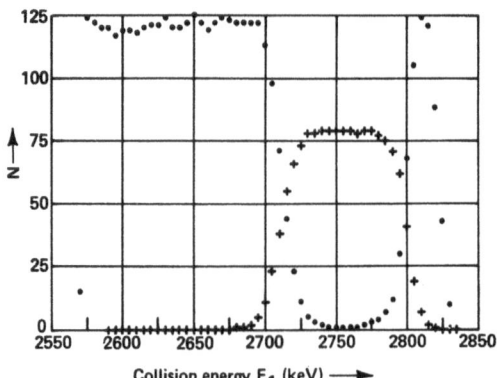

Fig. 5. Experimental spectrum from Fig. 4, transformed to the
form shown in Fig. 3c. The ordinate N is the atomic density
within each 5-kV interval of E_1, in units of 10^{15} atoms cm^{-2}.

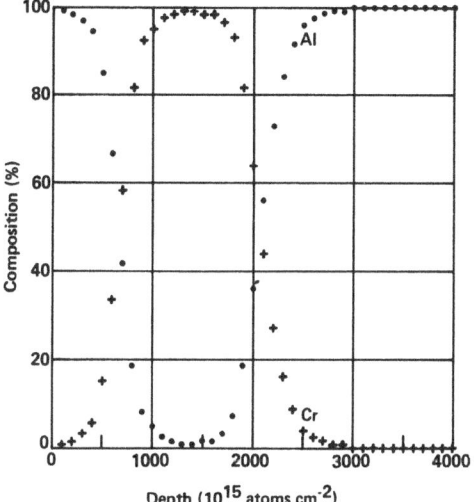

Fig. 6. Composition vs. depth profile, derived from Fig. 5 by integration and interpolation. The depth scale is in units of 10^{15} atoms cm^{-2}.

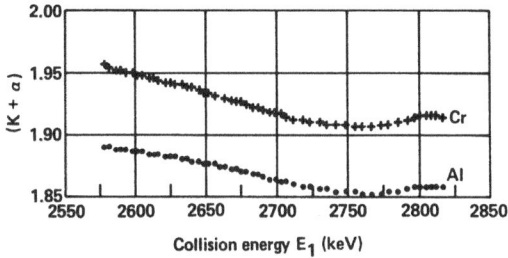

Fig. 7. Variation in $(K + \alpha)$ with depth for recoil from Cr and from Al. This is derived from the spectrum synthesis program, with Fig. 6 taken as input.

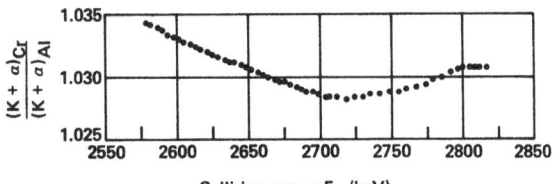

Fig. 8. Ratio of $(K + \alpha)$ values derived from Fig. 7. Note the small range of variation of this quantity.

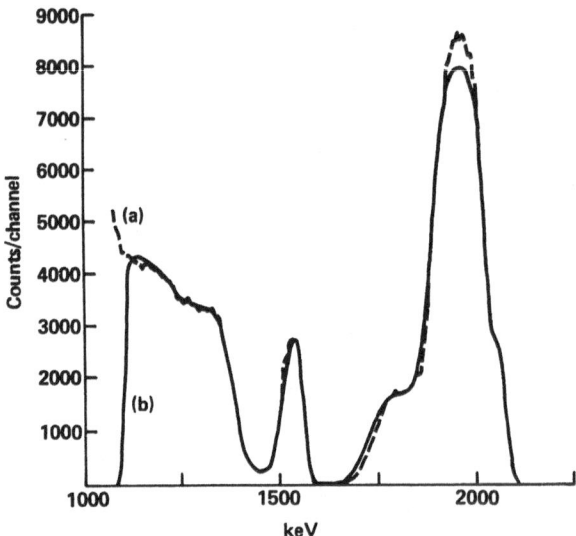

Fig. 9. Backscattering spectra for Cr-Al film heated at 400°C for 8 hours. (a) Experimental. (b) Theoretical, generated from calculated composition-depth plot (Fig. 11) by use of tabulated stopping cross sections. From Reference 1.

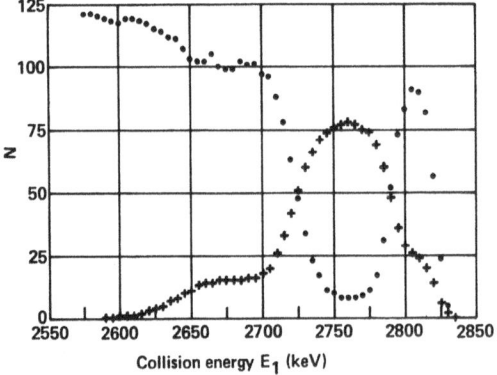

Fig. 10. Experimental spectrum from Fig. 9, transformed to the form shown in Fig. 3c. The ordinate N is the atomic density within each 5-kV interval of E_1, in units of 10^{15} atoms cm^{-2}.

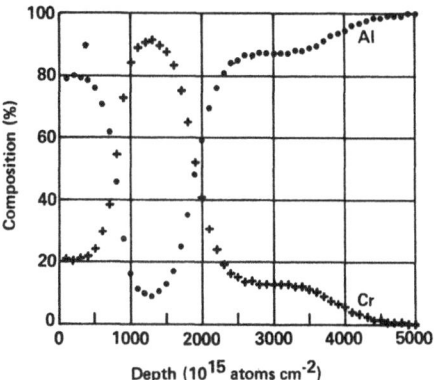

Fig. 11.   Composition vs.
depth profile derived from
Fig. 10 by integration and
interpolation.   The depth
scale is in units of 10^{15}
atoms cm^{-2}.

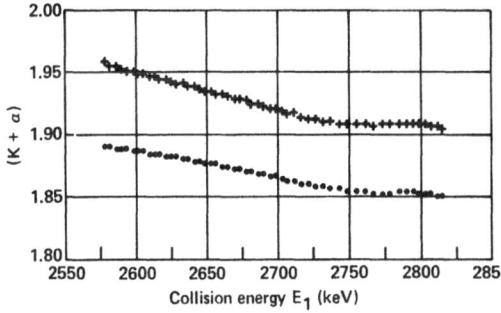

Fig. 12.   Variation in
$(K + \alpha)$ with depth for recoil
from Cr and from Al.   This
is derived from the spectrum
synthesis program, with
Fig. 11 taken as input.

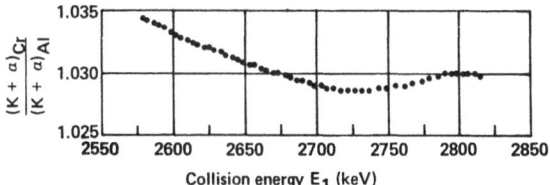

Fig. 13.   Ratio of $(K + \alpha)$
values derived from Fig. 12.
Though this curve is different
in detail from Fig. 8, the
range of variation remains
small.

Once the reconstructed spectrum had been resynthesized by use of the energy loss values tabulated by Ziegler and Chu [1], it became possible to calculate the values of $(K + \alpha)$ used in that calculation, and these are shown in Fig. 7 for both Al and Cr. The values of $(K + \alpha)$ are seen to increase by about 5% as E_1 decreases from 2820 keV at the surface of the film to 2580 keV at a depth of 5×10^{18} atoms cm^{-2}. Figure 8 shows that the ratio $(K + \alpha)_{Cr}/(K + \alpha)_{Al}$ varies by well under 1% with depth, showing that the Al and Cr spectra in Fig. 5 correspond with each other very closely indeed.

Figure 9 shows the spectrum obtained for the same Al-Cr-Al film after 8 hours of annealing at 400°C. The spectrum was then analyzed by the same procedure with the same values of α, and Fig. 10 shows the atom density vs. E_1 curve. Though the Al and Cr spectra fit together very well, it is clear that the fit is not as sensitive to the small changes in α_{Al} as in Fig. 5. Fig. 11 shows the resulting composition vs. depth profile. The composition at a depth of 3×10^{18} atoms cm^{-2} is 13% Cr, which strongly suggests the formation of $CrAl_7$ (which contains 12.5% Cr). The surface Al, in contact with excess Cr, appears to have formed $CrAl_4$. Figures 12 and 13 show the corresponding values of $(K + \alpha)$ and $(K + \alpha)_{Cr}/(K + \alpha)_{Al}$. Though these differ noticeably in detail from Figs. 7 and 8, the general conclusions are the same, namely that the variations in $(K + \alpha)$ with depth are small enough to confirm the validity of the method. The spectrum in Fig. 9, synthesized from the data in Fig. 11 for the Ziegler-Chu energy loss values, agrees well with the experimental spectrum.

ACKNOWLEDGMENT

I wish to acknowledge helpful discussions with W. K. Chu, without whose advice and encouragement this paper would not have been written.

REFERENCES

1. W. K. Chu, J. W. Mayer, M.-A. Nicolet, T. M. Buck, G. Amsel, and F. Eisen, Thin Solid Films 17, 1 (1973).
2. J. A. Borders and S. T. Picraux, Proc. IEEE 62, 1224 (1974).
3. J. F. Ziegler, J. W. Mayer, M. Ullrich, and W. K. Chu, in New Uses of Low Energy Accelerators, ed. J. F. Ziegler, Plenum Press, New York (1975).
4. D. K. Brice, Thin Solid Films 19, 121 (1973).
5. J. F. Ziegler and W. K. Chu, Atomic Data and Nuclear Data Tables 13, 463-489 (1974).
6. W. H. Bragg and R. Kleeman, Phil. Mag. 10, 5318 (1905).

7. An analytic expression for E_1 in terms of E_2, E_0, and energy loss information is given by W. K. Chu and J. F. Ziegler, J. Appl. Phys. <u>46</u>, 2768 (1975).

8. J. K. Howard, R. F. Lever, P. J. Smith, and P. S. Ho, J. Vac. Sci. Technol., to be published.

9. J. F. Ziegler, R. F. Lever, and J. K. Hirvonen, this conference.

COMPARATIVE ANALYSIS OF SURFACE LAYERS BY BACKSCATTERING

AND BY AUGER ELECTRON SPECTROSCOPY

J. K. Howard, W. K. Chu, and R. F. Lever

IBM Corporation, System Products Division, East Fishkill

Hopewell Junction, New York 12533   USA

ABSTRACT

The characteristics of backscattering and Auger electron spectroscopy
have been investigated by comparative analysis of a homogeneous
alloy (Al-Cu system); compound formation (Cr-Al system); and interdiffusion of two metals (Au-Ni system). The backscattering methods,
which are nondestructive, provide measurements of thickness and concentration, with an accuracy of 5% and a depth resolution of about
200 Å in the near surface region when a solid-state detector is
used. For thicker films, energy straggling of the beam decreases
the depth resolution. Backscattering provides quantitative information about the concentration of impurities and the depth distribution
of heavy elements in a light element matrix. It is not sensitive to
light impurities in a heavy element matrix.

Depth profiling with Auger electron spectroscopy (AES), which
requires ion sputtering, can detect nearly all elements. At the
surface it has a depth resolution of about 10 to 30 Å; with increasing depth, however, sputtering effects cause the resolution to
decrease.

INTRODUCTION

In the study of surfaces and thin films, Auger electron spectroscopy, AES [1], and ^{4}He ion backscattering [2] are two of the
methods most often used. AES is a surface method; to obtain depth
profiles, it must be combined with target sputtering. Sputtering
processes are reasonably well characterized for amorphous single-

element targets. For compounds, mixtures, or any other targets of
general interest, however, sputtering itself can disturb the surface
concentration; this fact complicates the interpretation of AES data,
making quantitative analysis difficult and less accurate. With appropriate standards, nevertheless, AES can provide quantitative
results. The major advantage of AES is that it is specific and
sensitive to nearly all elements; also, it has high depth resolution
near the surface region.

Backscattering yields depth profiles without sputtering. The
concentration analysis is performed without standards, because the
scattering and stopping cross sections for most materials systems
are known. Some weak points of backscattering are that it is not
sensitive to lighter atoms, and that it is mass-sensitive but not
specific; that is, materials of similar atomic number cannot easily
be analyzed.

These two methods, with their different capabilities, complement each other remarkably well. Several reports have described
a general comparison of analytical methods [3, 4]. This paper will
focus on a comparative analysis of specific materials systems by
both AES and backscattering methods, with emphasis on the quantitative features of each method. First, a detailed study of Cu
concentration measurements in the Al-Cu system (in a range from 1
to 12 at.%) is reported because of the importance of Al-Cu in
electromigration experiments [5]. Several AES procedures for
quantitative analysis are described, and the AES measurements of
relative concentration are directly compared to the absolute concentration data provided by backscattering and x-ray fluorescence.
Also, the influence of sputtering on the measurement of surface
concentration is discussed briefly. Second, the comparison of AES
and backscattering is extended to investigate compound formation
($CrAl_7$) in an Al-Cr thin-film couple. A paper to be published
elsewhere [6] will describe the kinetics and structure of the Al-Cr
reaction in greater detail. Finally, both methods are used to
study the interdiffusion of Au-Ni, with particular emphasis on the
quantitative analysis of the composition profile. Good agreement
for depth profiles was obtained under certain conditions.

EXPERIMENTAL PROCEDURE

Apparatus

Backscattering measurements with ^{4}He ions were performed at
2.0 to 2.8 MeV in a van de Graaff 3-MeV HVEC accelerator. The ^{4}He
beam was focused with a strong focusing quadrupole magnet. The
energy of the beam was analyzed by a magnetic analyzer, which held
the beam energy monenergetic to within 0.1%. The current was

typically 10-20 nA after the beam was collimated to 1/4 mm^2 on the
target. Backscattered particles were energy-analyzed by means of a
surface barrier Si detector with a preamplifier, a linear amplifier,
a biased amplifier, and a multichannel analyzer. The scattering
angle was 170°, and the solid angle was 4.11 msr. In most of the
runs the beam was normal to the target. In a few, the sample was
tilted to 60° to increase the effective thickness of the film.

AES measurements were performed on an Auger-ESCA spectrometer
(model 548) [7], manufactured by Physical Electronics Industries.
A double-pass cylindrical mirror analyzer was used without retard-
ation. The peak positions of the Auger signal were read directly
from a digital voltmeter to the nearest 0.1 eV. The AES spectra
were obtained when the primary electron beam was operated at 3, 4,
and 5 kV; the beam current was typically 40 μA. The depth profiles
of the films were measured by a sequential sputtering and multiplexing method [8]. For sputtering, the Xe ion gun was usually
operated at 1 kV and 30 mA with a Xe pressure of 5 x 10^{-5} Torr;
also, a 2-kV Xe beam and a current of 10 to 40 mA were used in a
study of sputtering effects. The position and focus of the ion
beam were adjusted by maximizing the charge collected in a Faraday
cup that had been pre-aligned with the electron beam.

Target Preparation

For the Al-Cu mixture film, samples were prepared by co-
evaporating Al and Cu onto oxidized Si wafers. The film thickness
was typically 5000 Å. The content of Cu in the film was controlled
by independently adjusting the current of the electron beam bombarding the Cu source, and thus changing the Cu deposition rate. The
Al evaporation was filament-heated at a constant rate of 30 Å/sec.
Both Al and Cu sources were thoroughly outgassed, and the system
was pumped for at least 16 hours to outgas the chamber walls. The
pressure of the system after outgassing was 1 x 10^{-8} Torr. A
liquid nitrogen cold trap was used around the evaporation chamber
and above the diffusion pump. The substrates were mounted on a
holder, which also acted as a heat sink during evaporation. No
cooling or heating was applied to the target during evaporation.
This minimized Al-Cu interdiffusion and compound formation, permitting
the fabrication of homogeneous alloy films.

The Al-Cr-Al layered films were prepared in the same evaporation system. About 5000 Å of Al was evaporated onto an oxidized
Si wafer by filament heating. About 1000 Å of Cr film was added by
electron beam evaporation; then a 1000-Å film was added to the
surface of the Cr. All three layers were deposited in a single
pumpdown. The as-deposited samples were cut into smaller pieces
and annealed in a quartz furnace that was flushed with dry nitrogen.
The temperature of the furnace was monitored with a thermocouple,
and controlled to within 1°C with a feedback system.

For the Au-Ni layered film, a thick Ni layer ($\leq$ 50,000 $\overset{o}{A}$) was electrochemically plated onto a substrate in a bath containing phosphoric acid. A thin Au layer (500 to 1000 $\overset{o}{A}$) was then evaporated onto the Ni film to form the diffusion couple.

Al-Cu MIXTURE

Ion Backscattering Analysis

In a backscattering analysis, the concentration of Cu and Al can be obtained from the spectrum height. The analysis is absolute if the scattering geometry and the electronics are calibrated. The concentration ratios can also be determined from the relative height of the backscattering signal. Let us define

$$H_1 = H_{Al}^{Al_{1-x}Cu_x} \tag{1}$$

$$H_2 = H_{Cu}^{Al_{1-x}Cu_x} \tag{2}$$

as the signal heights for Al and Cu from a target containing a uniform mixture of Al and Cu with x amount of Cu and (1 - x) amount of Al, where x < 1. The signal heights are related to the atom fraction x; to the differential scattering cross sections, σ_{Al} and σ_{Cu}; and to the energy loss and experimental conditions. Also,

$$H_1 = \frac{(1 - x)\sigma_{Al}\Omega Q \delta E}{[\varepsilon]_1} \tag{2}$$

$$H_2 = \frac{x\sigma_{Cu}\Omega Q \delta E}{[\varepsilon]_2} \tag{3}$$

where Ω is the solid angle of the scattering, Q is the total number of incident projectiles on the target, and δE is the energy channel for the detecting system, and where the scattering cross section factors $[\varepsilon]$ are defined as

$$[\varepsilon]_1 = [\varepsilon]_{Al}^{Al_{1-x}Cu_x} = K_{Al}\varepsilon(E) + \frac{1}{|\cos\theta|}\varepsilon(K_{Al}E) \tag{4}$$

$$[\varepsilon]_2 = [\varepsilon]_{Cu}^{Al_{1-x}Cu_x} = K_{Cu}\varepsilon(E) + \frac{1}{|\cos\theta|}\varepsilon(K_{Cu}E) \tag{5}$$

where K_{Al} and K_{Cu} are kinematic scattering factors for Al and Cu, respectively, and θ is the scattering angle, 170° for the present experiment. The stopping cross section for the mixture can be evaluated by using Bragg's rule:

$$\varepsilon = (1 - x)\varepsilon^{Al} + x\varepsilon^{Cu} \tag{6}$$

The stopping cross sections, ε^{Al} and ε^{Cu}, can be obtained from available tabulations.

A measurement of the spectrum height (Eq. 1 or 2) gives a solution of x. This is usually done by carefully measuring Ω, Q, and δE to obtain a height vs. x relation. An alternative is to take the relative height measurement. From Eqs. 1 and 2 we have the ratio

$$\frac{H_2}{H_1} = \frac{x}{1 - x} \frac{\sigma_{Cu}}{\sigma_{Al}} \frac{[\varepsilon]_1}{[\varepsilon]_2} \tag{7}$$

Three factors enter into this formula for height ratio: the concentration ratio, the scattering cross section ratio, and the inverse of the stopping cross section ratio. The last factor is nearly unity, because Eqs. 4 and 5 differ only by the scattering kinematics, and because the stopping cross section is in the same material. The ratio of Eqs. 4 and 5 is insensitive to the stoichiometry. For example, when $x = 0.01$,

$$\frac{[\varepsilon]_1}{[\varepsilon]_2} = 0.924$$

and when $x = 0.3$ the above factor becomes 0.911. The ratio of the scattering cross section σ_{Cu}/σ_{Al} can be approximated as $(29/13)^2$ -- that is, the square of the atomic number ratio. Corrected for recoil, it is 5.15. A height ratio measurement directly yields the concentration ratio.

The major advantage in relative measurements, as given in Eq. 7, is that any systematic error in ε, Q, and Ω does not influence the determination of concentration. The only disadvantage is that any low Z impurities that cannot be resolved in the spectrum will influence the height of the Al signal, and therefore will be overlooked in a relative calculation. One should attempt to make an absolute measurement to check the target purity and its consistency with the relative measurement. The relative target height does not have to come from the same spectrum; it could come from an independent run on a different target. For example, the spectrum height for pure Al should be

$$H_{Al}^{Al} = \frac{\sigma_{Al} \Omega Q \delta E}{[\varepsilon]_{Al}^{Al}} \qquad (8)$$

where $[\varepsilon]_{Al}^{Al}$ is the stopping cross section factor of pure Al. Any height ratio of Eq. 1 or Eqs. 2-8 will give information on the concentration x; and Ω, Q, and δE will be canceled in the calculation. Backscattering spectra for pure Al and for $Al_{1-x}Cu_x$ with x = 0.060 are given in Fig. 1. The 6% Cu value can be obtained from the height measurement, Eq. 1 or 2, or from relative height, Eq. 7, or by comparing Eq. 1 or 2 with Eq. 8. They all give the same answer, 6 ± 0.5%; Eq. 7 gives the best accuracy.

To check for consistency, a set of samples was run at 2.0 MeV and 2.8 MeV $^4He^+$. The results of both runs are in good agreement.

Equation 7 gives a nearly linear relation between height ratio and concentration ratio. As the above discussion shows, Eq. 7 can be reduced to

$$\frac{H_2}{H_1} = 4.7 \left(\frac{x}{1-x} \right) \qquad (9)$$

Four different Al samples of various Cu concentrations are measured at 2.0 MeV and 2.8 MeV $^4He^+$ backscattering, and their concentrations are determined by the analysis given above. They are 0.8%, 2.7%, 6%, and 11%.

The same four samples were analyzed by Auger electron spectroscopy. An example is given in Fig. 2, which gives the differentiated spectrum, N'(E) vs. Auger electron energy, for an Al sample containing 6% Cu. Before sputtering, the sample is partially covered with oxygen and carbon; their signals, along with the Cu and Al signals, can be seen in Fig. 2a. After sputtering for a few minutes, the C and O signals disappear and the Cu and Al signals are increased. A small amount of Xe is also observed (Fig. 2b), because some Xe has been implanted into the surface during sputtering.

AES Analysis

Three different methods are used in analyzing AES concentration measurements:

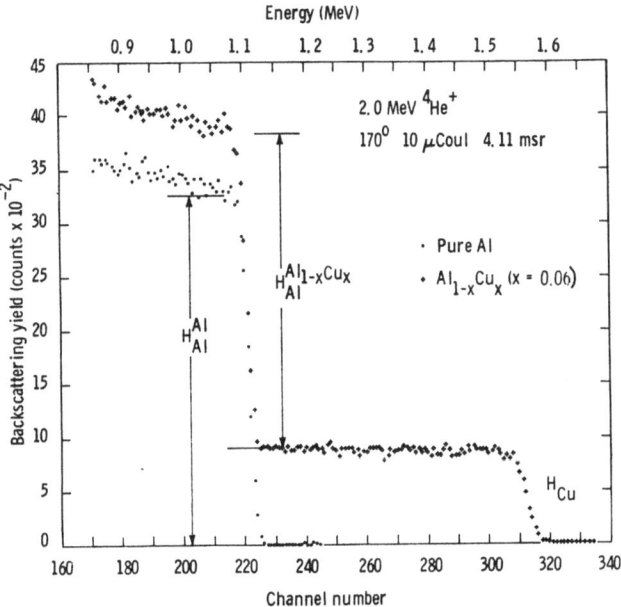

Fig. 1. Backscattering profile of Al-6% Cu sample; note that the Cu concentration is uniform throughout the film.

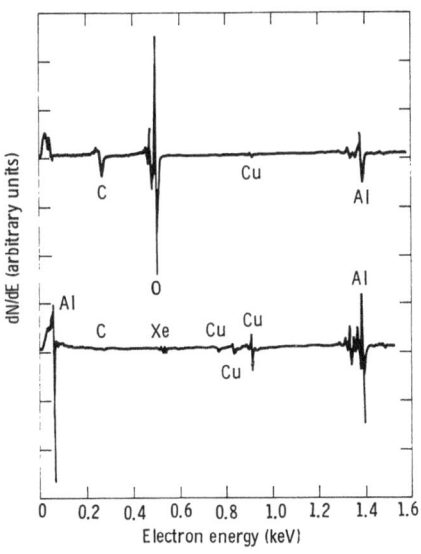

Fig. 2. AES spectrum of Al-6% Cu sample before (a) and after (b) sputtering.

(1)   Use of a standard sample (e.g. pure Cu) for cali-
      bration [9]
(2)   Use of secondary standards such as Ag [9]
(3)   A first-order (standard-less) approximation, suggested by
      Morabito [10]

We will describe the methods and summarize the analytical procedures required for quantitative AES measurements. The discussion
follows those of Palmberg [9], Morabito [10], and Chang [11].

The most important principle of quantitative AES is that the
Auger current, I_i for the i-th element, is proportional to the atom
fraction x_i of the i-th element in the sample over the range $0 <$
$x_i \leq 1$:

$$I_i = k_i x_i \qquad\qquad\qquad (10)$$

where I_i is the peak height in the N'(E) spectrum. The constant k_i
depends on the ionization cross section, the Auger transition
probability, the transmission of the cylindrical mirror analyzer,
the Auger escape depth, the backscattering correction, and other
experimental factors [11]. Equation 10 ignores matrix effects,
surface roughness, and variations in peak shape, and further
assumes that $E_p \gg E_i$ for all elements i, where E_p is the primary
electron energy and E_i is the atomic binding energy for element i.

Since in the present state of theoretical knowledge k_i cannot
be calculated for any arbitrary system, k_i must be measured by use
of a standard sample from which x_i is known. In general we may
write

$$x_i = (I_i / I_i^S) x_i^S \qquad\qquad\qquad (11)$$

where I_i^S is the signal height of the i-th element in the standard,
and x_i^S is the atom fraction of i in the standard. If a pure
elemental standard is used, $I_i^S = I_i^0$ and $x_i^S = x_i^0 = 1$. A potential
disadvantage of using pure standards is that the escape depth of
minor constituents, such as Cu in Al, may be considerably different
from the standard, Cu in Cu.

Often it is not convenient to use primary standards (pure
element i or mixtures with a known concentration of i). Secondary
standards can then be used. A good example is given in the handbook [12] provided by Physical Electronics Industries, Inc., which
depicts an extensive series of Auger spectra for an incident electron energy of 3 kV. The spectra were taken under similar experimental conditions, and the peak heights were referenced to Ag,
which exhibited the best signal-to-noise ratio. Using the relation-

ship in Eq. 11 for elements i and j, and ignoring matrix effects, we can write

$$(x_i/x_j) = (I_i/I_j) \cdot \left(I_j^0 \Big/ I_i^0\right) = (I_i/I_j) \cdot \left(I_j^s \Big/ I_i^s\right) \tag{12}$$

where I_i^s and I_j^s are normalized signal heights of elements i and j in standard samples. There are two ways of using this relation. The first involves the use of a secondary standard, Ag^{12} or Si^{13}, so that $x_j = x_j^0 = 1$ and

$$x = \left(I_i \Big/ I_j^0\right) \cdot \left(I_j^s \Big/ I_i^s\right) = \left(I_i \Big/ I_j^0\right) \alpha_i \tag{13}$$

where α_i is the inverse sensitivity factor for element i referenced to standard sample j. Provided our system is operated at $E_p = 3$ kV, then

$$\alpha_i = I_{Ag}^H \Big/ I_i^H$$

(relative to Ag) and

$$x_i = \left(I_i \Big/ I_{Ag}^0\right) \cdot \left(I_{Ag}^H \Big/ I_i^H\right) \tag{14}$$

where I_{Ag}^H and I_i^H are handbook [12] peak heights normalized by the proper scaling factor, and I_{Ag}^0 is the signal from a pure Ag standard. The second method involves the case in which a Ag standard is not available and requires that Eq. 13 be summed over all k elements in the sample so that $\sum_k x_k = 1$. It follows that

$$x_i = I_i \alpha_i \Big/ \sum_k^k I_k \alpha_k \tag{15}$$

Equation 15 has the advantage that no standard is required; the α_i's, however, are known only for $E_p = 3$ kV [12] or must be determined experimentally [13] and may vary with the experimental parameters. Chang [11, 13] has generalized the treatment of Eq. 15 to include matrix effects; the reader is referred to his work for details.

Morabito [10] has derived a first-order approximation for the quantitative analysis of Auger spectra obtained by use of a CMA. Using the assumptions implicit in the derivation, we can write

$$I_i/I_j = (x_i/\sqrt{E_i})/(x_j/\sqrt{E_j}) \qquad\qquad (16)$$

which, from Eq. 10, implies that $k_i \simeq 1/\sqrt{E_i}$ and $k_j \simeq 1/\sqrt{E_j}$, where E_i and E_j are the kinetic energy values of AES peaks for elements i and j. The atom fraction of element i is obtained from Eq. 5 by using $\alpha_i = \sqrt{E_i}/\sqrt{E_M}$, where E_M is the kinetic energy of an Auger line from the matrix. This relation requires that $\alpha_M = 1$ and assumes that Eq. 10 is strictly true for all components; it ignores matrix effects, etc.

The distribution of Cu throughout the film thickness was determined by means of AES combined with ion sputtering. The experimental conditions were E_p = 5 kV, I_p = 40 mA; the Xe sputtering conditions were E_{Xe} = 2 kV, I_{Xe} = 40 mA. The sputtering was interrupted at various times to obtain an entire spectrum for more accurate peak measurements and to detect the presence of new impurities not included in the multiplexing experiment. Survey scans were made at 5 eV/sec with a time constant of 0.1 sec; the modulation voltage was usually 6 volts peak to peak for the high-energy Al and Cu peaks. Detail scans of individual peaks were performed at a sweep rate of 1 eV/sec, with a time constant of 1 sec. The modulation voltage was 2 V. The depth profiles of all but one sample, Al-11% Cu, showed a uniform Cu profile throughout the film. The peak height ratio (Cu/Al) for the 11% sample was 0.95 for the first few minutes, and then dropped to 0.793 and remained constant at that level. The normalized peak heights are listed in Table 1. Segments from the same wafers were then mounted on the sample holder with a Cu standard in the form of a 5000-Å evaporated film of Cu. In order to compare the various quantitative AES methods discussed in the preceding section, depth profiles and spectra scans were obtained for E_p = 3 kV, I_p = 40 μA, and for E_{Xe} = 1 kV, I_{Xe} = 30 mA. The Cu standard was examined immediately after an Al-Cu sample to minimize the effect of multiplier drift with time. The Auger signal for the Cu standard (column 3 in Table 1) increased by 53% during the same day with all other experimental factors held constant. The Cu standard was sputtered about half as long as the Al-Cu samples so that the effect of roughness on AES yield would be comparable.

The AES concentration data are summarized in Table 2. Several statements can be made about the relationship between the absolute Cu levels and the relative Cu measurements. At E_p = 3 kV, the data from the Cu standard and the secondary (Ag) standard differ on the average by only 22%, but are greater than the absolute Cu levels by a factor of about 1.98 to 2.4. At the same E_p, the Cu concentration

Table 1. Auger Peak Height Data (Normalized).

E_p (kV)	Sample	I^{Cu}_{std}	I^{Cu}	I^{Al}	I^{Cu}/I^{Al}
3	A	6.55	0.105	1.11	0.095
	B	6.55	0.31	0.8	0.387
	C	4.29	0.55	0.7	0.785
	D	4.29	1.075	0.675	1.59
5	A		0.3	6	0.05
	B		0.6	3.87	0.155
	C		1.775	5.0	0.355
	D		3.61	4.55	0.793

obtained from the standard-less method of Morabito [10] is greater
than the absolute measurements by a factor of 5 to 9. At E_p = 5 kV
the Cu/Al ratio is lower than at E_p = 3 kV, by about 2.1; the
standard-less data are still greater than the absolute Cu concentration, on the average by a factor of 4.2. Finally, the backscattering and x-ray data for absolute concentration differ by only
5%.

Plotting the Cu/Al Auger peak ratio against the Cu/Al atomic
ratio, as measured by ion backscattering, yields a calibration
curve for E_p = 3 kV. The 3-kV calibration curve is plotted in
Fig. 3 along with an unpublished calibration curve for 5 kV [14].
Inspection of Table 2 or Fig. 3 shows that the 5-kV experimental
data from this study are in excellent agreement with the previously
determined data [14]. The agreement is probably no better than
±5-10%, allowing for measurement errors. Note that the initial
5-kV calibration curve was established with a single-pass analyzer
and Ar sputter etching; thus the peak ratio calibration method
appears to eliminate operational variables.

The minimum Cu concentration detectable in Al was estimated to
be less than 0.1 at.% for the conditions used in the current experiments. The signal-to-noise ratio was 22 for the 0.8% sample
(E_p = 3 kV, I_p = 40 μA); the background level corresponded to 0.04
at.%, and a signal twice the background height could be easily
distinguished. The error limits for the E_p = 3 kV study were

Table 2. AES Concentration Data (at.% Cu).
          Cu (920 eV)/Al (1396 eV).

(a) AES: E_p = 3 keV, E_{Xe} = 1 keV. Absolute concentration
calibrated by ion backscattering.

Sample	Cu/Al	Cu standard	Ag standard* (Refs. 9, 12)	Standard-less[†] (Ref.10)	Absolute concen-tration
A	0.095	1.6	1.9	7.1	0.8
B	0.387	4.7	7.3	24	2.7
C	0.785	13	14	39	6.0
D	1.59	25	25	56	11.3

(b) AES: E_p = 5 keV, E_{Xe} = 2 keV. Absolute concentration
calibrated by x-ray fluorescence.

Sample	Cu/Al	Calibration standard (Ref. 14)	Standard-less (Ref. 10)	Absolute concen-tration
A	0.05	0.8	3.9	0.8
B	0.155	2.3	11	2.5
C	0.355	5.1	22	5.3
D§	0.793	11.4	39	11.3

*Inverse sensitivity factors: Cu (920 eV), 0.313; Al (1396 eV),
 1.527 (valid only for E_p = 3 keV).

[†]Inverse sensitivity factor: Cu, 0.812.

§Spectra recorded at peak-to-peak modulation voltages of
 2, 4, and 6 V, with a maximum difference of 11% for Cu/Al.

established to be ±5% or better for the Cu/Al ratio from the same
surface.

     The factors that affect the Cu/Al ratio also directly affect
the Cu concentration measurement as obtained by secondary standards
[9, 12] and by Morabito's approximation method [10]. The effects
that can influence the Cu/Al ratio are summarized in the following
paragraphs.
     The effect of E_p. The relative ionization cross section is
known to be a function of the reduced energy [15] E_p/E_w, where E_w
is the atomic binding energy for ionization. If backscattering is
included, then the maximum ionization occurs when $E_p/E_w \simeq 6$ [11, 16].

Table 3. E_p/E_w, for Cu and Al transitions

Transition	E_p (kV)	E_w (kV)	E_p/E_w
Cu (LMM)	5	0.9311	5.369
	3	0.9311	3.222
Al (KLL)	5	1.5996	3.206
	3	1.5996	1.923

Table 3 contains the values for E_p/E_w for Cu and Al transitions. In Table 2 the Cu/Al ratio is reduced by a factor of 2 when E_p is increased from 3 to 5 kV. From the general shape [16] of the curve of ionization cross section vs. E_p/E_w, and the information in Table 2, one can speculate that the Cu/Al ratio (Table 2) is reduced because the ionization cross section increases more for Al than for Cu. The sputtering energy was also increased from 1 to 2 kV as E_p was increased from 3 to 5 kV. To estimate the effect of primary energy on the Cu/Al ratio, E_{Xe} was held constant at 1 kV and E_p was increased from 3 to 5 kV, in increments of 1 kV. The Cu/Al ratio decreased by about 35% as E_p increased from 3 to 5 kV. For fixed values of E_p (3, 4, and 5 kV) the Cu/Al ratio increased only about 15%, on the average, as E_{Xe} was decreased from 2 to 1 kV. The best agreement with absolute concentration values would appear to occur for higher values of E_p, for more efficient ionization, and at higher values of sputtering energy. Note, from Tables 2 and 3, that higher values of E_p/E_w would tend to bring Morabito's data into better agreement with the absolute Cu values. This is equivalent to satisfying Morabito's condition that $E_p >> E_w$.

   The effect of preferential sputtering and knock-on. Preferential sputtering and knock-on are sputtering effects [17] that can cause the elemental composition of the surface to be different from the bulk value. The extent to which sputtering changes the Cu/Al ratio was estimated in the following experiment. First, an Al-11% Cu sample was sputtered in the conditions E_{Xe} = 1 kV, I_{Xe} = 30 mA for 5 minutes to remove the accumulation of oxide and hydrocarbon from the surface. The AES yields for Cu, Al, and Xe are plotted against sputtering time in Fig. 4. At 10 minutes, we increased the

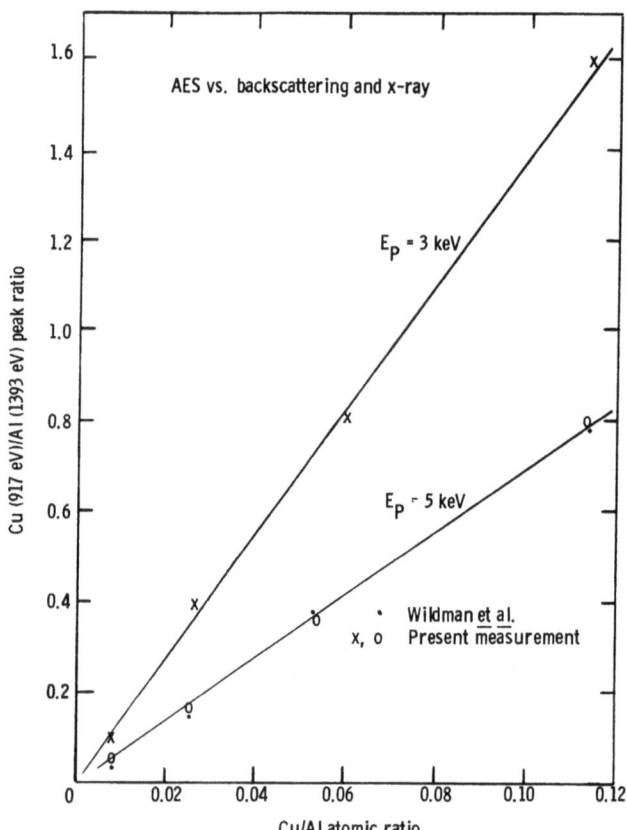

Fig. 3.   AES calibration curves
for E_p = 3 and 5 kV.

current to 40 mA, but no significant effect was observed. At 13.5
minutes we suddenly changed the sputtering conditions to 2 kV and
30 mA. After 1.5 minutes more, the Cu yield decreased and the Al
yield increased, to establish a new equilibrium surface concentration. The Cu/Al ratio decreased by 19% in the region of maximum
change (15 min), but an equilibrium decrease of 13% was soon
obtained. It appears that the signal height ratio (Cu/Al) is
influenced more by the energy of the sputtered ions than by the ion
dose (current). The amount of Xe retained in the Al film seems to
behave similarly.

The effect of escape depths in a homogeneous alloy. A homogeneous alloy of Al and Cu can produce conditions in which the
escape depths affect the Cu/Al signal ratio in a complex
manner [11, 13]. As Palmberg pointed out [18], the escape depth
depends critically on the valence band structure of the matrix.
The escape depth for Cu in an Al-rich matrix may be very different
from the escape depth of Cu in a Cu matrix. Such a difference
could certainly affect the results of the Cu standard method
(Table 2). To compensate for the dependence of escape depth on the
host material, it would be desirable to use Al-Cu standards with
known impurity concentrations [18], or else resort to a calibration
curve of the type displayed in Fig. 3. Since AES is more sensitive
to the top few monolayers, the alloy component with the shorter
escape depth may produce the greater yield [11]. The magnitude of
these effects on the Cu/Al ratio is unknown, but as a source of
potential error they must be considered.

The effect of chemical environment, peak shape change, and
backscattering. For co-deposited Al-Cu, these matrix effects are
not expected to have a significant effect on the Cu/Al signal
ratio. Analysis of Al/Cu films that were deposited cold and not
annealed should prevent any compound formation or bonding effects
that might affect peak shapes. The effect of energetic backscattered
electrons on the high-energy transitions of Cu and Al is also not
expected to change the Cu/Al ratio significantly.

The effect of sputtering on the surface concentration. All of
the relative AES measurements overestimate the Cu concentration
except those for E_p = 5 kV (Table 2, Fig. 3). The calibration
standards were obtained from x-ray measurements, which average the
Cu concentration over the thickness of the film and therefore are
insensitive to surface effects. Further, for E_p = 3 or 5 kV the
Cu/Al ratio increased by 13 to 19% as the sputtering energy, E_{Xe},
was decreased from 2 to 1 kV. These results suggested that
low-voltage sputtering could enrich the surface with Cu,
and might partially account for the overestimate of the Cu concentration. Since AES must use sputtering to obtain elemental
depth profiles, it could not be used to examine the effect of
sputtering on surface composition. Backscattering is usually not
very sensitive to the top few monolayers, and therefore a standard
analysis of backscattering data might not detect composition changes
in the near surface region. For that purpose, a more sensitive

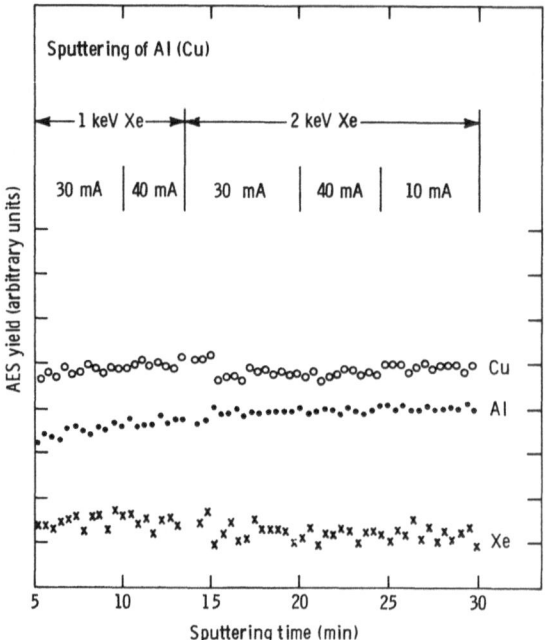

Fig. 4. AES depth profile of Al-11% Cu sample, showing the effect of sputtering variables on Al, Cu, and Xe yields.

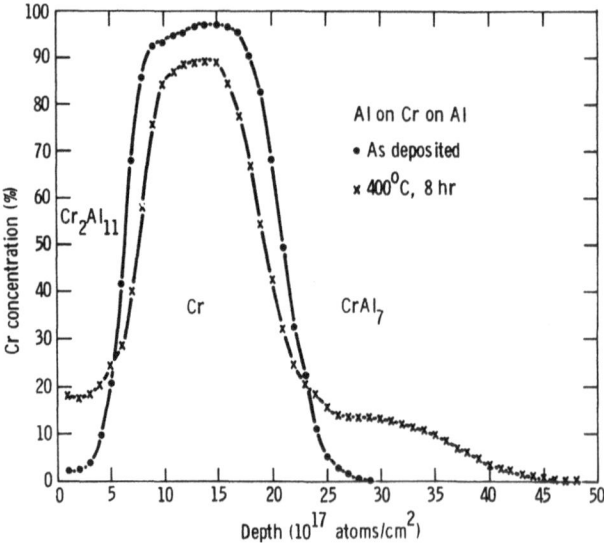

Fig. 5. Concentration-depth profile of Al-Cr-Al layer film, derived from backscattering spectra, shows compound formation after an 8-hour anneal at 400°C.

method was used by which the difference of two Cu spectra was taken
before and after sputtering [19]. The present analysis confirms
that after sputtering, excess Cu is found on the surface of samples
C and D. This enrichment is probably the result of enhanced diffusion and preferential segregation of Cu at the surface. At any
rate, it implies that the factor of 2 agreement between the data
for E_p = 3 kV (Table 2) for the relative Cu concentration and the
absolute (average) Cu level is a worst case. Elimination of the
enrichment or an independent measurement of the surface concentration should produce better agreement.

Al-Cr COMPOUND FORMATION

     Compound formation in thin-film couples of Al and Cr has been
investigated with backscattering to determine the phases formed,
Cr_2Al_{11} and $CrAl_7$, and the growth kinetics of $CrAl_7$ [6]. In this
section we compare information about metal-metal reactions (Al/Cr)
that can be obtained by means of AES and backscattering.

     The Al-Cr film was deposited in a layered sequence--1000 Å of
Al, 1200 Å of Cr, and 5000 Å of Al--on an oxidized Si wafer. Figure
5 depicts a composition-depth profile, derived from backscattering
spectra [20], before and after annealing at $400°C$ for 8 hours. A
measurement of the moment change in the backscattering spectra
yields the amount of compound formed [21]. By the moment method,
one can measure a thin layer of compound whose thickness is of the
order of the depth resolution.

     Quantitative analysis of the Al to Cr ratio in the AES
spectrum is possible but more complicated than for a homogeneous
film of Al-Cu. In the layered system, the AES signal ratio must be
converted into a concentration ratio, and also the sputtering time
must be converted into a depth scale. Figure 6 depicts an AES
depth profile of the $CrAl_7$ region in the sample. The starting
point for analysis was arbitrarily selected at a point at which Al
and Cr have the same signal heights, and should correspond to a
location at which the concentration ratio of Cr to Al is about one
to four. Measurements are included for the layer as deposited, and
after heat treatment at $400°C$ for 0.5, 2, 4, and 8 hours. The
upper half of the curve shows the height of the Al peak, and the
lower half is a plot of the corresponding Cr peak heights. During
the sputtering experiment, the multiplexing was stopped to record
the entire spectrum for a more accurate measurement of peak shape
and position. The Al spectra are superimposed on Fig. 6 for the
sample annealed at $400°C$ for 8 hours (identical with Fig. 5). The
spectra were recorded at the moments when the sputtering time
corresponded to 6, 12.5, 25, and 41 minutes. Note that the Al peak
was shifted from 66.1 ± 0.1 eV to 66.9 ± 0.1 eV; however, no corres-

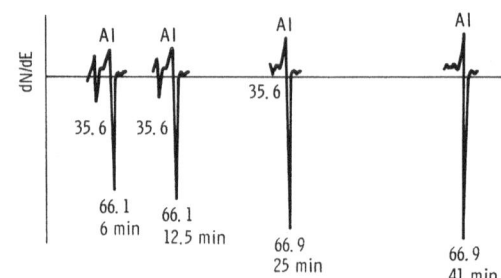

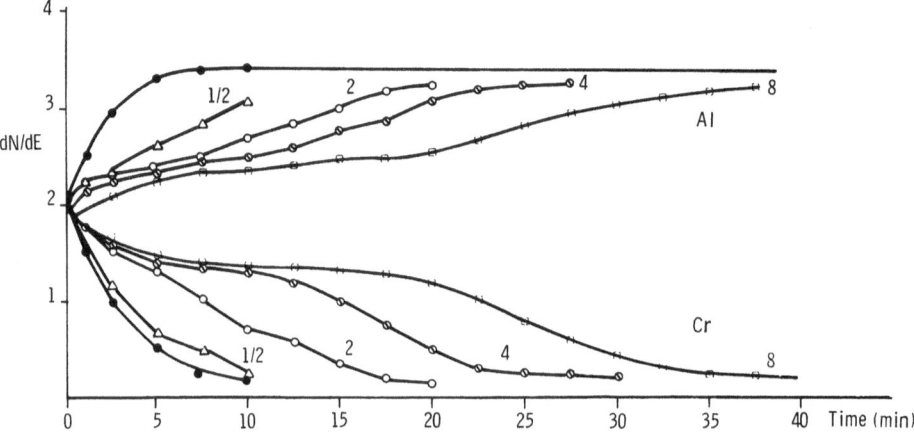

Fig. 6.   AES depth profiles showing CrAl₇ compound formation after anneals at 400°C. The Al spectra are superimposed to show an apparent chemical shift; the Cr peak positions did not change at the various depths in the film.

ponding shift of the Cr peak was observed (527.5 ± 0.2 eV). The identical behavior was observed for the Al and Cr peaks after 0.5, 2, and 4 hours; the as-deposited sample yielded peak positions of 67 eV for Al and 527 eV for Cr. Within experimental error, the low-energy Cr peak at 35.6 eV remains unchanged, as does the high-energy Cr peak. If the peak shift can be used to differentiate Al in $CrAl_7$ from elemental Al mixed with Cr, then a more exact value for the thickness of $CrAl_7$ can be determined by combining the AES and backscattering data. The midpoint of the tangent line drawn through the sloping section of the Al-Cr profile (Fig. 6) corresponds to about 25 minutes on the sputter-time axis. If this point is selected as the interface separating $CrAl_7$ from the unreacted film (66.9 eV Al peak in Fig. 6), then beyond this point Cr-Al interdiffusion can be considered to occur. The equivalent interface in Fig. 5 occurs at a depth of about 37.5×10^{17} atoms/cm^2. If we use the same zero point as in Fig. 6, the $CrAl_7$ layer contains about 16.8×10^{17} atoms/cm^2. This corresponds to a thickness of about 2658 Å obtained for a density of 3.18 gm/cm^3 for $CrAl_7$ [22]. By combining the AES and backscattering data, a Xe sputter rate of 106 Å/min is obtained for $CrAl_7$. The time axis in Fig. 6 can be directly converted to distance by use of the sputter rate measurement, if the sputter rate does not vary with composition.

Analyzing $CrAl_7$ by both AES and backscattering demonstrates how the two methods complement each other. Since backscattering infers the stoichiometry of compounds from elemental ratios (Al-to-Cr), it is insensitive to the location of the interface separating reacted and unreacted film. For $CrAl_7$ samples annealed 8 hours, the moment method yielded a total thickness of 3225 Å (2040×10^{17} atoms/cm^2). If based only on the backscattering data, the thickness calculated would be in error by about 20%.

Au-Ni INTERDIFFUSION

We have demonstrated (Fig. 2a) that AES is sensitive to low Z contaminants on the surface; backscattering cannot detect small amounts of contaminants if they have an atomic mass lighter than that of the material under study. A different example for inter-diffusion, is described here.

A 500-Å Au film was deposited on an electroless Ni deposit. The electroless Ni is known [23] to contain P and Ni_3P; thus interdiffusion in the Au-Ni-P system can lead to problems for backscattering. Figure 7 shows AES depth profiles of the Au-Ni(P) film, both as deposited and after being annealed at $400°C$ for 2

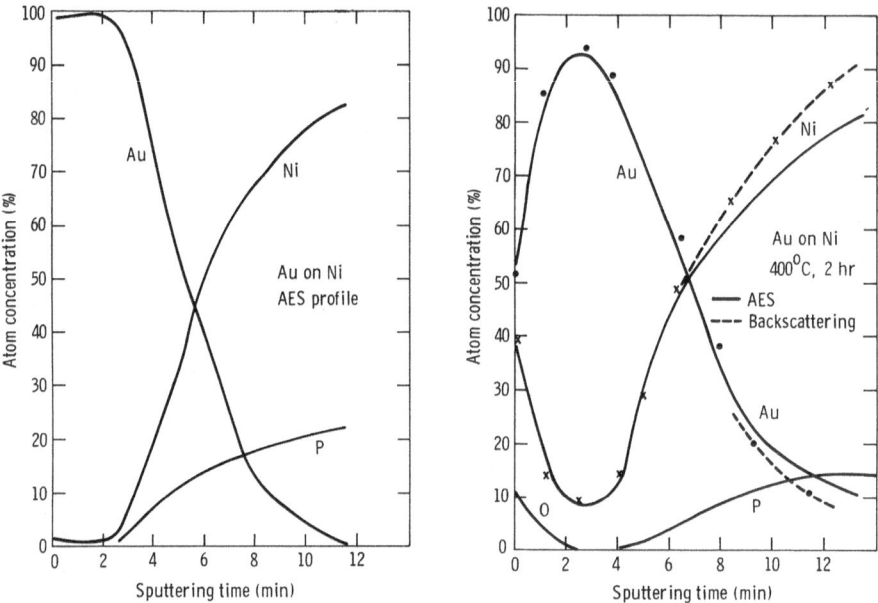

Fig. 7. AES depth profile of Au-Ni(P) thin-film couple before anneal (a) and AES-backscattering profiles after 400°C for 2 hours (b).

hours. The values for relative concentration were obtained by using Eq. 15 and the appropriate sensitivity values [12]. The peak heights were measured directly from the AES spectrum by interrupting the sputtering at various depths in the film. After the heat treatment, large amounts of Ni and O are observed at the surface and considerable Au-Ni interdiffusion has taken place. Backscattering spectra results are given in Fig. 7b as individual points. The conversion from a backscattering depth scale (atoms/cm^2) to AES sputter time (min) was made by use of a single conversion factor, which forced the 50% Au signal in the backscattering and AES profiles to pass through the same point. When the P signal is low, the AES and backscattering data are in good agreement. In the P-rich part of the Ni film, however, they begin to diverge. After about 10.5 minutes, the AES results are 12% P, 16% Au, and 72% Ni; the backscattering data are 12% Au and 80% Ni, which does not add up to 100% as required. The deviation between the two profiles could be due to increased surface roughening with depth, and to composition-dependent sputter rates. The sputtering effects are expected to be minimal, however, because the interface was reached in only a few minutes. The major source of error between AES and backscattering is attributed to the inability of backscattering to detect P in a Au-Ni matrix. The 8% not accounted for in the backscattering data can probably be assigned to P. The AES P measurement, 12%, is off by 50%, an acceptable error for the relative standards method.

SUMMARY

Homogeneous Al-Cu Film

Quantitative AES measurements were performed on a series of Al-Cu alloy films by various published methods [10-13]. The relative concentration values were obtained by using primary (Cu) and secondary (Ag) standards, and were typically within a factor of two of absolute Cu levels, as determined with backscattering and x-ray fluorescence. However, Cu-to-Al peak height ratio measurements were in excellent agreement ($\pm$5-10%) with calibration curves [14]. Deviations from the absolute Cu levels were attributed to a combination of matrix effects and surface enrichment of Cu due to sputtering [19]. Backscattering provides accurate (average) concentration values without standards and is an excellent calibration standard for AES.

Sequence of Al, Cr, and CrAl$_7$

Backscattering produces a quantitative depth profile of compound layers that can be converted to thickness if bulk density values are available. Although the profile is obtained without sputtering, only the atomic concentration ratio at a given depth is available; thus it is not sensitive to the interface separating the compound layer from interdiffusion of Al and Cr. AES sputtering experiments yield a semiquantitative profile of the compound layer, but thickness measurements depend on an independent determination of sputter rate. An apparent chemical shift in the AES spectrum was used to locate the compound interface in the backscattering profile, thus providing a potential improvement in the evaluation of compound thickness.

Au-Ni Interdiffusion

The interdiffusion of Au-Ni thin films was investigated with AES sputtering and backscattering. The effect of low Z impurities such as P on the composition profile was investigated. The AES and backscattering profiles were in good agreement for low P concentrations, but diverged appreciably when the P concentration reached about 12%. The error was attributed to the insensitivity of backscattering to low Z impurities in a high Z matrix, as well as surface roughness effects in AES sputtering.

Acknowledgment

The authors would like to thank Dr. H. H. Andersen of the University of Aarhus, Denmark, for discussions on the surface enrichment of Cu observed in this study.

REFERENCES

1. P. W. Palmberg, J. Vac. Sci. Technol., 9, 160 (1972); J. Vac. Sci. Technol., 10, 274 (1973).
2. W. K. Chu, J. W. Mayer, M.-A. Nicolet, T. M. Buck, G. Amsel, and F. Eisen, Thin Solid Films, 17, 1 (1973).
3. J. W. Mayer and A. Turos, Thin Solid Films, 19, 1 (1973).
4. R. Behrisch, B. M. U. Scherzer, and P. Staib, Thin Solid Films, 19, 57 (1973).
5. F. d'Heurle and R. Rosenberg, Physics of Thin Films, 7 (Academic Press, NY, 1973), p. 257; P. S. Ho and J. K. Howard, J. Appl. Phys., 45, 3229 (1974).
6. J. K. Howard, R. F. Lever, P. J. Smith, and P. S. Ho, J. Vac. Sci. Technol. (to be published).
7. P. W. Palmberg, J. Vac. Sci. Technol., 12, 370 (1975).
8. P. W. Palmberg, J. Vac. Sci. Technol., 9, 160 (1972).
9. P. W. Palmberg, Anal. Chem., 45, 549 (1973).
10. J. M. Morabito, Surf. Sci., 49, 318 (1975).
11. C. C. Chang, in Characterization of Solid Surfaces, eds. P. F. Kane and G. B. Larrabee (Plenum Press, New York, 1974), ch. 20.
12. P. W. Palmberg, G. E. Riach, R. E. Weber, and N. C. MacDonald, Handbook of Auger Electron Spectroscopy (Physical Electronics Industries, Inc., Edina, Minnesota, 1972).
13. C. C. Chang, Surf. Sci., 48, 9 (1975).
14. H. S. Wildman, J. K. Howard, and P. S. Ho, J. Vac. Sci. Technol., 12, 75 (1975).
15. H. E. Bishop and J. C. Riviere, J. Appl. Phys., 40, 1740 (1969).
16. J. Gallon, J. Phys. D., 5, 822 (1972).
17. J. W. Coburn and E. Kay, CRC Critical Reviews in Solid State Sciences, 4, 561 (1974).
18. P. W. Palmberg, in Electron Spectroscopy, edited by D. A. Shirley (Elsevier, New York, 1972), p. 857.
19. W. K. Chu, R. F. Lever, and J. K. Howard, to be published.
20. R. F. Lever (this conference).
21. R. F. Lever and W. K. Chu (this conference).
22. M. J. Cooper, Acta Cryst., 13, 157 (1960).
23. A. H. Graham, R. W. Lindsay, and H. J. Read, J. Electrochem. Soc., 112, 401 (1965); L. Spencer, Metal Finishing, 72, 35 (1974).

DISCUSSION

Q:(W.L. Brown) You suggest that Auger measurements prove good
relative values of concentration. Is this still the case taking
account of the effects of ligand atoms in the Auger spectra
themselves? As the Au/Cu concentration changes for example or
as compounds are formed?

A: (W.K. Chu) As the Cu/Al concentration changes so does the Cu/Al
AES signal ratio. Up to 12 % Cu concentration, we have a linear
relation as shown in one of the figures. We know at some higher
concentration this linear relation will break down but we have not
studied where it breaks down. Probably at low concentration. The
effects of ligand atoms in Auger yield are corrected in a same way
where "relative" measurement is made. That means the effects being
taken care of by normalization.

Q: (R. Behrisch) What was the diameter of your ion beam for sputtering compared to the electron beam for the Auger analysis in
order to avoid fringe effects?

A: (W.K. Chu) The diameter of the sputtered area is about 4 mm.
The electron beam size is roughly 1/4 of that of the ion spot.

Q: (J.M. Poate) Where do you think the Cu is sitting inside the
Al films?

A: (W.K. Chu) Cu, Al are co-evaporated. The amount of Cu is exceeding the solid solubility and the sample is not heat treated.
Therefore I would expect that Cu will precipitate in very small
size inside the Al matrix.

ANALYZING THE FORMATION OF A THIN COMPOUND FILM BY TAKING

MOMENTS ON BACKSCATTERING SPECTRA

R. F. Lever and W. K. Chu

IBM Corporation, System Products Division, East Fishkill

Hopewell Junction, New York 12533

ABSTRACT

The use of backscattering in the analysis of compound formation in a thin-film reaction becomes difficult when the thickness
of the compound layer is comparable to the depth resolution of
the instrument. This paper describes a method of circumventing
this difficulty by using the change that takes place in the
moment of the distribution of a given element. The growth of the
compound layer is proportional to the square root of the change
of moment between a reacted sample and an unreacted one. Alternative analytic methods are discussed, and an empirical cubic
correction is found to give results in good agreement with those
obtained by the moment method. The formation kinetics of a thin
film of Pd_2Si is taken as an example.

INTRODUCTION

Nuclear backscattering has often been used in studying thin-
film interactions [1-3]. It has proved particularly useful in
studying compound formation. In the early stages of compound
formation, however, the precise thickness of the compound layer
is very difficult to determine because it is of the order of the
depth resolution limit, 100 to 300 Å, of conventional semiconductor
particle detectors.

We will demonstrate this difficulty in a study of Pd_2Si
formation, and propose a method to circumvent it. The method is
to use the change of moment of a given element distribution to
determine the small amount of element that changes into compound.

149

The Pd_2Si system was chosen because (a) it has been studied extensively[4-8], and is well understood; and (b) it forms at temperatures around $200°C$, with relatively sharp interfaces.

We also studied Al-Cr interaction by means of backscattering moment analysis, sheet resistance measurement, Auger electron spectroscopy, and transmission electron microscopy; the results will be reported elsewhere [9].

EXPERIMENTAL PROCEDURE

Samples for this study were prepared by evaporation of Pd onto <100> Si wafers at a pressure of 1×10^{-7} Torr. Surface oxides were removed by dipping the substrates in HF and then cleaning them with deionized water just before pumpdown. The evaporated samples were heat-treated in a vacuum furnace at $200°C$ for 10 to 160 minutes, at a pressure of 10^{-6} to 10^{-5} Torr.

Backscattering measurements with ^{4}He ions were performed at 2.0 and 2.8 MeV with a 3-MeV van de Graaff accelerator. An ion beam of 10 nA was collimated and focused to $1/4$ mm^2. The scattering angle was $170°$, and the solid angle was 4.11 msr. Samples were run at a tilt of $0°$, and also at $60°$ to increase the effective thickness. At $60°$, because of the experimental setup, the solid angle was reduced to 2.127 msr. Energy spectra were obtained by using a surface barrier Si detector with a preamplifier, a linear amplifier, a biased amplifier, and a multichannel analyzer.

ANALYTICAL METHOD

Backscattering analysis has been well documented [10,11]. Here we will not go into details on the conventional method. Rather, we will look at the difficulty of applying it to a compound film whose thickness is of the order of the depth resolution. We will then propose a moment calculation as a method of circumventing the difficulties, and demonstrate this method on the same example.

Figure 1 is a schematic diagram of a Pd portion of a back-scattering spectrum. The dashed-line rectangular shape represents the ideal for a Pd film before reaction. The dashed curve, a typical error function curve, takes into consideration the detector resolution and the energy straggling; the former produces the slope on the front surface, and both produce the slope on the interface of Pd and the Si substrate. The solid line represents the spectrum for a target in which two-thirds of the Pd film has been converted into Pd_2Si after reaction with the Si substrate in heat treatment. The solid curve takes into consideration the detector resolution and the energy straggling. For simplicity we ignore the change in the energy loss over the film thickness and the change in the differential scattering cross section.

The thickness of the compound film formed can be calculated from the energy width ΔE, which is defined in Fig. 1 as the distance between the two half-heights. It is obvious that when the resolution functions of both steps do not interfere with each other, ΔE is an accurate measurement of the amount of the compound formation. This analytical method has been used in nearly all the metal-silicon interactions in backscattering analysis.

For a thinner film of compound, the precision of ΔE becomes uncertain. This is illustrated in Fig. 2, a schematic diagram of a backscattering spectrum of Pd. Only the interface region is shown. The dashed curve represents the signal from the unreacted Pd, and the solid curve represents the early stage of Pd_2Si formation. If we define ΔE_1 from the half-height separation of the reaction front, just as we defined ΔE in Fig. 1, we will overestimate the thickness of the compound layer. This is because, even for the unreacted Pd film, there is a finite spread ΔE_0. To remove the contribution of ΔE_0 from ΔE_1 is not easy.

In Fig. 1, where the two error functions are well separated, ΔE accurately reflects the thickness of the measurements. In Fig. 2, the error function at the Si/Pd_2Si interface and that at the Pd_2Si/Pd interface overlap. There is no simple direct relation between ΔE and the thickness of the film.

Figure 3 gives the real spectra of our backscattering measurements on Pd-Si interaction at $200^\circ C$ for 10, 20, 40, 80, and 160 minutes. The spectra are plotted with different notations; the spectrum of the as-deposited film is given as a dashed curve.

It is obvious that at the early stage of compound formation the ΔE separation is due mainly to the contribution of the resolution function ΔE_0, defined in Fig. 2. The uniformity of the compound layers has been verified for much thicker films. Several suggested ways of removing the contribution of ΔE_0 from ΔE for thin films have been considered. Simple subtraction $(\Delta E - \Delta E_0)$ and orthogonal subtraction $[(\Delta E^2 - \Delta E_0^2)^{1/2}]$ would seem to be the obvious methods, but neither works well.

We propose, by taking the change of moment on a given element distribution, to determine the small amount of compound formed. The moment P for a given element is defined as

$$P = \int yx \, dx = \delta x \sum yx \qquad (1)$$

where y is the elemental concentration (in atoms/cm^3) per unit thickness dx or δx at a given depth x, and x is expressed in

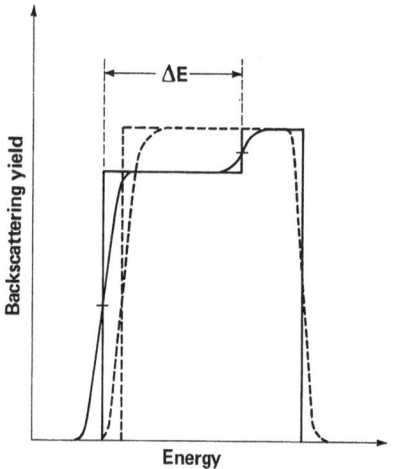

Fig. 1. Schematic Pd spectra for pure Pd (dashed line) and for thick Pd_2Si (solid line), where ΔE is large.

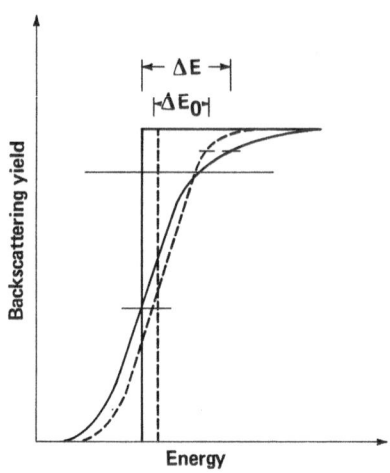

Fig. 2. The low-energy part of the Pd spectrum for pure Pd (dashed line) and for thin Pd_2Si (solid line), where ΔE is comparable with ΔE_0.

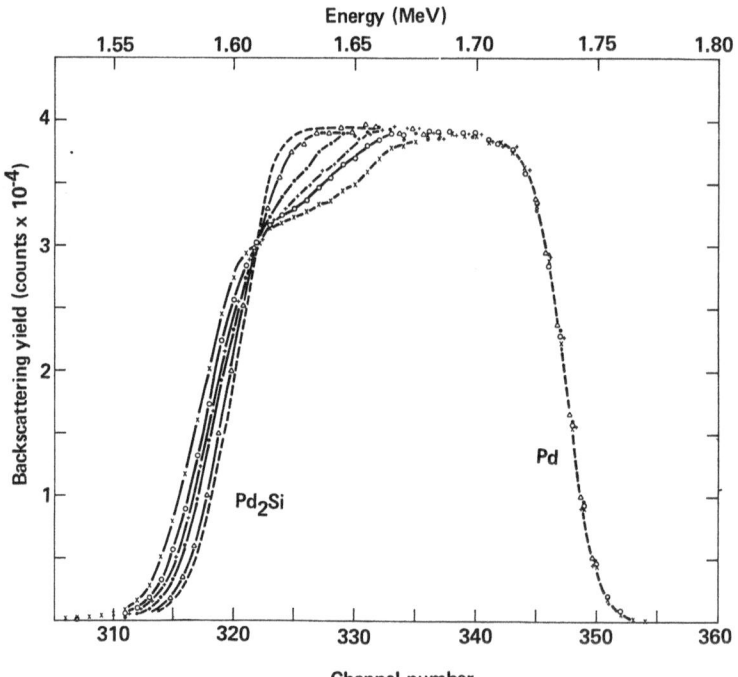

Fig. 3. The 2-MeV backscattering spectrum for Pd-Si reaction at 200°C, for 0, 20, 40, 80, and 160 minutes.

atoms/cm^2. In a backscattering energy spectrum, y can be treated
as the spectrum height (backscattering yield) at a given energy
x in the element profile. The latter case is an approximation,
usable   only for a film thin enough that the scattering cross
section and the energy loss do not change over the film thickness.
Where the scattering cross section and the stopping cross section
change appreciably, the energy spectrum can be normalized before
the moment is calculated.

The concept of moment is derived from the observation that
the total amount of material is conserved during compound formation.
The only change is the redistribution of material. In the
energy spectrum, the integrated counts for different growth
spectra are conserved after correction for the energy dependence
on the stopping and differential scattering cross sections.
Spectra taken at various stages of growth differ only in the
distribution of y, and this difference can be measured from the
shift of the interface, as given in Fig. 1 for thicker films, or
from the change of the moment. The effect is analogous to the
case of leverage, in which redistribution of mass changes the
torque of the system while the total mass is conserved.

We will demonstrate and verify the moment method from the
superposition of error functions and also apply it in analyzing
the formation of a thin Pd_2Si film (Fig. 3).

The application of the moment method to an error function is
described and justified in the appendix. In this section, we
will apply the method to the early stage of the growth of Pd_2Si.
The Pd parts of the Pd_2Si spectra for different stages of growth
are given in Fig. 3. This figure gives backscattering yield vs.
energy detected, and the spectra can be translated into the
energy just before scattering. The relation of the detected
energy, the incident energy, and the energy just before scattering
is complicated and can be handled by approximation [12] or by an
analytical method [13].

After Fig. 3 is translated into backscattering yield vs.
energy just before scattering, the yield can be normalized by
making a $(1/E^2)$ correction for the energy dependence of the
scattering cross section. For example, for a film that produces
an energy loss of 100 keV for 2-MeV ^{4}He ions in the incident
path, the maximum correction the yield is $(1.90/2.00)^2$, i.e.
10%. The scattering yield is also related to the energy loss in
both the incident path and the outgoing path in the Pd_2Si, and to
the kinematics of scattering [10]. The variation of the scattering
yield due to energy loss change is only 1%. For the purpose of
taking moment differences between different spectra, this 1%
error is partially canceled. In fact, for practical application
one can ignore the variation of yield due to energy loss change.

Before calculating the moment (Eq. 1), one has to check whether the total area under the spectrum is the same after different heat treatments, by determining whether each film contains the same number of Pd atoms per unit area. A small nonuniformity in the film thickness can change the moment of a given spectrum so much that no meaningful result can be obtained. To avoid this difficulty, one can adjust the total number of Pd atoms in the system by arbitrarily clipping the surface part of the spectrum, in which no compound forms. The amount clipped depends on the variation of the film thicknesses among the targets studied. The effect of this operation is to specify an artificial surface position under which there is no variation in the total amount of Pd. For different samples, different artificial surface positions are specified.

Then, by use of Eq. (1), the moment of a given system is calculated from the artificial surface (taken as the origin) down to the underlying Pd. For this calculation, y still has the original definition, but x is already offset because the starting points (artificial surfaces) are different for different spectra.

The description of the method is repeated in the schematics given in Fig. 4. Figure 4a shows a given amount of material uniformly distributed in depth, or in an energy spectrum that gives counts per channel uniformly distributed over the energy scale. This is the case for the spectrum of an as-deposited film. The total area under the rectangular shape is 64 units, and the total moment is 168 units, as calculated from Eq. (1) and given in Fig. 4a. For the moment calculation the origin is arbitrarily taken to be the surface of the film. As Fig. 4b shows, after a given heat treatment there is compound formation, and the amount of material is redistributed at a greater depth. For convenience, the height of the compound is arbitrarily selected to be half of the original height. Now an artificial surface is drawn at x = 0; thus the total area on the left-hand side of the origin is identical with that of Fig. 4a, i.e., 64 units. The amount of material on the right-hand side of the artificial surface is due to nonuniformity of the film, and will be ignored in the moment calculation. For this application, comparisons of moment are meaningful only for systems containing the same amount of material. In Fig. 4b, the total moment of the system is 184 units and the moment change from the as-deposited system is 16 units. The amount of compound formed, judging from the energy shift, is $\Delta x = 4$ units.

For Fig. 4c the sample has been further heat-treated to form a thicker compound film. The line representing the artificial surface is drawn such that the total area is conserved. The moment is calculated to be 232 units, the moment change from the as-deposited system is 64 units, and the compound growth is $\Delta x = 8$ units.

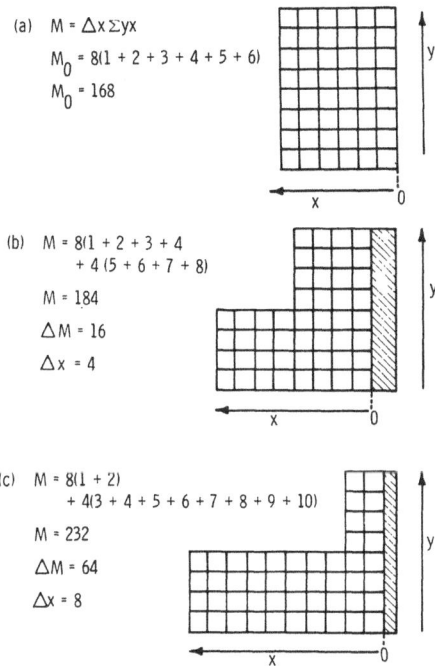

Fig. 4. Description of the moment method. (a) Unreacted film.
(b) Compound layer 4 units wide. (c) Compound layer 8 units wide.
The shaded areas in (b) and (c) represent additional thickness due
to nonuniformity in the film. The moment change is proportional to
the square of the compound thickness.

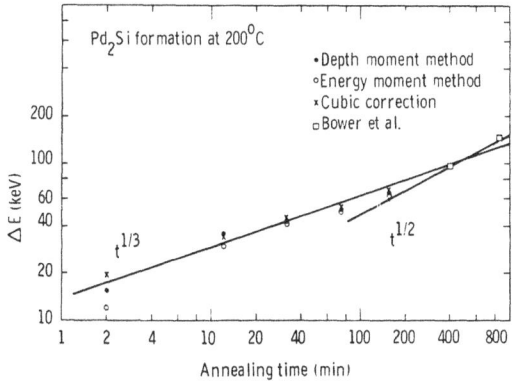

Fig. 5. The time dependence of the Pd_2Si thickness, as estimated
by various methods. The magnitude of the growth rate agrees with
that given by Bower __et al.__[6]. The early stages of growth appear
to obey a $t^{1/3}$ law.

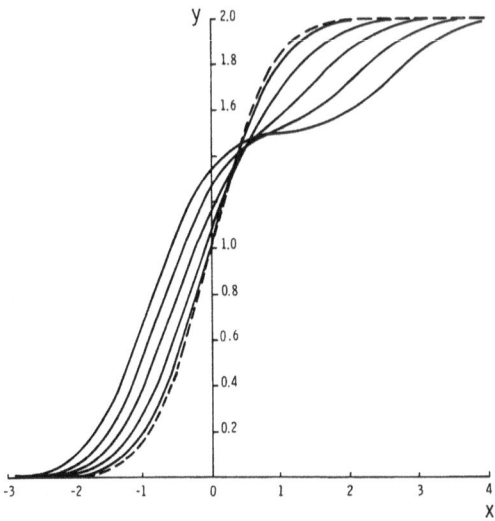

Fig. 6.   Simulated thin compound film formation according to Eq. (A4).

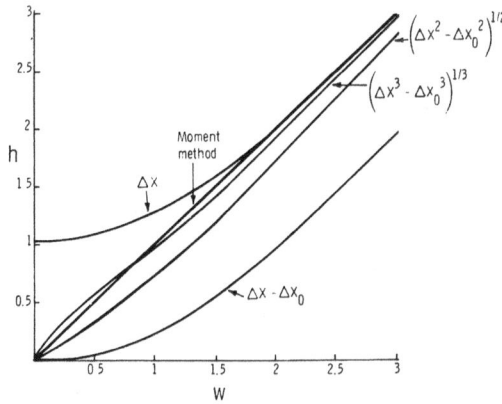

Fig. 7.   Calculated film thickness h from Fig. 6 vs. the real thickness w for varous methods.   The moment method is exact, and the
cubic method gives an excellent approximation.

In this schematic example we have shown how to select the artificial surface so as to conserve the total area, and how to calculate moment changes. Also, we have demonstrated that the moment change is proportional to the square of the thickness change. In the Pd_2Si analysis, corrected spectra are then shifted by 0.3 to 1 channel to give the same area.

Next the moment of a given system is calculated for the shifted spectrum, and the amount of compound formed is calculated from the moment changes by the following relation:

$$w = \sqrt{2\Delta M/[y_0 f(1 - f)]} \qquad (2)$$

With respect to the energy domain, y_0 is the spectrum height for the unreacted part of the spectrum, f is the spectrum height ratio of the reacted part to the unreacted part, and w is the number of channels or the energy width of the compound layer. With respect to the depth profile, y_0 is the atomic density volume of the unreacted elemental film, $y_0 f$ is the atomic density for the reacted compound, and w is the thickness of the compound film in atoms per unit area.

Equation (2) can be derived simply from the definition of moment and its change due to compound formation. The moment method can be used in studying the kinetics of the formation of Pd_2Si. When the moment change is plotted against annealing time, an intercept at 8 minutes is found for zero moment change. We define this intercept as the incubation time, and offset all the time scales by this 8 minutes. In Fig. 5, the time dependence of the Pd_2Si growth at $200^\circ C$ is given. We use the energy shift ΔE to indicate the growth. The closed circles give the growth from the moment change, calculated from the depth profile. The conversion from depth to energy is $[\varepsilon] = 460 \times 10^{15}$ eV/molecule cm^{-2}. The open circles give the moment change as calculated directly from the energy spectrum. The x's represent the direct measurement of energy shift after a cubic correction such as that given in the appendix. The squares are measurements made by Bower et al. [6]. The good agreement between the results of the current analysis and those of Bower et al. is encouraging. The overall slope of the formation kinetics at the early stage of the growth appears to follow a $t^{1/3}$ relation rather than the $t^{1/2}$ relation reported in Ref. 6. We do not understand the reaction kinetics at the beginning of the growth. Rather, we use the growth of Pd_2Si at a very early stage as the vehicle to demonstrate the analytical method.

If the data analysis is computerized, the moment method is straightforward, and unique in detecting a small amount of compound growth. If the analysis is done by hand, the moment method is not very practical because of its complexity. Then a cubic

correction, as given in the appendix, will give a very accurate
value for the amount of compound formed. This cubic correction
is an empirical approximation, and cannot be derived or justified.
Its accuracy can be estimated from the last figure in the appendix.

   The moment method may also be applied directly to certain
diffusion problems, as long as the diffusion coefficient D is
independent of concentration. In this case the moment change
will be a simple function of $\sqrt{Dt}$.

CONCLUSIONS

1.   In the backscattering analysis of compound formation, a
     change of moment in the profile of the energy spectrum makes
     possible an accurate and sensitive detection of an amount of
     compound formation too small to be evaluated correctly by
     the conventional method of analysis.
2.   The moment method assumes that
     (a) The resolution function is symmetric and independent
          of depth over the region of compound film.
     (b) Only a single compound phase, with well defined
          stoichiometry, forms during reaction.
     (c) The compound is uniformly distributed laterally, and
          interfaces between compound and elements are sharply
          defined.
3.   In the earlier stage, the growth kinetics of Pd_2Si follows
     the $t^{1/3}$ rule. For a thicker compound, the reaction rate
     agrees with the published data.
4.   A cubic correction method is also introduced as an alternative
     method for evaluating the formation of a thin film of
     compound. This empirical correction gives excellent results.

APPENDIX

   Under idealized conditions, the backscattering spectrum of a
compound may be considered to result from the convolution of a
Gaussian of standard deviation σ with a step of height y_o. The
change of moment ΔM on such a step is given by

$$\Delta M = \frac{1}{2} y_0 \sigma^2 \tag{A1}$$

Since any profile having $y \to y_0$ as $x \to \infty$ and $y \to 0$ as $x \to -\infty$ can
be regarded as being made up of a series of small steps, and
since the moment change for each step does not depend on the
point about which moments are taken, Eq. (A1) applies to any step
of height y, irrespective of the shape of the step. In practice,

the moment calculation has to include the whole part of the profile in which variation is <u>appreciable</u>. Although, in theory, y varies somewhat over the whole range $-\infty < y < +\infty$, in practice it is sufficient to stay within $\pm 3\sigma$ of any interface. Similarly, the moment change of peaks may be ignored if the moment is taken from a point at least 3σ away from the peak.

Let us consider two "ideal" spectra having moments M_1 and M_2, so that according to Eq. (2),

$$w = \left\{ [2(M_1 - M_2)] / [y_0 f(1 - f)] \right\}^{1/2} \qquad (A2)$$

On convolution with a Gaussian of standard deviation σ, the new moments will be

$$M_1' = M_1 + \frac{1}{2} y_0 \sigma^2$$

$$M_2' = M_2 + \frac{1}{2} y_0 \sigma^2$$

Hence, $M_1' - M_2 = M_1' - M_2'$, and we have

$$w = \left\{ [2(M_1' - M_2')] / [y_0 f(1 - f)] \right\}^{1/2} \qquad (A3)$$

Hence it is not necessary to obtain the ideal spectra by de-convoluting the real spectra, since the moment difference between spectra does not vary on convolution. This will be generally true for any symmetric resolution function.

We now examine possible approximation methods based on measuring differences between individual points on spectra, for cases in which integration is not convenient. Consider the series of curves:

$$y = 0.75 \text{ erfc } (0.25 \ w - x) + 0.25 \text{ erfc } (0.75 \ w + x) \qquad (A4)$$

for $w = N/\sqrt{2}$ and $N = 0, 1, 2, 3, 4, 5$. These represent the back-scattering spectra produced by a compound whose spectral height is 3/4 of that for the pure compound and whose energy spread before convolution with the Gaussian detector junction is $N\sigma$. This family of curves is shown in Fig. 6. The problem is to measure w from the curves. We have examined various functions h $h(\Delta x)$, where Δx corresponds to the quantity ΔE shown in Figs. 1 and 2. The value of Δx for a given curve in Fig. 6 is the difference between half heights, i.e., the intersections of the curve with the lines $y = 1.75$ and $y = 0.75$. The most widely used method of analysis is to write

$$h = \Delta x \tag{A5}$$

This is known to be very accurate when the error functions are separated widely enough; Eq. (A5) is shown as the top curve in Fig. 7. It is seen that h is very close to w when w > 2, but becomes rapidly worse for lower values. Equation (A5) gives $h = (\Delta x)_0 = 1.04$ when w = 0—that is, when the method is applied to the dashed curve of Fig. 6. This curve is simply y = erfc (-x), representing an abrupt interface with no compound present. To avoid this problem it might appear intuitively reasonable to write

$$h = \Delta x - (\Delta x)_0 \tag{A6}$$

This expression, shown as the bottom curve in Fig. 7, correctly gives h = 0 at w = 0. For all other values of w, however, it gives results that are much too low. Clearly, a method is required which somehow subtracts $(\Delta x)_0$ but gives it less weight.

The other curves shown in Fig. 7 are:

$$h = [(\Delta x)^2 - (\Delta x)_0^2]^{1/2} \tag{A7}$$

$$h = [(\Delta x)^3 - (\Delta x)_0^3]^{1/3} \tag{A8}$$

It can be seen that (A8) represents an excellent approximation, which does not detract from the virtually exact answer given by (A5) at w > 3 and yet gives a reasonable fit down to w = 0.5 σ.

ACKNOWLEDGMENT

We are indebted to R. E. Bischoff for his assistance in preparing the Pd_2Si samples, and to J. K. Howard for several stimulating discussions.

REFERENCES

1. J. W. Mayer, Proc. International Conference on Applications of Ion Beam to Metals, Albuquerque, New Mexico, Oct. 1973, p. 141.
2. J. A. Borders and S. T. Picraux, Proc. IEEE 62, 1224 (1974).
3. J. W. Mayer and K. N. Tu, J. Vac. Sci. Technol. 11, 86 (1974).
4. R. W. Bower and J. W. Mayer, Appl. Phys. Lett. 20, 359 (1972).
5. C. J. Kircher, Solid State Electron. 14, 507 (1971).
6. R. W. Bower, D. Sigurd, and R. E. Scott, Solid State Electron. 16, 1461 (1973).

7. W. D. Buckley and S. C. Moss, Solid State Electron. 15, 1331 (1972).
8. D. H. Lee, R. R. Hart, D. A. Kiewit, and O. J. Marsh, Phys. Status Solidi (a) 15, 645 (1973).
9. J. H. Howard, R. F. Lever, P. J. Smith, and P. S. Ho, to be published in J. Vac. Sci. Technol.
10. W. K. Chu, J. W. Mayer, N-A. Nicolet, T. M. Buck, G. Amsel and F. Eisen, Thin Solid Films 17, 1 (1973).
11. J. F. Ziegler, J. W. Mayer, M. Ullrich, and W. K. Chu, New Uses of Low Energy Accelerators, edited by J. F. Ziegler, Plenum Press, New York (1975).
12. R. F. Lever (this conference).
13. W. K. Chu and J. F. Ziegler, J. Appl. Phys. 46, 2769 (1975).

DISCUSSION

Q: (S.T. Picraux) What are constraints on the moments method? For example what if one has a change in the compound being formed or the presence of multiple compounds ?

A: (W.K. Chu) The boundary condition for applying moment method should be determined before the analysis. That means one should know the stoichiometry of the compound. In our application we restrict ourself to a single compound formation system. The analysis can be extended to the diffusion analysis where no compound is formed.

COMPUTER ANALYSIS OF NUCLEAR BACKSCATTERING

J. F. Ziegler
IBM-Research
Yorktown Heights, New York 10598, USA

R. F. Lever
IBM S.P. Div.
Hopewell Junction, New York, 12533, USA

J. K. Hirvonen
Code 6672
Naval Research Lab
Washington, D.C., USA

ABSTRACT: The potential use of computers in analyzing nuclear
backscattering spectra is discussed in respect to simplifying
necessary input so spectra can be easily generated by non-experts.
Such spectra can be used for the analysis of merged peaks, the
removal of isotope effects from high resolution spectroscopy, the
deconvolution of spectra to minimize effects of detector resolution,
and the parameterization of thin film interdiffusion. Both experimental and theoretical spectra are shown illustrating the power
and difficulties of advanced analysis.

INTRODUCTION

One of the advantages of nuclear backscattering is its
simplicity of interpretation. This has been true until recently
because targets were selected with consideration of the limitations
of the technique. Except for the study of oxides, there have been
few papers written which involved the untangling of merged peaks.

The formalism of the calculation of backscattering spectra has
been detailed before by various authors.(1-3) These calculations
are straightforward, and merely calculate detailed numerical integration of the energy loss into a target, the elastic scattering of
the projectile, and energy loss back out of the target. Brice (4)
added to the calculation the effects of projectile straggle for
thick targets, and also proposed a method for untangling merged
peaks by assuming Bragg's rule (the additivity of stopping cross-

sections in multi-element layers). Recently an analytic solution
to backscattering spectra has been described (5) which eliminates
the numerical integrations and allows such spectra to be calculated
quickly on very small computers.

We have added to these calculations the effects of isotopes,
the effects of detector resolution, the energy loss of projectiles
in the detector surface metal layer, and the interdiffusion of two
or more layers. We have not included projectile straggling as the
calculation of this effect cannot be accurately predicted (6,7).

BASIC CONCEPT

We have based our computer program on the premise that backscattering spectra calculations can be packaged so that spectra
may be generated by persons with only limited knowledge of backscattering. Such persons can use simple spectrum generating
facilities to match backscattering spectra with models of their
targets. This approach allows convincing evaluation of layer
thicknesses, compound stoichiometry, and contamination identification. Also, the non-expert can identify many second order effects
such as layer interdiffusion and thin surface oxides. This approach
is possible because the physical processes of backscattering are
well understood, and the experimental equipment is very stable.
Only for a few common elements (such as Zr and Pb) are there troubles
because energy loss data appear inaccurate, and for some compounds
(such as SiO_2) where Bragg's Rule appears to be in error by at
least 10%. And once this spectrum generator program is written,
much of the effort in backscattering analysis disappears, and more
time can be devoted to the physical experiment. The data base
needed for backscattering spectra calculations includes:

Target Information
- Table of Chemical Symbols
- Table of Atomic Numbers
- Table of Atomic Masses
- Table of Isotope Masses
- Table of Isotope Relative Abundances
- Table of Normal Densities (Atoms/cm^3)

Projectile Information
- Table of Stopping Cross Sections

Detector Information
- Table of Energy Loss in Detector Dead Layer

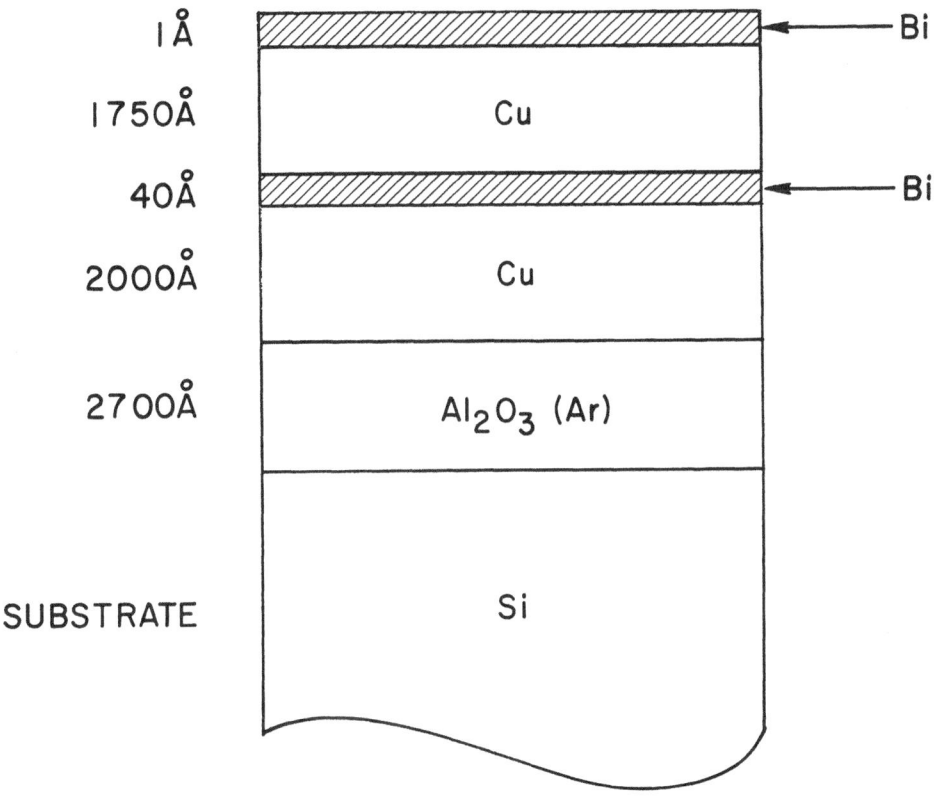

Figure 1.   Schematic of a complicated sample of several layers.
            Backscattering from such a sample is shown in Figures 2
            and 3 where both experimental and theoretical spectra
            are superimposed.

Experimental
Information
{
Projectile Atomic Number, Mass and Energy
Target Tilt to the Beam
Detector Angle, Solid Angle, and Resolution (FWHM)
Total Projectile Coulombs
Multichannel Analyzer (keV/Channel)
}

        The above information allows the generation of backscattering
spectra once the target is identified. The user must only input the
target, specifying the elements as a function of depth. For example,
consider the target shown in Figure 1. This may be specified to
the computer as:

LAYER	ELEMENT	PERCENTAGE	THICKNESS
1	BI	100	1 Å
2	CU BI	99.9 0.1	1750 Å
3	BI	100	40 Å
4	CU	100	2000 Å
5	AL O AR	35 55 10	2700 Å
6	SI	100	20000 Å

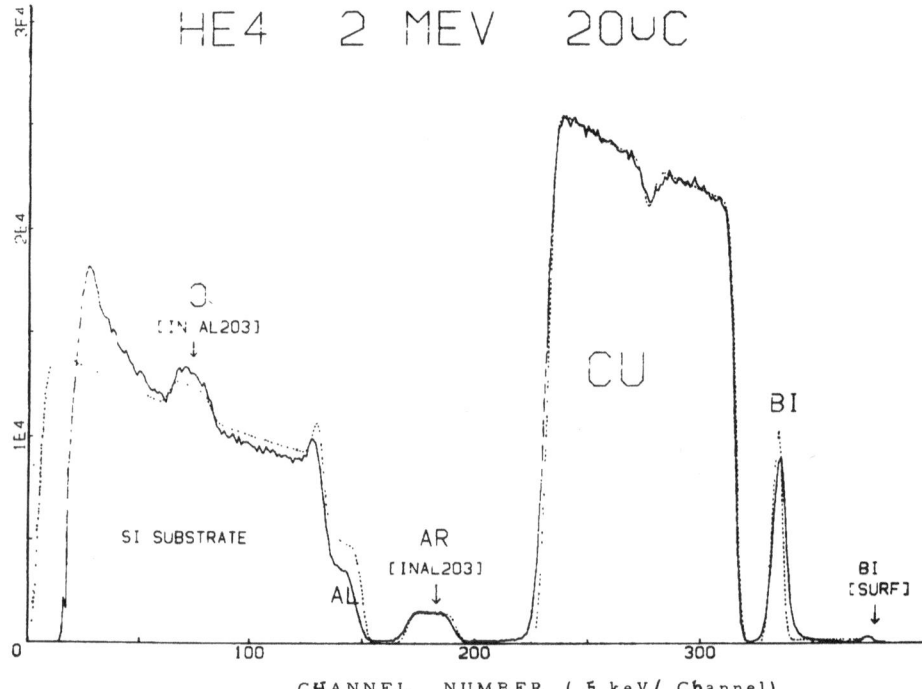

HE4   2 MEV   20UC

Figure 2. Backscattering spectra are shown for the sample illustrated in Figure 1. The solid line is experimental data,
and the dotted line is a calculated spectrum. The calculation includes all target isotopes, detector resolution, and the detector dead layer. No account is made
of beam straggle in the target. The experimental conditions are: ^{4}He; 2MeV; 20μC; Target tilt = 0°; Detector
resolution = 22 keV (FWHM); Multichannel analyzer = 5 keV
/channel. The ordinate is in channel numbers, and the
abscissa is counts.

This simple language input is all that is needed to generate a spectrum once the data base is in the computer. This sample is complex, and several iterations of input are necessary to match the various thicknesses to the observed spectrum.

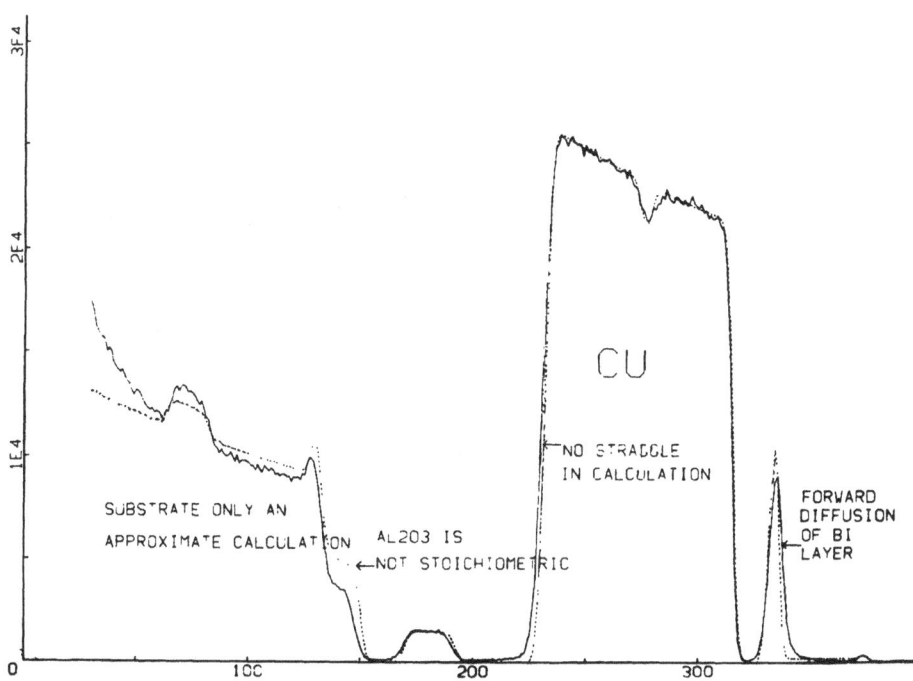

Figure 3.   The same spectra as in Figure 2, with notation about spectrum differences. The calculated thin Bi layer peak matches the back edge of the Bi data peak, but it appears that there has been Bi diffusion into the upper Cu layer because of the skewness of the Bi data peak. The extent of straggle is shown by the gap between the two spectra at the left hand side of the Cu peak. The Al_2O_3 layer is not stoichiometric and is rather oxygen rich (the argon in this layer occurs because it was sputtered in an argon ambient). The substrate calculation is only approximate and its accuracy becomes poor at very low energies.

The results of the calculation are shown in Figure 2 along with data from a backscattering run on such a sample. The various peaks are identified and can be correlated with the target schematic of Figure 1. In Figure 3 various differences between the spectra are identified and commented about in the figure captions. Of particular interest is the degree of straggle after 3791 Å of target material. The back edge on the <u>calculated</u> Cu peak is

significantly skewed because of the Cu isotopes, and the straggle
appears rather small. The Si substrate calculation is of limited
accuracy (to minimize computer time) as it has mostly esthetic
purpose. This substrate calculation assumes no changes in energy
loss with depth, but does include the change in scattering cross-
section with depth. The profile generated by such an approximation
usually is accurate (∿5%) for the first 5000 Å of the substrate.
(This approximation is discussed in the Appendix, and shown
explicitly in Figure 12.) The overall vertical deviation of the
substrate data from the calculation for the substrate may be due to
partial channeling of the He beam in the Si substrate.

   Typically such a theoretical calculation takes about 16 seconds
of computer time on a small computer with a cycle time of 24 nano-
seconds.

HIGH RESOLUTION BACKSCATTERING

   The use of magnetic spectrometers to obtain high resolution
backscattering spectra has in it a necessary commitment to computer

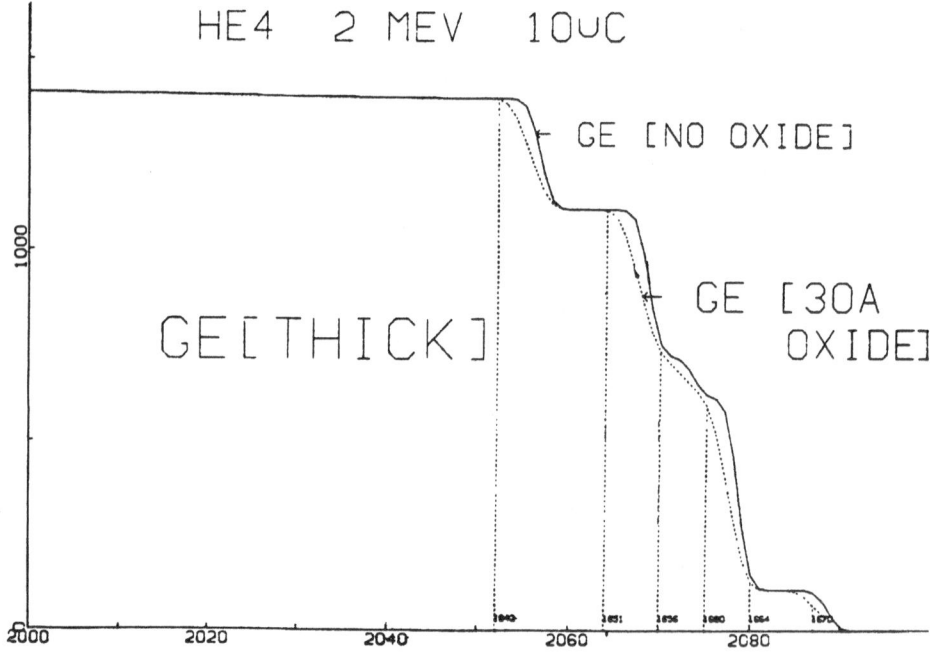

Figure 4. Calculated spectra for He ions onto a thick Ge target. Ge
         has five significant isotopes which can be seen as sepa-
         rate edges. If the target is given a surface layer of
         germanium oxide, the spectrum becomes significantly skewed.

analysis because of isotope effects. The 15 keV resolution of solid
state particle detectors is very happy. This resolution manages to
blur together the isotopes of each element to the degree that they
could be ignored. However, high resolution backscattering almost
always shows every isotope separately, and what might have been two
peaks (say of Mo and Sn) now becomes 14 to 16 identifiable layers
partially merged together. Also, thin surface layers (such as
native oxides) are very evident in the skewness they give to surface
edges.

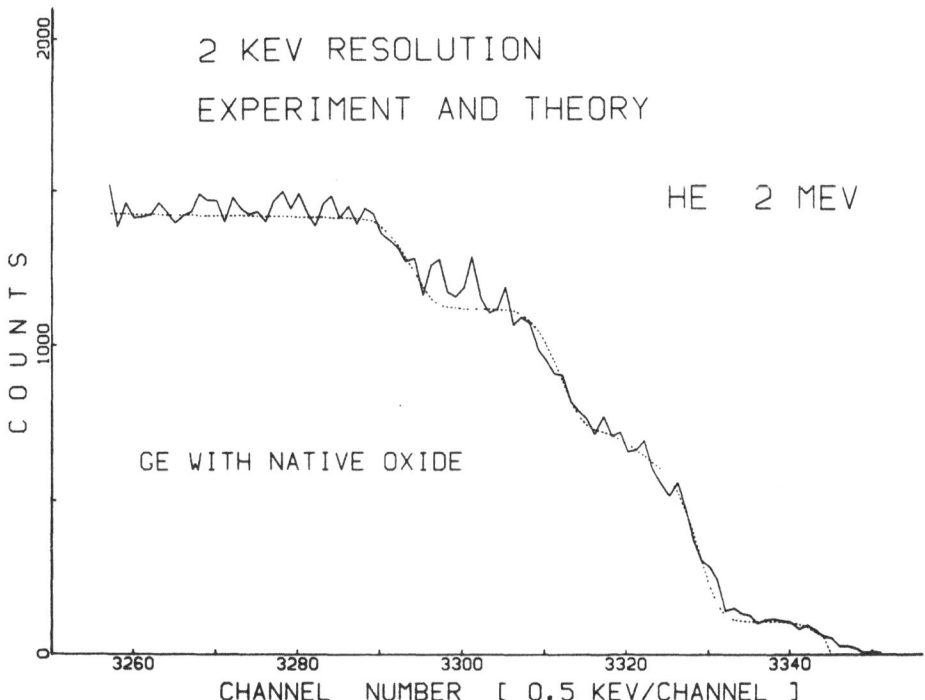

Figure 5. Experiment (solid line) and theory (dotted line) for He
           ions onto Ge. The data fit nicely the calculation of
           Figure 4 which includes the native oxide. The experi-
           mental parameters are (see Appendix for symbols):
           E_o = 2 MeV, Q = 20μC, θ = 135°, φ = 0°, Ω = 2.5 msr.,
           Det. FWHM = 2KeV, 0.5 keV/channel.

     Figure 4 shows a calculated spectrum for He projectiles on a
thick Ge sample. The curves show clear differences for scattering
from pure Ge, and from Ge with a typical 30 Å oxide. The necessity
for including this oxide is shown in Figure 5 where actual data is
superimposed on the calculation (experimental details for the spectra in Figs. 5 and 6 can be found in the paper, "Applications of a
High Resolution Magnetic Spectrometer" by J. K. Hirvonen, in this

book). The data clearly demands detailed accounting of the actual
surface oxide layers.

A similar and more evident example of isotope effects can be
seen in Figure 6 where we backscatter with high resolution from a
pure Mg target. Because the Mg isotopes differ in mass by 4%, each
isotope is clearly seen. If one were to estimate the magnetic
analyzer resolution from the experimental data, one might conclude
that the resolution was ∿ 5 keV. But Mg forms a thick native oxide
MgO, and this oxide skews the front edge of each isotope to give
the appearance of poor resolution. The actual calculated spectrum
shown had a resolution of 2 keV, and it fits the data adequately.

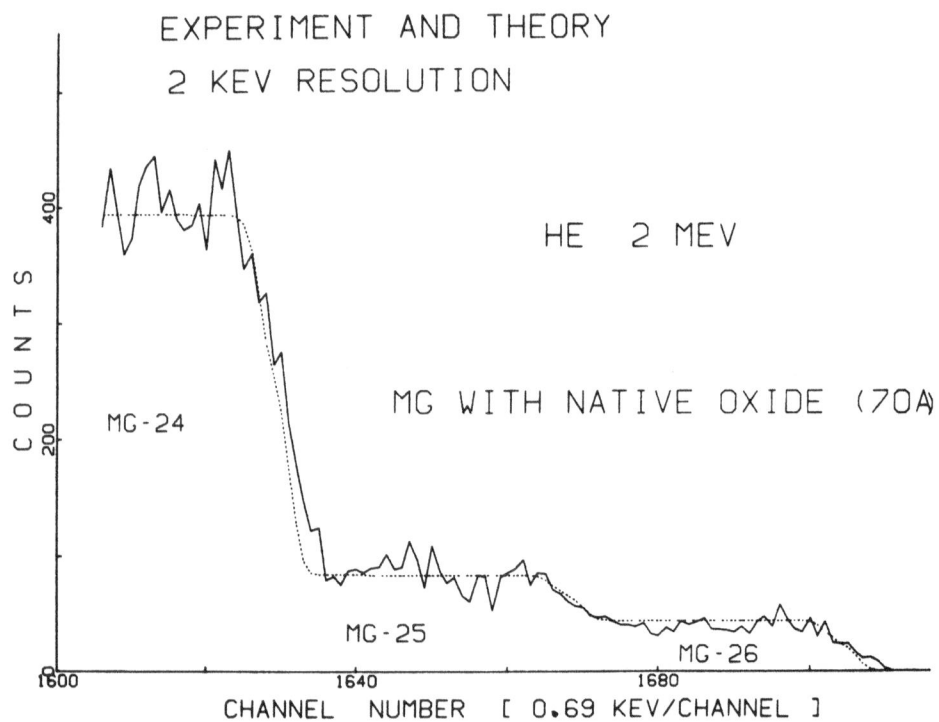

Figure 6. Experimental data and theoretical calculations for He
           ions onto Mg. The data (solid lines) was taken with a
           magnetic spectrometer with a resolution of 2 keV. The
           three isotopes of Mg are clearly seen because they differ
           in mass by 4%. The calculation parameters are (see
           Appendix for symbols): E_o = 2 MeV, Q = 15μC, θ = 135°,
           φ = 0°, Ω = 2.5 msr, Det FWHM = 2 keV, 0.69 keV/channel.

DIFFUSION EFFECTS

The interdiffusion of thin films is so complex that one cannot
hope to make a reasonable model which is simple enough to be quickly
set up and calculated. However, in most cases we have seen, it is
possible to obtain a set of first-order diffusion constants which
are adequate to fit a few backscattering spectra. These few constants, then, can be called the 'data parameters' even though they
may have no physical significance. This technique has been used by
nuclear physicists for many years to reduce complex data to a
few simpler constants. For the case of thin film interdiffusion
this may be accomplished to first order by assuming Fick's Law and

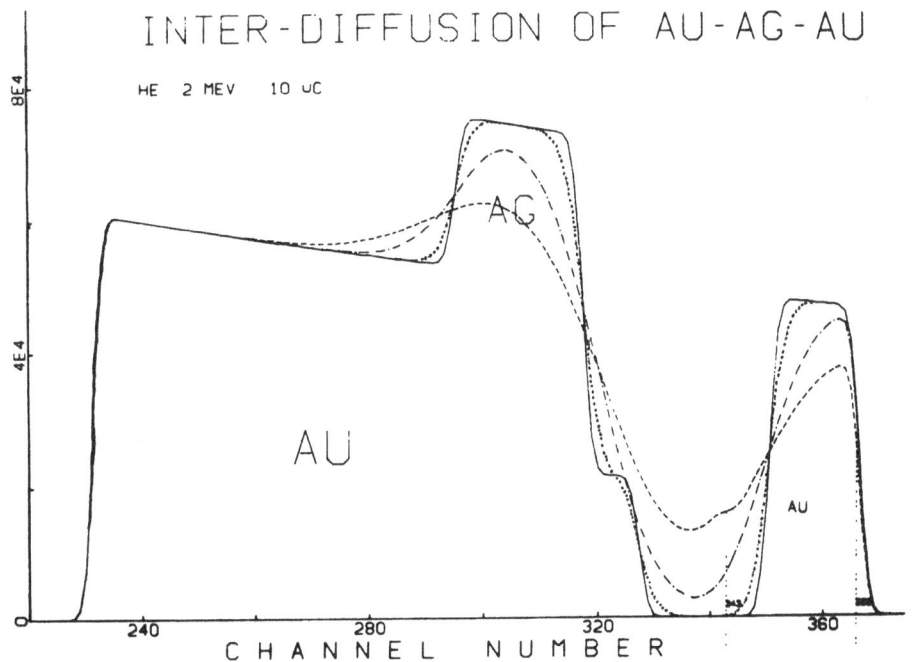

Figure 7. Calculations of backscattering spectra from the inter-
           diffusion of Au and Ag layers. The target is assumed
           to be a sandwich with Au above and below the 1000 Å Ag
           layer. The activation energy for both Ag and Au were
           taken from the literature as 2.0 eV, with diffusion
           constants of 0.2 cm^2/sec. The layers are assumed to
           have been heat treated at 475°C for times of 10 minutes
           (solid line), 1 minute (dotted line), 10 minutes (dot-
           dash line), and 30 minutes (dashed line). The experi-
           mental parameters are: E_0 = 2 MeV, θ = 170°, ϕ = 0°,
           Q = 10μC, Det. FWHM = 15 keV, Ω = 4.11 msr, and 5 keV/
           channel.

assigning an activation energy, Ea, and a diffusion constant, Do,
to each layer. Just these two parameters for each layer seem to
be able to reproduce spectral data which appears to be bulk interdiffusion such as illustrated in Figure 7. Experimental data
(which fit the curves reasonably well), are not shown to prevent
confusion. For films with grain-boundary diffusion, in addition to
bulk diffusion, there appears to be less chance of a simple model
to parameterize the data. This is even more true of thin film
reactions where a new compound is formed. Until there is a model
of how these compounds are formed, there is little possibility of
parameterizing data. However, some valuable information is provided by a simulated spectrum. In particular, it allows the extraction
of the <u>gradient</u> of the moving species through the compound to be
determined.

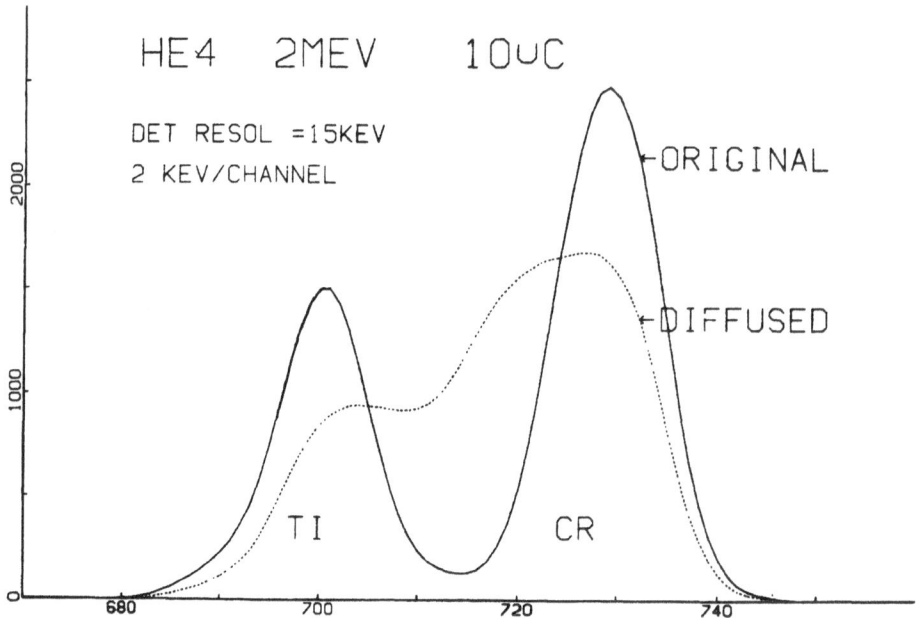

Figure 8.   Calculated spectra are shown from 100 Å layers of Cr and
Ti, and from such films after interdiffusion. The
spectra are typical of those from surface barrier
detectors which blur together the isotopes and give the
appearance of a single peak for each element.

DIFFUSION AND HIGH RESOLUTION BACKSCATTERING

The use of high resolution backscattering to obtain the fine
details of interdiffusion can be a can of worms. The isotope effects

become almost intractable when each element's concentration profile
(after diffusion) must be guessed. To illustrate this we show the
interdiffusion of two layers of Cr and Ti, each 100 Å thick. Using
conventional surface barrier detectors with a resolution of 15 keV
we would see spectra as shown in Figure 8. The original layers
appear like Gaussians, with the interdiffusion appearing as smooth
inter-mixing. One has some hope of untangling these smooth peaks.

With high resolution (FWHM = 2 keV) we would see a spectrum
as shown in Figure 9 for the same Ti-Cr films (the 15 keV resolution
spectrum is shown in dots for comparison). The many isotopes make
ragged curves out of the two thin layers. The interdiffusion of
the films is shown in Figure 10 with high resolution. Although the
final spectrum is somewhat smooth, it will be only with the greatest
difficulty that a smooth interdiffusion gradient can be guessed
which will untangle all 9 separate isotopes from that final shape.

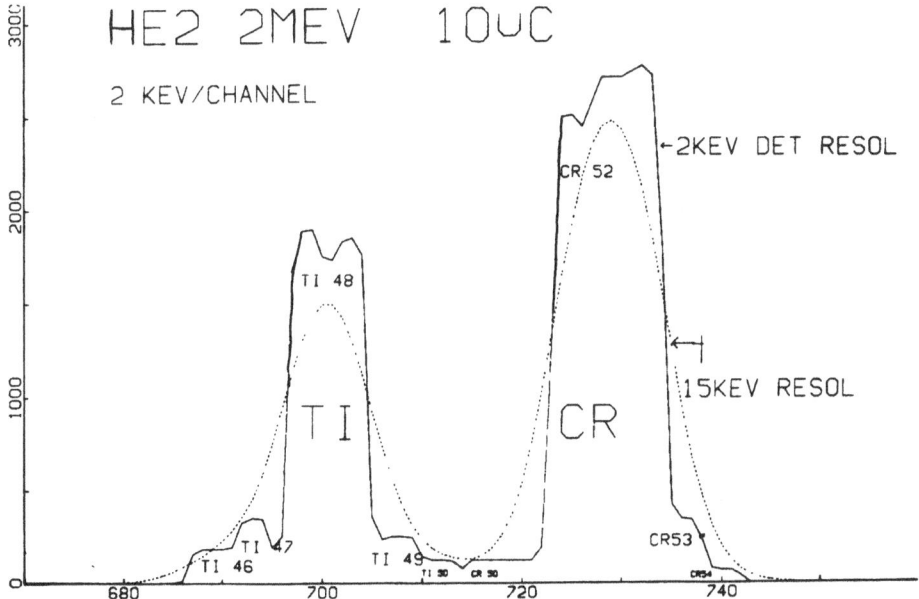

Figure 9. Spectra identical to those of Figure 8, but assuming 2
           keV detector resolution. The five isotopes of Ti and
           the four of Cr are distinctly seen. The dotted line
           shows how these layers appear with a detector of 15 keV
           resolution.

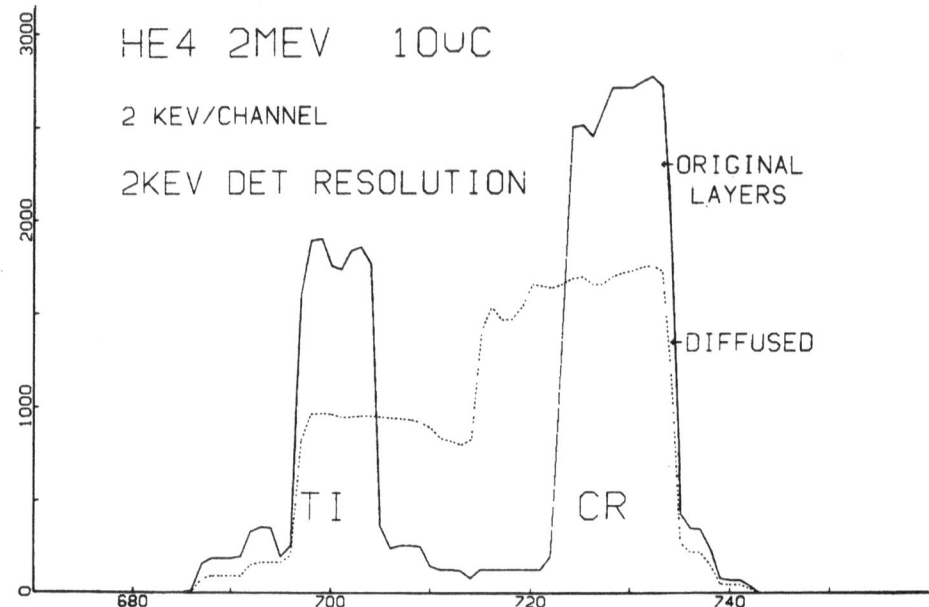

Figure 10. The interdiffusion of Ti-Cr layers is calculated for
            2 keV detector resolution. To unfold such a spectrum
            one has the difficult task of assuming concentration
            distributions which, when distributed among the nine
            isotopes of the element, will give the interdiffused
            spectrum (dotted line). The experimental parameters
            are the same as for Figure 7 except detector resolution
            is 2 keV, and 2 keV/channel.

DECONVOLUTION

     At intervals, highly complex papers have appeared which discuss
the deconvolution of spectra. (8,9) That is, the removal of equipment resolution from spectra so that one has the semblance of
physical data without any equipment blurring. We find that these
papers have never proven applicable to a wide range of data. The
problem is simple to state: the scatter of data in Poisson statistics
cannot be described by the convolution of any smooth function with
a Gaussian. If one thinks of deconvolution using Fourier trans-
formations, then the process with 'normal' statistics is to take
the transform of the data, and the resolution function, divide them,
then take the inverse transform. The difficulty arises because the
Poisson statistics contain significant higher frequency components
than the resolution function. Thus one divides the high frequency
coefficients of the data by a Gaussian transform whose high frequency coefficients are almost zero. This blows-up the high fre-

quencies, so that the inverse transform contains unrealistic oscillations.

A far simpler approach is to convolute. That is, simulate the target and the backscattering, then add in equipment resolution at the end. A few iterations of comparing calculations with data and the equipment resolution can be determined. One then has the 'deconvoluted' spectra by a process of successive simulations.

CONCLUSIONS

The use of computer simulation of backscattering spectra can be packaged so that non-experts can easily obtain detailed analysis of data. By overlaying data with calculations, one can quickly determine layer thicknesses, non-stoichiometry, interdiffusion effects, and contaminants. The use of high resolution magnetic spectrometers demands an equal commitment to computer analysis because of isotope effects. For samples with many isotopes, extraction of diffusion profiles from high resolution spectra appears a formidable task.

APPENDIX - SPECTRUM SYNTHESIS PROCEDURE

The spectra are based on Coulombic scattering theory and on the Ziegler-Chu energy loss polynomials. (10) Input variables characterizing the backscattering experiment are the incident energy E_o, the angle between the incident beam and the detector, θ, the solid angle subtended by the detector at the target Ω, and the total incident beam charge Q. The normal to the target is assumed to be coplanar with the beam and the center of the detector, and to make an angle ϕ with the backward beam direction so that $\phi=0$ represents normal incidence. The detector response is characterized by the spectrum characteristic of a single incident energy and is usually calculated from a Gaussian of known width (FWHM).

Input variables characterizing the film are as follows. The specimen whose spectrum is to be calculated is divided into layers sufficiently thin to accurately represent variations in composition with depth. Since the input layers determine the integration step size in the calculation, the spectrum will present a stepped appearance if the energy spread corresponding to each layer is much greater than the detector resolution. Typically, the layer depths (in Angstroms) are three times the detector FWHM (in KeV). The components of the film are specified by the atomic numbers Z_α and the masses M_α of the α-th elements present, and the composition of each layer by the relative proportions of each atomic mass in the layer. These layers are combined into a composition matrix C and the total number of atoms in each layer represented by a vector N, in units of 10^{15} atoms cm^{-2}. The procedure is best understood by reference to Table 1 which gives the input composition matrix C, and the Z,M and N vectors for 1500 Å of SiO_2 on silicon, where the density of SiO_2 is taken as 2.25×10^{22} molecules cm^{-3}.

TABLE 1. INPUT DATA FOR 1500 Å of SiO_2 ON SILICON

Z =	[8	14	14	14]		N ↓
M =	[16	28	29	30]		
C =	200	92.2	4.7	3.1		200
	200	92.2	4.7	3.1		200
	200	92.2	4.7	3.1		200
	200	92.2	4.7	3.1		200
	200	92.2	4.7	3.1		200
	0	92.2	4.7	3.1		100
	0	92.2	4.7	3.1		400
	"	"	"	"		"
	"	"	"	"		"

As can be seen, for light elements where the relative isotope separation is large but the number of isotopes is small, isotopes are handled by treating them as separate components having different M values for the same Z. Elements with many isotopes are handled as described later. Given input data as described above, the generation of the backscattering spectrum proceeds by the following steps.

(1) Reference 10 provides a table of polynomial coefficients $a_{\alpha i}$ such that the energy loss per unit areal density of element α is given by

$$\epsilon_\alpha = \sum_{i=0}^{5} a_{\alpha i} E^i$$

where ϵ is in units of eV/10^{15} atoms cm^{-2} and E is in keV. This dependence will simply be written as $\epsilon_\alpha(E)$ in what follows.

(2) The composition matrix is converted into a matrix of atom fractions F_α for each atomic species present, and Bragg's rule is assumed so that the composite energy loss in eV/10^{15} atoms cm^{-2} is simply

$$\sum_{\alpha} F_\alpha \, \epsilon_\alpha (E)$$

(3) For a given incident energy E_o, the energy of the incident beam E_r is calculated as it exits from each layer r (Figure 11).

$$E_r = E_o - \sum_r \delta_r \qquad\qquad (1)$$

$$\delta_r = N_r \, \sec\phi \sum F_\alpha \, \epsilon_\alpha (E_{r-1}) \qquad\qquad (2)$$

The energy loss values for a layer r is calculated for the incident energy E_{r-1}. The set of energies E_r then serve as collision energies for subsequent calculations.

(4) The desired end result of this step is the set of final energies $E_{r1\alpha}$, the energy of a particle exiting from the top of layer 1 at angle θ having previously backscattered from a nucleus of type α at the bottom of layer r with incident energy E_r. If $E_{rs\alpha}$ denotes the energy of a particle exiting from the top of layer s having previously backscattered from the bottom of layer r, we have (see Figure 11).

$$E_{rs\alpha} = K_\alpha \, E_r - \sum_{\gamma=0}^{\gamma=r-s} \delta_{r,r-\gamma} \qquad\qquad (3)$$

where

$$\delta_{rs} = -N_s \, \sec(\theta-\phi) \sum_\alpha F_\alpha \, \epsilon_\epsilon (E_{r,s+1,\alpha}) \qquad\qquad (4)$$

as before, the energy loss coefficient for layer s is calculated for the _incident_ energy ($E_{r,s+1}$ in this case.)

In equation (4) the initial backscattered energy $E_{r,r+1,\alpha}$ is simply

$$E_{r,r+1,\alpha} = K_\alpha \, E_r$$

where K_α, the kinematic energy recoil coefficient is given by

$$K_\alpha^{\,2} = \frac{m \cos\theta + (M_\alpha^{\,2} - m^2 \sin^2\theta)^{1/2}}{m + M_\alpha} \qquad\qquad (5)$$

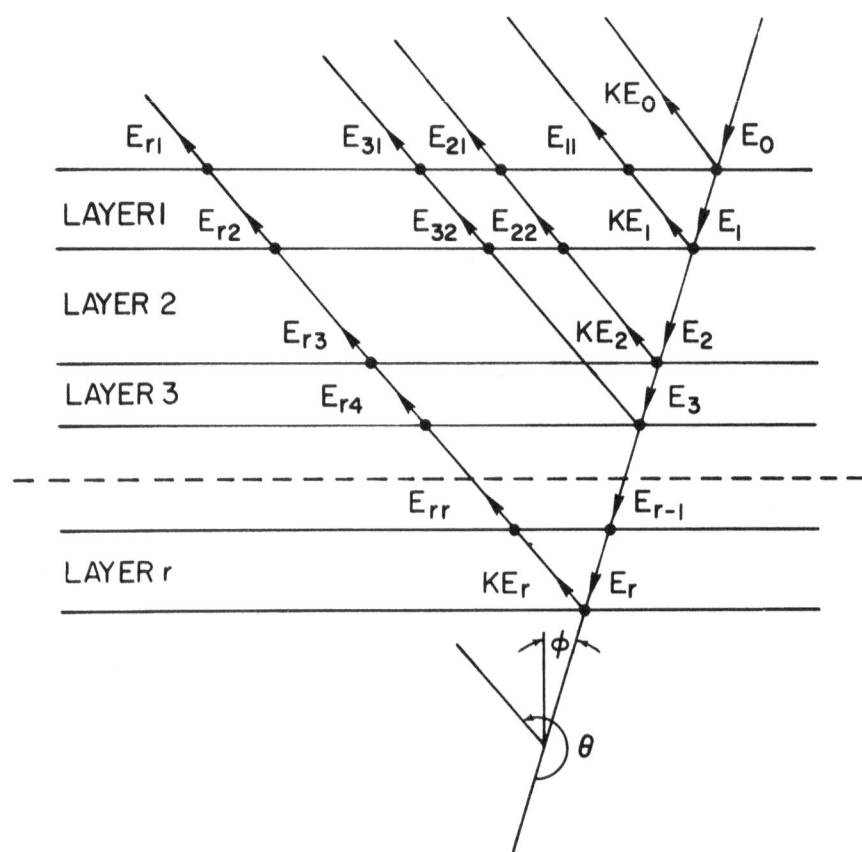

Figure 11. Schematic showing the various parameters used in the
            calculation of backscattering spectra. The layer
            thicknesses are dependent on the desired calculation
            resolution. The incident ion energies are found for
            each layer and are called $E_0, E_1, E_2, \ldots E_r$. The energy
            after recoil from each layer edge is identified as
            $KE_0, KE_1, \ldots,$ and separate tallies must be kept for
            each element of the target. Then the energy of the
            backscattered ion is determined as it passes through
            each layer edge. The final energies are identified
            as $E_{01}, E_{11}, E_{21}, \ldots, E_{r1}$ (the first subscript tags
            from which layer the backscattering occurred).

where m is the particle mass (4 for He) and M_α is the mass of
species α.

     We now have a complete set of beam exit energies corresponding
to the layer edges for each element present in the total film. For

light elements where there are only a few isotopes but fractionally
large mass differences, each isotope may be treated as a separate
element having the same Z but different M as illustrated in
Table 1 for silicon. This means calculating a whole series of ϵ_α
values for each isotope even though the $a_{\alpha i}$ are the same. For
heavy atoms having many isotopes this would be extremely wasteful,
particularly since the energy shifts between isotopes are quite
small. In this case one calculates the values of $E_{rl\alpha}$ for the
heaviest isotope only and then assumes that $(K_\alpha E_r - E_{rl\alpha})$ has the
same value for all the other isotopes for which the values of K_α
are slightly different.

(5) We may now calculate the number of backscattered particles
at the detector corresponding to each interval in the final energies.
The scattering cross section is assumed to be constant for each
layer, and characteristic of the average energy for that layer. Thus
the collision energy for the rth layer is taken as $1/2(E_{r-1} + E_r)$.
This will result in a stepped appearance in the spectrum if the
layers are too thick, which is advantageous when running the calculation for the first time. This feature may be avoided however
by taking the cross-section corresponding to an incident energy E_o
and correcting later for the energy dependence of the cross section.
The cross section is calculated from the standard coulombic
scattering formula (11). The number of counts at the detector in
the energy range $E_{rl\alpha}$ to $E_{r-1,1,\alpha}$ from atoms of type α is
given by

$$n = 6.25 \times 10^{24} \, Q \, \Omega \, \sigma \, F_{r\alpha} \, N_r$$

where $F_{r\alpha}$ is the fraction of atoms of type α in layer r, N_r
is the total number of atoms in layer r in units of 10^{15} atoms
cm^{-2}, Q is in microcoulombs, Ω in millisteradians and σ in
cm^2/steradian.

(6) A backscattering spectrum has now been obtained in that
the number of counts for each nuclear species present has been found
for each layer r and the corresponding exit energies have been
calculated. It is now necessary to interpolate to a standard channel
width, say 5kV, so that the various partial spectra may be summed
to yield the final spectrum. This is accomplished by summing
the total counts in each spectrum from the high energy end, performing a linear interpolation on the sum at 5 kV intervals and then
differencing the result to give the number of counts per channel for
the new 5 kV channels. If in step (5) the cross-sections had been
determined for a collision energy of E_o, it is necessary at this
point to correct to the actual cross section. This is done by
making a table of $E_{rl\alpha}$ versus E_r for each element, and relating
the exit energy to the collision energy. One then takes the new
channel energies, E_{out}, and calculates the corresponding collision

energies, E_{hit} , by interpolation on the above table. One then multiplies the just calculated counts/channel by the factor $(E_o/E_{hit})^2$ to allow for the increase in cross section with decrease in collision energy.

(7)   It is possible to introduce a faster approximate isotope calculation instead of the detailed approach discussed in section (4) above.

As we indicated earlier, there are two detailed ways to include isotope effects. These treat each isotope as a separate layer components, and a spectrum calculation is done in entirity for each isotope. A coarser, but fast approach, is to do the spectrum calculation for the heaviest isotope only. In this case, one may replace each spectrum for each element, by a repeated series of this spectrum offset by the value of KE_o for each isotope of the element, and normalized by the respective isotopic abundance for each isotope. That is, for a Si spectrum we calculate a Si^{30} spectrum, then displace it to the relative positions for back-scattering from Si^{28}, Si^{29}, and Si^{30}, then normalize each by multiplying by .94, .04, and .02, and make our final spectrum the sum of these three.

This technique is reasonable for thin layers. For thick layers, the trailing edge of the combined spectra may become erroneous because this calculation approximation assumes the energy loss of the ion to the various recoiling isotopes does not change with depth. The error introduced will be approximately proportional to $\Delta E/E_o$, where ΔE is the energy loss on the inward path through the layer.

(8)   At this point it is convenient to allow for detector spread, by convolving the spectrum with the detector resolution. While it should be possible to vary the spread function with depth and thereby allow for energy straggle, this has not yet been included. The final step is to sum all the spectra together to obtain the final spectrum. The various components of the final spectrum are, of course, available separately if needed. Figure 12 shows, by way of illustration, the spectrum generated from Table 1 where 14 layers of silicon were included to provide overlap between the silicon and oxygen spectra. Other assumed conditions were Q=10 microcoulombs, Ω = 4.11 millisteradians, E_o = 2000 keV, θ = 170, ϕ = 0, and channel width = 5 keV/channel. Also shown is the fast substrate approximation that was discussed in the text in reference to Figure 2 and 3. This is obtained by introducing only one silicon layer containing 10^{19} atoms in place of the 14 layers used in Table 1.

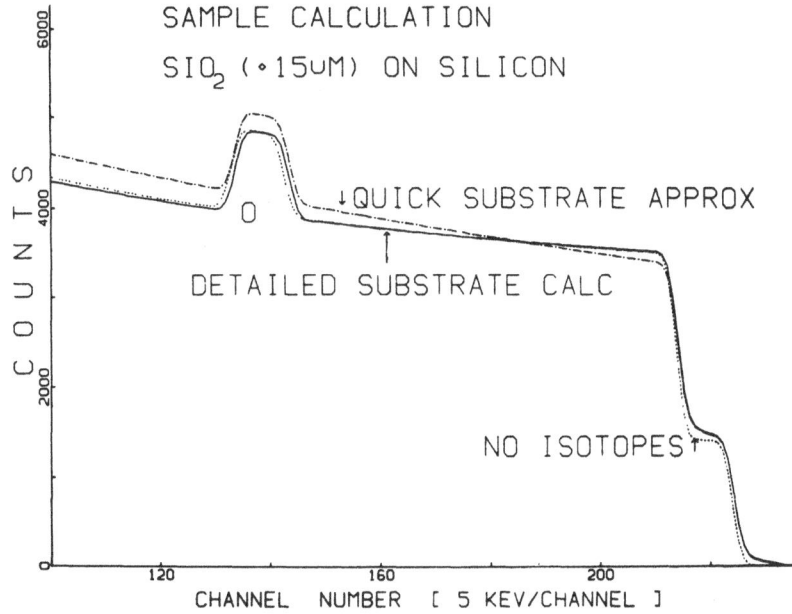

Figure 12.   The solid line spectrum was generated from the data of
Table I, with the experimental parameters itemized at
the end of the Appendix. Also shown (in dot-dash lines)
is the approximate substrate calculation discussed in
the text where it is assumed that the energy loss of the
projectile is constant throughout the layer, and only
changes in scattering cross-section are calculated in
detail. Since a substrate detailed calculation can
triple spectrum computation time, it is desirable to
have a quick approximate solution if the substrate is
included only for esthetic reasons. A calculation
which does not include isotopes is shown in dotted
lines.

REFERENCES

1.   D. Powers and W. Whaling, Phys. Rev. **126**, 61 (1962).

2.   W. A. Wenzel, Ph. D. Thesis, California Institute of Technology,
     1952 (unpublished).

3.   E. I. Siritonin, A. F. Tulinov, A. Fiderkevich, and K. S.
     Shyskin, Vestnik MGU, **12**, 541 (1971).

4.   D. K. Brice, Thin Solid Films 19, 121 (1973).

5.   W. K. Chu and J. F. Ziegler, Journal of Applied Physics, 46,
      2768 (1975).

6.   J. M. Harris, W. K. Chu, and M. A. Nicolet, Thin Solid Films
      19, 259 (1973).

7.   J. Rickards, "Energy Straggling of Protons in Carbon" (to be
      published in Nuclear Inst. and Methods).

8.   B. C. Cook, Nucl. Inst. and Methods, 24, 256 (1963); and
      A. Barcy, B. Cichochi, A. Turos, L. Wielunski (unknown
      publication).

9.   J. F. Ziegler and J.E.E. Baglin, J. Appl. Phys. 42, 2031 (1971).

10.   J. F. Ziegler and W. K. Chu, Atomic and Nuclear Data Tables,
      13, 463 (1973).

11.   J. F. Ziegler and R. F. Lever, Thin Solid Films, 19, 291 (1973).

DISCUSSION

Q: (W. Brown) Since backscattering never really proves information
on actual mechanical thickness, an input specification of Å seems
poor. It surely implies introduction of a set of densities which can
be very uncertain. Do you insist on film thickness or depth in real
distance?

A: (J.F. Ziegler) The input to the illustrated example (Figure 1 and
2) was used by a physical chemist who preferred depths denoted in Å.
Other possible inputs are in units of gm/cm^2 or $atoms/cm^2$. Inside the
computer program all calculations are based upon units of gm/cm^2. But
the important point is that one let the computer do the conversion
work and let the user use the units with which he is most comfortable.

Q: (R. Behrisch) Generally the ion dose and the opening angle of the
detector are not known very accurately in an experiment. Did you fit
your spectra in height in order to get this excellent agreement between the calculated and the measured spectra ?

A: (J.F. Ziegler) There are no free parameters or normalization in the
program. One must enter all experimental data. But for use by non-experts, one would usually use standardized values so that these do not
have to be entered by the user.

Q: (B. Scherzer) How thick are the layers of constant composition you use in the computer code and how many layers can be calculated?

A: (J.F. Ziegler) The depth layers can be made as thin as desired, however computer time increases linearly with the number of layers. Usually, the program selects the number of layers so that there are several layers included in each channel of the final spectrum. For diffusion effects the layers may become quite narrow in order to calculate correctly the interdiffusion of layers.

SOME PRACTICAL ASPECTS OF DEPTH PROFILING GASES IN METALS BY PROTON

BACKSCATTERING: APPLICATION TO HELIUM AND HYDROGEN ISOTOPES[*]

Robert S. Blewer

Sandia Laboratories

Albuquerque, New Mexico 87115

ABSTRACT

Reports of the application of the proton backscattering technique to determine the depth distributions of helium and hydrogen
in metals have appeared in the published literature for almost two
years, but a detailed discussion devoted to the practical considerations in its use has not yet been presented. Maximum detection
sensitivity for hydrogen isotopes and helium is achieved by employing thin foil samples backed by a "beam trap" to prevent any portion
of the incident beam which is not backscattered in the foil sample
from reaching the detector. Measurements have been made to determine the magnitude and origin of "tail" or "background" counts which
appear in parts of the spectra which ideally should contain no
counts and which, by their presence, reduce the detection sensitivity that should theoretically be achievable for light elements.
For copper foil targets, the ratio of the counts per channel of the
Cu peak to the counts per channel for the background is $\sim 5 \times 10^3$.
The energy distribution of the background has been measured and
leads one to the conclusion that for the apparatus used, these counts
arise principally from decollimated incident beam protons which
strike the sample holder instead of the foil sample. The enhancement in elastic scattering cross section for protons incident on ^{4}He
is discussed and the enhancement factor for a number of other low Z
elements is tabulated both for $E_0=2.5$ MeV and for energies at which
sharp resonances exist. Example spectra will be drawn from fusion
reactor first wall studies presently being conducted which illustrate the effect of applying the principles discussed.

[*]This work was supported by the U. S. Energy Research and Development Administration.

INTRODUCTION

Rutherford Ion Backscattering Spectrometry (RIBS) has in the
last several years been shown to be a remarkably powerful near-
surface analytical technique, providing quantitative information on
composition and concentration with depth up to several thousand
Angstroms in solid surfaces.[1] Detection sensitivity has been shown
to increase rapidly with atomic number; thus the technique is most
useful for studies involving high Z impurities in low Z hosts.
Indeed, because of the unfavorable ratio of Rutherford cross-sections
for low Z atoms in metals, detection of helium and hydrogen isotopes
was, until recently, generally regarded to be beyond the capabilities
of RIBS.[2] However, a specialized form of proton backscattering has
been shown to be capable of sensing 0.5 at.% He in Cu with a depth
resolution of less than 350 Å.[3] Though this technique has been used
in several investigations in the recent literature, there has been
little discussion of the more practical aspects of its use. This
paper is addressed to some of the topics of concern.

PROTON BACKSCATTERING TECHNIQUE

Proton elastic scattering as a surface analytical technique is
essentially identical in concept to helium ion backscattering. The
appropriate experimental setup is shown in Figure 1. A monoenergetic,

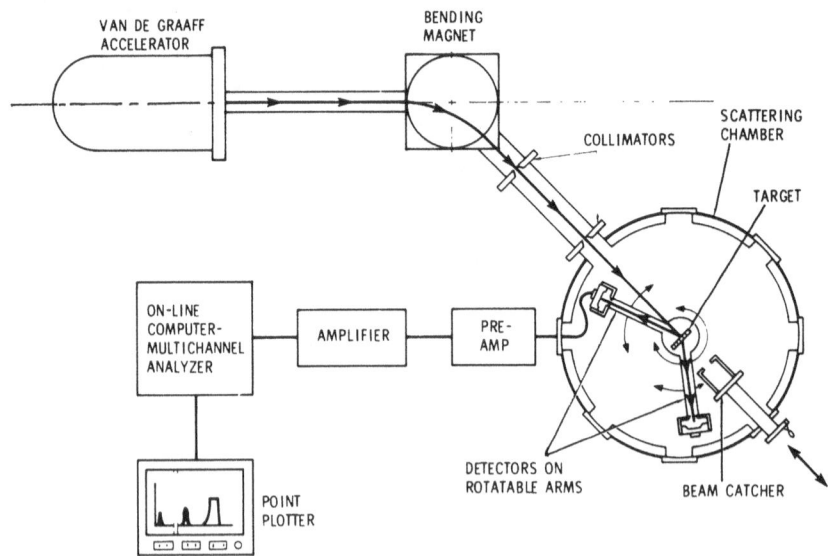

Figure 1.   Schematic of experimental apparatus used in ion backscattering experiments (not to scale).

mass analyzed proton beam of $\sim$ 2.5 MeV energy is collimated to a
size of $\sim$ 1 mm^2 and directed at normal incidence onto a suitable
target. Protons which strike target atoms (and impurity species
within the target) and which are backscattered at a fixed angle
> 90° are collected and energy analyzed by means of a silicon surface
barrier detector. The associated electronics equipment is standard
for backscattering experiments[1] and will not be discussed here.
The data are presented as spectra in which the number of protons
backscattered per incident energy increment (counts per channel) is
registered on the ordinate axis versus contiguous equal energy intervals (channels) from zero to the incident energy, E_0, which appear
on the abscissa.

     The forte of elastic backscattering techniques lies in the fact
that the energy E retained by a given incident ion of mass m after
a collision at an angle θ with a target surface atom of mass M is
unique for each of the possible target element species in the
periodic table. Thus if one measures the energy of each of the backscattered ions for an incident beam of fixed composition and known
energy at a given angle θ, one can determine the species of atoms
in the surface using the relation given in Figure 2(a).

$$k = \frac{E}{E_0} = \left\{ \frac{\cos\theta + [(M/m)^2 - \sin^2\theta]^{1/2}}{1 + M/m} \right\}^2$$

where k is known as the "fractional retained energy" or the "kinematic recoil factor" of the incident ion. Note that the percentage
change in k from element to element in the low Z range is much
greater for incident protons than for more commonly used $^4He^+$ ions.
In Figure 2(b), the same relationship is plotted, with the axes
interchanged to illustrate the (energy) position at which backscattered protons and $^4He^+$ ions would appear on the multi-channel
analyzer display. For instance, helium ions incident on a target
containing a surface oxide would, if backscattered at 164° (laboratory),
retain $\sim$ 40% of their initial energy when scattering from oxygen
atoms, whereas protons incident on the same target would retain $\sim$ 80%
of their initial energy. If one is primarily interested in observing
low Z elements in the target, kinematics dictates that a larger part
of the spectrum will be available to register the detail of backscattered protons from elements below fluorine than will be the case
for backscattered $^4He^+$ ions. Therefore one can expect better mass
resolution among low Z elements using protons, in addition to removing
the kinematic limitation of obtaining no backscattering at all for
$^4He^+$ ions incident on elements lighter than ^6Li.

     The superior energy dispersion between low Z elements (and even
isotopes) for proton backscattering is more vividly illustrated in

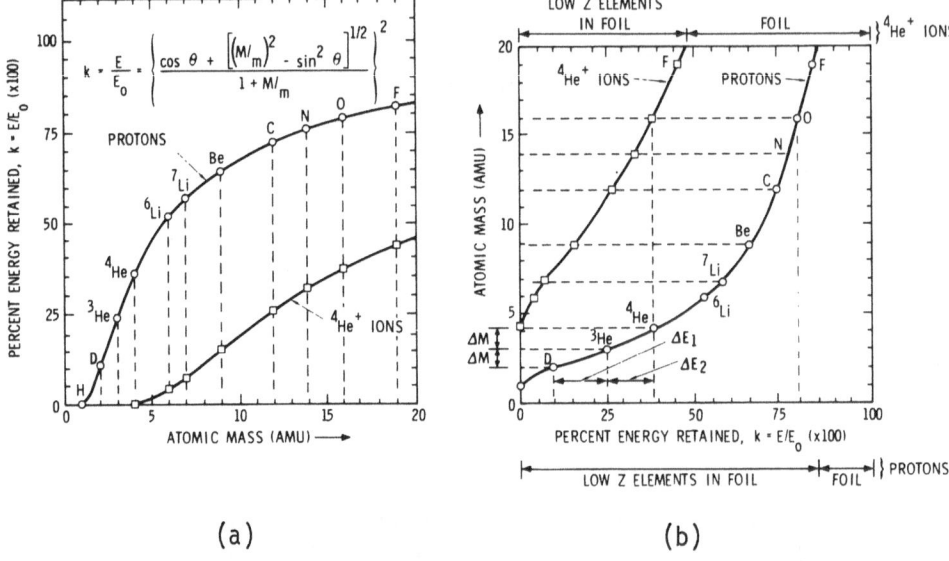

(a)   (b)

Figure 2.   (a) Variation of the kinematic recoil factor, k, with
                target atom mass for a fixed backscattering angle of
                164° (lab).

             (b) Replot of Fig. 2(a) with axes interchanged. Note the
                 large fractional energy change for protons back-
                 scattered from D, ^{3}He and ^{4}He, atoms which differ in
                 mass by only one atomic mass unit.

the plots of derivatives of the k factor with respect to target
atom mass shown in Figure 3 for both incident protons and ^{4}He$^+$ ions.
For target elements up to oxygen, the dispersion is more than twice
as great for protons as for ^{4}He$^+$ ions. Also the proton dispersion
curve is peaked toward the lower end of the mass range where ^{4}He$^+$
backscattering from low Z target atoms is not possible.

    The influence of the scattering angle θ on the energy separation
of protons backscattered from various elements is depicted in Figure
4 (lower part). The dispersion increases as one detects backscattered protons at larger angles and is greatest at 180°. The
upper part of the figure illustrates the appearance of a measured
spectrum from protons scattered from a target composed of equi-atom
density monolayer species (with backscattering from the fictional
substrate eliminated).

    According to Rutherford theory, the elastic backscattering cross
section should increase as (Z^2) target. As can be seen in the spec-

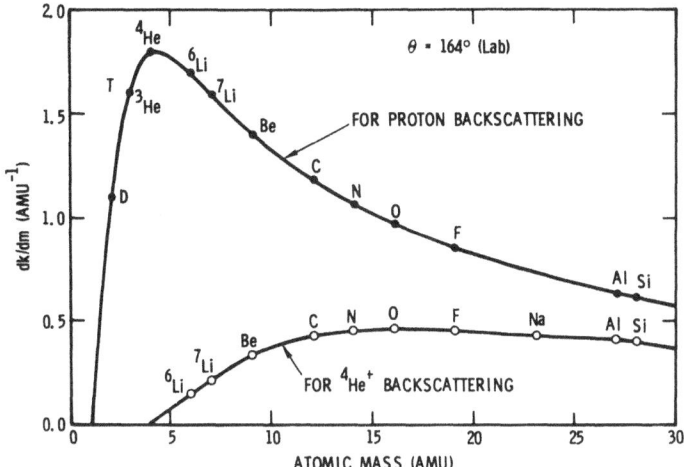

Figure 3. A plot of the derivative of k with respect to target atom
           mass versus target atom mass for both incident protons
           and incident $^4He^+$ ions.

trum in Figure 4 the scattering cross sections of low Z elements in
the target are orders of magnitude smaller than for common metals
such as titanium or silver. Moreover, as shown in Fig. 5, the thick
target yield from a metal host would completely bury the weak backscattering peak expected from the Rutherford cross section value for
~1 at. % of an implanted low Z impurity such as helium.

     In order to optimize chances for observing the helium peak, the
thick target yield should be suppressed by some means. This can be
accomplished by using a thin foil (1-10 μm) as the host and trapping
all protons which do not backscatter in the foil itself. This
trapping can be accomplished by use of a beam trap like that illustrated in Fig. 6. Note that all protons which are incident normally
and penetrate any part of the foil, are prevented by the geometrical
placement of the detector from being counted. If the beam trap is
tilted or otherwise aligned improperly, the thick target background
will be only partially suppressed, as shown by the dashed line in
Figure 5.

     The thickness of the foil, the depth distribution and abundance
of all the elements present in the foil (except ^1H) can be determined
from the spectra obtained using proton backscattering. A discussion
of the relevant formulae is given in Ref. 4 and will not be repeated
here.

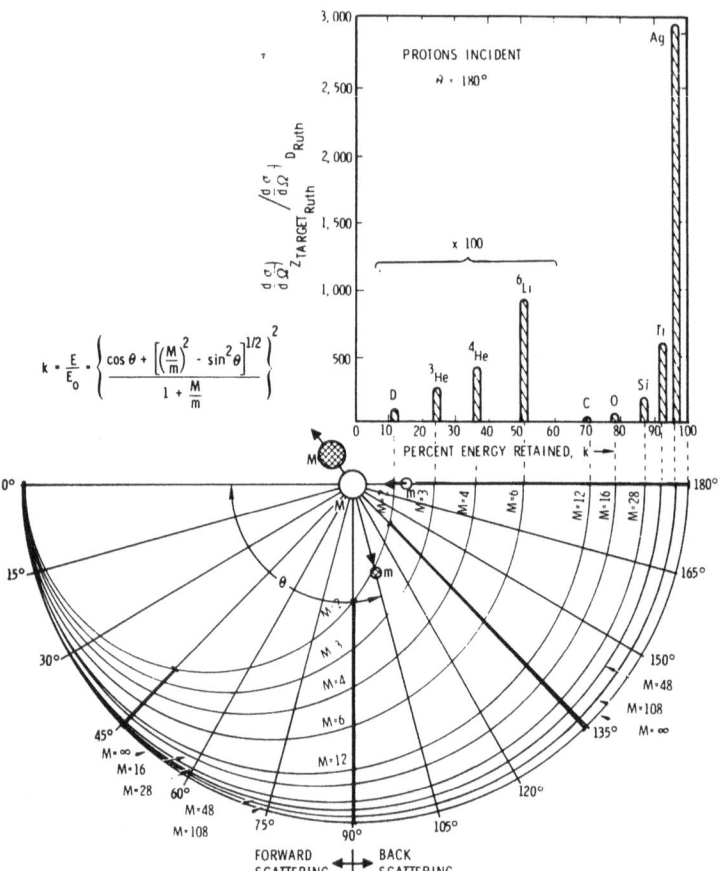

ENERGY DISPERSION OF SCATTERED PROTONS
AS A FUNCTION OF SCATTERING ANGLE

Figure 4. Schematic representation of the change in proton fractional energy retention, k, with scattering angle θ for several common elements.[1] The upper part of the figure illustrates the relative cross section of the same elements and the energy position a monolayer of each would occupy on a multichannel analyzer spectrum.

SCATTERING CROSS SECTION ENHANCEMENT

Precautions discussed up to this point which can be taken to optimize the chances for detection of helium buried in metals would be fruitless were it not for large scattering cross section

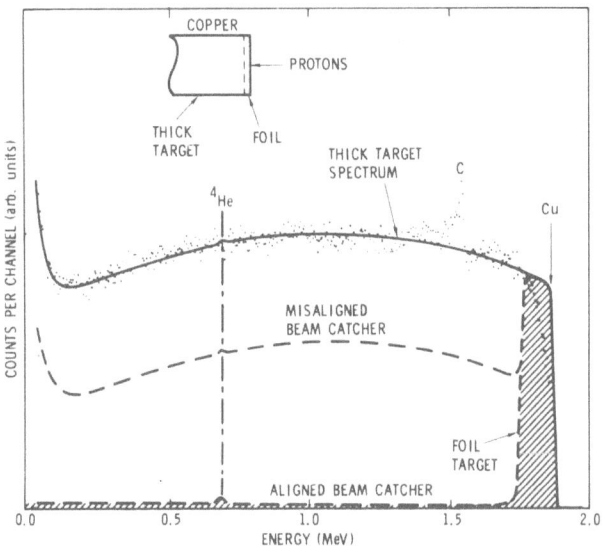

Figure 5. Backscattering spectrum of a thick copper target implanted with 3 x 10¹⁷ He/cm² at 50 keV.

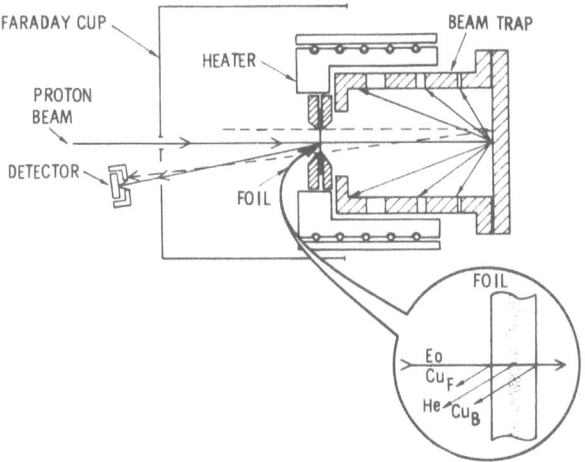

Figure 6. Schematic representation of a helium implanted copper foil mounted in a "beam trap" to prevent detection of any protons except those scattered in the foil.

enhancement first discovered by Chadwick in 1921 for the collision between high energy alpha particles and hydrogen atoms.[5] Over the years, accurate cross section measurements for the $^4He(p,p)^4He$ reaction have been made for a variety of incident angles and for energies ranging from $\sim$ 1 to 40 MeV (Fig. 7). The departure of $d\sigma/d\Omega$ from the Rutherford dependence is marked for the energies and angles shown, such departure being least at 0.95 MeV. A comparison between the magnitude of $d\sigma/d\Omega)_{Ruth}$ and $d\sigma/d\Omega)_{exp}$ is indicated for $E_0 = 2.53$ MeV by the shading in Figure 7. An enhancement is seen to exist for $\theta_{cm} \sim 25°$ for this energy.

For comparative purposes the ratio between the experimentally measured and the calculated Rutherford cross sections at 2.53 MeV is exhibited in Fig. 8. The enhancement is > 200% at 40° and increases to more than two orders of magnitude for scattering angles greater than $\sim$ 135°. For the experiments performed by this author, $\theta_{lab} = 164°$ and $E_0 = 2.53$ MeV at which angle and energy $\sigma_{meas}/\sigma_{Ruth} = 212$, making possible the detection of as little as 0.5 at.% ^4He in copper. Enhancements of a similar nature have been observed for protons incident at 2.5 MeV on other low Z elements such as D, ^3He, ^6Li, ^7Li, ^9Be, ^{12}C, ^{14}N, and ^{16}O, though the measured magnitude of the enhancements are usually less than for ^4He. The ratios of measured values to calculated Rutherford values are shown in Table I for several low Z elements of general interest whose scattering cross sections have been measured in detail.

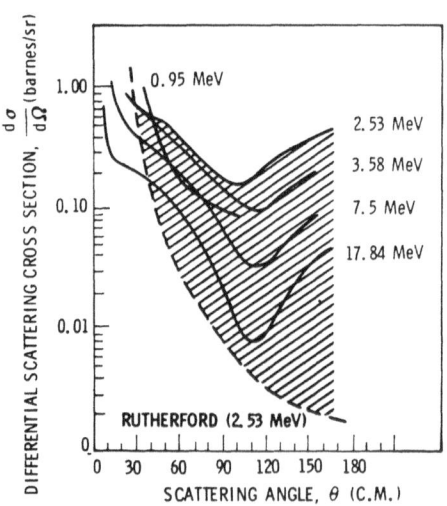

Figure 7.   Experimentally measured differential scattering cross section values for the $^4He(p,p)^4He$ reaction as a function of scattering angle (center-of-mass system) for incident proton energies (lab) of 0.95 MeV, 2.53 MeV, 3.58 MeV (Ref. 6), 7.5 MeV (Ref.7) and 17.84 MeV (Ref. 8).

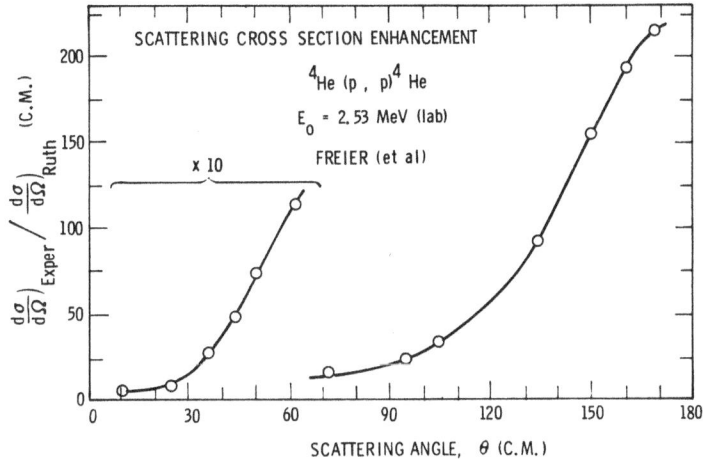

Figure 8. Plot of the ratio of the measured differential elastic
scattering cross section for the ^{4}He(p,p)^{4}He reaction
(Ref. 6) to the calculated Rutherford value assuming
θ = 168(CM), E_0 = 2.53 MeV(lab).

TABLE I

Reaction	Scattering Angle (CM)	$\frac{\sigma}{\sigma_{Ruth}}$ at E_0 = 2.5 MeV	$\frac{\sigma}{\sigma_{Ruth}}$ at Resonance Energy	Ref
D (p, p)D	172.5°	360	375 (E_0 = 2.8 MeV)	9
^{4}He(p, p)^{4}He	168°	212	181 (E_0 = 2.22 MeV)	6
^{9}Be(p, p)^{9}Be	146°	31.4	33.6 (E_0 = 2.52 MeV)	10
^{12}C(p, p)^{12}C	169.2°	7.6	46.1 (E_0 = 1.7 MeV)	11
^{14}N(p, p)^{14}N	160.9°	9.9	12.5 (E_0 = 2.35 MeV)	12
^{16}O(p, p)^{16}O	164°	13.6	97.5 (E_0 = 3.47 MeV)	13
^{16}O(α, α)^{16}O	168°	90.0	7.8 (E_0 = 3.04 MeV)	14

EFFECT OF SUBSTRATE THICKNESS ON DETECTION SENSITIVITY

Maximum sensitivity for detecting He and hydrogen isotopes in solids depends on minimizing the background on which these back-scattering peaks ride. A comparison of the magnitude of background counts for helium implanted in thick-target and foil-target titanium deuteride is shown in Figure 9. The upper spectrum is that of a helium implanted titanium deuteride foil while that below is from a helium implanted titanium deuteride film deposited on a thick copper substrate. The elastic scattering cross section enhancement for 2.5 MeV protons incident on ^{4}He and D is sufficient in both cases to observe backscattering peaks for these low Z elements in the titanium deuteride host.[15] The principal difference in the two spectra is the high thick target background in the lower spectrum (which must be subtracted from the light element peaks to obtain quantitative depth profiles) and the convoluted metal peak spectra at high energy which occurs in the lower spectrum because the energy position for front surface titanium overlaps that of the copper sub-strate. In the upper spectrum (foil substrate) the helium and deuterium peaks are much more prominent, and the background counts have been almost entirely eliminated.

An example of the role that thick target background plays in the spectra of the light elements is more clearly illustrated in Fig. 10. Exhibited in the upper part of the figure is the helium peak in the foil spectrum of Fig. 9 after the foil has been sequentially annealed at two elevated temperatures. It can be seen from the near perfect overlapping of the points (counts per channel) that the helium depth distribution remains unchanged after annealing. The lower spectrum is that of the helium peak in the copper backed titanium deuteride film after sequential anneals under the same conditions. Not only is the character of the helium profile unclear, it is also non-repeatable between the two runs so that few conclusions can be drawn about the helium behavior. Though an average can be taken of the thick target background counts (per channel) and used as a baseline against which to observe the desired helium distribution, there is no means to eliminate the statistical fluctuation the background adds to each of the points in the distribution in question. Only in the case where the background has been reduced at the outset to zero (or near zero) through the use of a thin foil substrate and a beam trap can variations in height be regarded as being related solely to variations in the helium depth concentration profile itself.

The situation is even less favorable for the deuterium depth profile on the thick target substrate. As can be seen in the lower spectrum in Fig. 9, the position of the deuterium peak falls just where the thick target background is changing most rapidly with energy and thus it is even more difficult to select the proper level

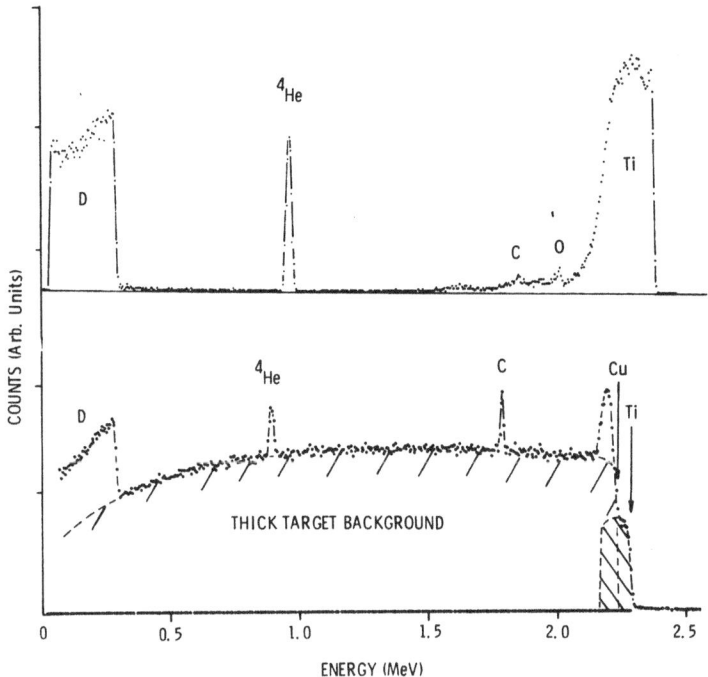

Figure 9.   Comparison between proton backscattering spectra of a
~ 1.5 μm titanium deuteride foil (above) and a titanium
deuteride film deposited on a thick copper substrate
(below). Both samples were implanted with ^{4}He at 50 keV
to a dose of 4 x 10^{17} He/cm^2. The energy scale of the
upper spectrum is slightly expanded with respect to that
of the lower one.

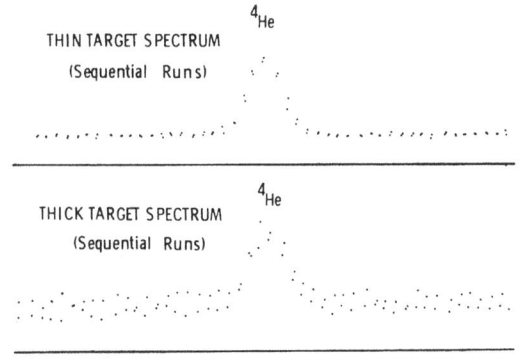

Figure 10.   Implanted helium peaks from the samples whose room temperature spectra appear in Figure 9.

to subtract from the total yield to obtain the deuterium profile.
This problem is minimized by using foil samples.

SOURCES OF BACKGROUND COUNTS

Detection sensitivity for impurities depends principally on the
cross section for elastic scattering events between incident protons
and atoms under study in the host lattice. As described in previous
sections, in detecting low Z elements in foils by the use of the
proton backscattering technique, one avoids counting scattering events
which occur on surfaces of the scattering chamber other than the foil
being analyzed. This produces the desired reduction in thick target
yield to allow more sensitive detection of isotopes such as ^{4}He,
^{3}He and D, but even for a well aligned beam-sample holder arrangement, some counts accumulate where, according to kinematics, there
should be no counts. Such a tail or background can be observed in
the upper spectrum of Figure 9 at energies between the deuterium
peak and the ^{4}He peak and also above the ^{4}He peak.

For the proton backscattering technique being described the
magnitude of the tail is generally small ($\sim$ 1-5 x 10^{-3} times as
large as the host metal peak) and is almost constant with energy
so that it presents no real problem in analyzing impurity concentrations as low as 0.5 at.% for the case of ^{4}He in Cu. However the
presence of background counts remain the principal limit in detection
sensitivity. Moreover this problem needs to be minimized to improve
the technique, in any case.

A beam trap of the type shown in Fig. 6 has been used by this
author to check possible sources of background counts. Since the
reduction of counts per channel from the full thick-target value
to the level observed in the upper spectrum of Fig. 9 is dependent
on the design and alignment of the beam trap, one naturally suspects
that inefficiency of the trap may be the major source of background
accumulation. To check this possibility, an empty titanium foil
holder (no foil) was mounted on the (copper) beam trap and an experimental run made in the normal manner by allowing the incident protons
to enter the beam trap without first scattering in a foil. The
spectrum obtained is shown in the lower part of Fig. 11. Despite a
charge of 10 µC incident into the beam trap, fewer than 10 counts
per channel were recorded at the detector over most of the energy
range, with the front shoulder of the distribution occurring at an
energy characteristic of titanium. The energy dependence is approximately that expected from a proton backscattering spectrum of a
thick Ti target. An identical run performed with ^{4}He ions produced
a similar spectrum whose energy dependence was characteristic of
scattering from thick target titanium. No counts were observed at
higher energies, where protons backscattered from copper (the beam

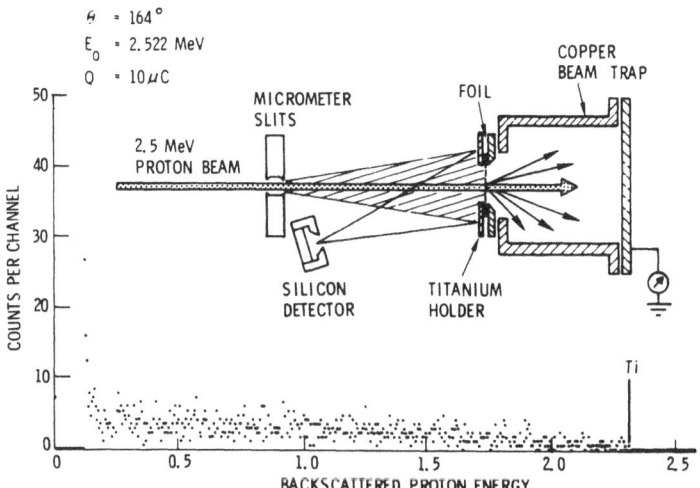

Figure 11. A schematic diagram of slit scattering and a repre-
            sentative spectrum using a Ti holder (with no foil
            sample) and a beam trap.

trap material) would be expected to occur. Double scattering events
within the beam trap, would result in energies lower than the position of the Ti shoulder, and may contribute to the spectrum to a
small extent at lower energies, but the probability for such events
is small.

     For this particular experiment, the most convincing explanation
is that the incident beam is accompanied by a "halo" as it approaches
the beam trap which arises from some protons being decollimated by
the slits from the main beam trajectory as shown by the shaded area
of Fig. 11. These protons strike the holder instead of passing into
the beam trap and are registered by the detector as titanium thick
target counts. This effect can be reduced by installing double or
triple detector collimation so that only the central part of the
foil in its holder can be "seen" by the detector, or by use of
different beam collimator design. The problem can also be reduced
by placing the final collimating slits close to the target. In
other experimental systems, halo or slit scattering may be of only
minor importance.

     Double scattering within the beam trap may be a more serious
problem when foil samples are mounted on the beam trap, for a
portion of the incident beam need be deflected only $\sim$ 30° in order
to provide a $\sim$ 180° backscatter path from the interior side of the
trap back through the foil to the detector as shown by the dashed
line in Fig. 11. However, this probability is also low, though
increased over the similar case for no foil in place.

Other factors may also contribute to a greater or lesser extent to the observed background, including pulse pile-up and similar electronics problems, double or multiple scattering in the sample foil and the like. Others have performed much more extensive research in this area using $^4He^+$ ions incident on thick targets at MeV[16] and sub-MeV[17] energies. No dependence on count rate (dead time) was found. However, the number of background counts per channel was found to decrease with increasing energy. One of the researchers[17] has observed a target film thickness dependence which, perhaps, can be explained in terms of plural scattering, which increases at lower energies. A similar effect has been observed by this author.

The problem of residual background counts warrants continued investigation. The beam trap plus foil arrangement seems to be an efficient way to study this phenomenon since one can isolate various contributory factors independently with and without foils in place. Using thin foils or no foils the energy dependence of the whole available spectrum can be observed simultaneously, without interference from thick target counts. Also with test foils mounted in place, the change in background with foil thickness and Z-number can be tested unambiguously.

REFERENCES

1. For a recent review paper, see W. K. Chu, J. W. Mayer, M-A. Nicolet, T. M. Buck, G. Amsel and F. Eisen, Thin Sol. Films 17 (1973) 1; M. A. Nicolet and W. K. Chu, American Laboratory, March 1975, pp. 22-34, see Figs. 3 and 5.

2. J. W. Mayer, L. Eriksson, and J. A. Davies, Ion Implantation in Semiconductors, Academic Press (New York) 1970, p. 16.

3. R. S. Blewer, Appl. Phys. Lett. 53 (1973) 593; L. R. Mervine, et.al., LLL Report UCRL-73087, 1971.

4. R. S. Blewer, Advances in Chemistry (to be published).

5. J. Chadwick and E. S. Bieler, Phil. Mag. 42 (1921) 923.

6. G. Freier, E. Lampi, W. Sleator and J. W. Williams, Phys. Rev. 75, 1345 (1949).

7. M. Putnam, J. E. Brolley, and L. Rosen, Phys. Rev. 104, 1303 (1956).

8. K. W. Brockman, Phys. Rev. 108, 1000 (1957).

9.   R.S. Langley, Bul. Amer. Phys. Soc. II, 20 (1975) 499.

10. U. Rohrer and L. Brown, Nuclear Physics A210, (1973) 465.

11. H.L. Jackson, A.I. Galonsky, F.J. Eppling, R.W. Hill,
     E. Goldberg, and J.R. Cameron, Phys. Rev. 89 (1953) 365.

12. S. Bashkin, Phys. Rev. 114 (1959) 1552.

13. R.A. Laubenstein, M.J.W. Laubenstein, L.J. Koester,
     and R.C. Mobley, Phys. Rev. 84 (1951) 12.

14. J.R. Cameron, Phys. Rev. 90 (1953) 839.

15. The first observation of deuterium and helium in thick target
     samples was made by R.A. Langley, see Ref. 9.

16. W.K. Chu, I. Mitchell, M.-A. Nicolet and J.W. Mayer, private
     communication, November 1974.

17. B.M.U. Scherzer, private communication, May 1975.

DICUSSION

Q: (J.P. Biersack) How sensitive is your method of using a resonance energy for detection of He in metals?

A: (R.S. Blewer) The detection sensitivity for helium in metals
using the proton backscattering technique depends on the residual
background level (tail), the incident beam energy (strength of
the ^{4}He(p,p)^{4}He resonance) and the host metal of the target (dE/dx
dependence). A typical value for 2.5 MeV incident protons is 0.5
at.% ^{4}He in copper.

Q: (G. Dearnaley) Is there any possibility that the unusual stress
conditions present in a thin foil ($\sim$5 μm) will interfere with
migration of mobile species such as hydrogen, by comparison with
that in the bulk metal specimens?

A: (R.S. Blewer) Because of their very small thickness (1-10 μm
for these experiments), foils should be sensitive self-indicators
of intrinsic stress. No obvious manifestations of unusual stress
(such as wrinkling, sagging, etc.) have been observed in about 15
different foil materials we have investigated. Conversely, intrinsic
stress in evaporated thin films is known to approach the bulk yield
strength in many metals (K.L. Chopra, Thin Film Phenomena, McGraw
Hill (New York) 1969, p. 266). In such cases, stress enhanced diffusion of mobile species such as hydrogen may be more evident.

DEPTH PROFILING OF DEUTERIUM AND HELIUM IN METALS BY ELASTIC PROTON
SCATTERING: A MEASUREMENT OF THE ENHANCEMENT OF THE ELASTIC SCATTER-
ING CROSS SECTION OVER RUTHERFORD SCATTERING CROSS SECTION*

R. A. Langley

Sandia Laboratories

Albuquerque, New Mexico 87115

I. INTRODUCTION

The characterization of solids implanted and diffused with impurity atoms is of concern to a variety of fields. Both nuclear
microanalysis and ion backscattering of energetic light ions have
proven to be useful tools in the study of thin films and surfaces.
Light atom impurities in heavier atom substrates have not been
readily detected by ion backscattering because the elemental Rutherford cross section at a given energy decreases as the square of the
target atomic number. Thus, the backscattered ion energy is low
for light impurities and usually is superimposed on the larger yield
from the heavier substrate atoms. Nuclear reactions, therefore,
have been used to study various phenomena associated with light atom
impurities. Many of the nuclear reactions used in nuclear microanalysis have been summarized and reviewed in previous articles,[1]
but this technique is fraught with certain difficulties. The reaction cross sections are relatively small so that in order to obtain
good statistics considerable fluences of the probing ion must be
used. This results in both long counting times and perturbation of
the systems studied because of implantation of the probing beam.

Recently the use of ion backscattering to observe deuterium
and helium in thin films on thick substrates has been successfully
accomplished.[2] Using this technique the elastic scattering cross
section for protons on deuterium at 170°C (lab) was found to be
approximately 140 times greater than the Rutherford cross section
at 2.0 MeV and approximately 260 times greater at 2.8 MeV, with

*This work was supported by the United States Energy Research
 and Development Administration, ERDA.

the enhancement increasing linearly with energy. In addition it
was found that for protons incident on ^{4}He the elastic cross section
is increased over the Rutherford cross section by approximately 300
at 2.8 MeV. Previously ion backscattering has been used to observe
deuterium and helium in thin free-standing films.[3,4]

The remainder of this paper contains a discussion of the experimental setup, results of the measurement of the reaction cross section, a demonstration of the technique on specially-prepared films,
and finally a description of an analytical technique to provide
impurity concentration vs depth from yield vs energy spectra.

II. EXPERIMENTAL SETUP

The source of energetic ions was a 2.5 MeV Van de Graaf accelerator which with its associated analyzing magnet provided a collimated monoenergetic beam of particles. The energy of the ion beam
was determined by measuring the magnetic field of the analyzing
magnet. The entire experimental system (consisting of the accelerator, magnet and backscattering chamber) was energy-calibrated by
observing resonances of ^{19}F$[p,\alpha]^{16}$O* $\rightarrow$ hν + ^{16}O. The resonances
which occur at beam energies of 340.5, 872.5, 935, 1347 and 1381 keV
were studied by observing the emitted gamma rays.[5] The maximum
possible error in this calibration was less than 0.5%.

The target chamber has been described in a previous paper and
will not be fully described here.[6] The geometry of the backscattering setup is such that the ion beam was incident normal to the surface of the substrate and the backscattered beam was detected at an
angle of 170° with respect to the incident ion beam by a silicon
barrier detector. The signal from the detector was processed using
a charge-sensitive preamplifier, a linear amplifier with a base-line
restorer and a multichannel analyzer. A precision pulser was used
to calibrate the counting system by supplying a pulse which was
capacitatively coupled to the preamplifier input. The quoted integral nonlinearity of the pulser was less than 0.002%, the integral
nonlinearity of the analyzer was found to be better than ± 0.025%
for 2048 channels by using the pulser. A series of collimating
slits defined the beam to a 0.76 mm x 0.76 mm square with a full
angular divergence of 0.02°. The energy resolution of the system
was about 8 keV. The beam current was kept sufficiently low for the
average count rate to be < 0.01 of the maximum count rate of the
analyzer; therefore, dead time of the analyzer was negligible and
pulse pileup was minimized. A typical backscattered proton spectrum is shown in Fig. 1. The target for this particular spectrum
was an erbium deuteride film on a kovar substrate. The film thickness of the erbium deuteride film was 2.59 x 10^{18} molecules/cm^2 as
determined by ^{4}He ion backscattering. The stoichiometry of the film
was nominally erbium dideuteride. The erbium and deuterium peaks

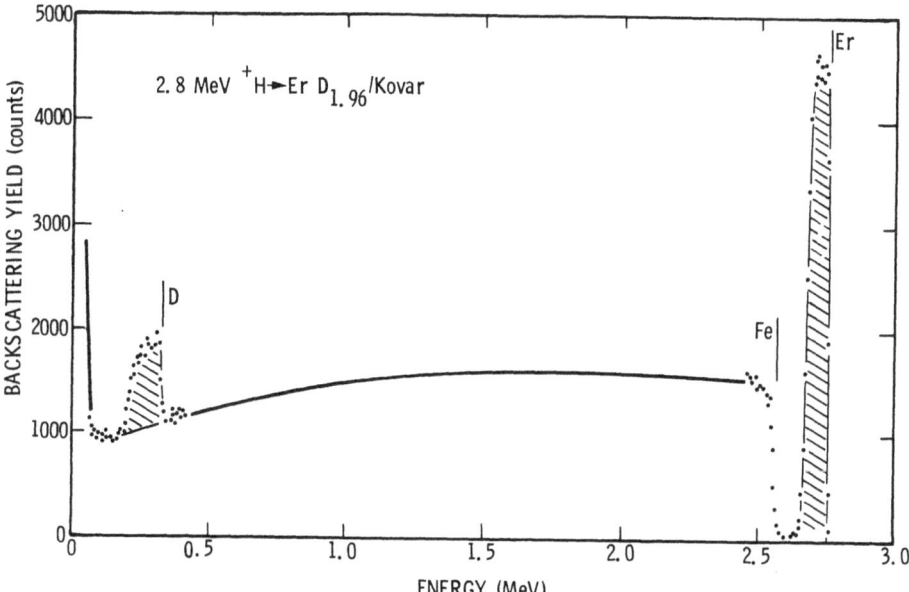

Fig. 1. Energy spectrum of 2.8 MeV protons backscattered from a
vacuum-deposited Er film about 8000 Å thick. The Er film was
deuterided to [D]/[Er] = 1.96. The energy edges for Er, Fe and D
are indicated.

are labeled and the iron edge from the kovar substrate is labeled.
The equations for the kinematics of the reaction are given in Ref. 7
and will not be discussed further here, although it should be noted
that even though the cross section for the reaction is enhanced the
reaction remains elastic.

III. MEASUREMENTS OF THE REACTION CROSS SECTION FOR H$^+$ ON D, ^{4}He

        Erbium deuteride films of appropriate thicknesses were pre-
pared by deposition of erbium on various substrates. The substrates
were held at 400°C during deposition. After deposition these films
were in situ hydrided by exposing the samples to a deuterium atmosphere of 5 Torr at a substrate temperature of 300°C. To terminate
the hydriding procedure the chamber was pumped free of deuterium
before the substrate temperature was returned to room temperature.
This procedure precludes formation of erbium trideuteride. It was
known from previous work that there was essentially no interaction
between kovar or alumina substrates and the erbium deuteride film
during the deposition process. This was further substantiated by
^{4}He ion backscattering spectra.

Substrates of chemically polished and vapor-blasted kovar and sapphire were used. Films were deposited on four substrates of each type. From each group of four, one substrate was analyzed for the amount of erbium present and for the amount of occluded deuterium. The amount of deuterium was determined by mass spectrometric determination on outgassing of the substrate.[8] The amount of erbium was measured by weight gain during deposition and by dissolution of the erbium from the substrate and subsequent analysis using atomic absorption spectroscopy.[9]

The elastic scattering cross section was determined by measuring the area under the deuterium peak and the area under the erbium peak by assuming that the elastic scattering of protons from erbium is Rutherford-like and independently measuring the loading ratio. Loading ratio is defined as the total amount of deuterium to the total amount of erbium, i.e., [D]/[Er]. Rutherford scattering is considered valid for $\alpha \gg 1$ where $\alpha = 2\ Zz/137\beta$ and Z is the target atomic number, z is the incident ion atomic number and $\beta = v/c$. For protons on erbium at 2.8 MeV $\alpha \sim 13$ and at 2.0 MeV $\alpha \sim 15$. For protons on deuterium at 2.8 MeV $\alpha \approx 0.4$ and at 2.0 MeV $\alpha \approx 0.45$, so that one can legitimately assume that the elastic scattering cross sections of protons scattered from erbium are Rutherford-like but not from deuterium. In determining the cross section it is necessary to account for the difference in solid angle subtended in the center-of-mass system by the solid state detector for protons scattered from erbium and from deuterium. The results for the elastic cross section obtained are shown in Fig. 2. In the usual notation the elastic scattering cross section is given in the center-of-mass system in mB/sr and energy in the lab system in MeV. Protons scattered from deuterium at 170° in the lab frame are equivalent to scattering at 172.5° in the center-of-mass frame. Shown in comparison in the same figure are the recent results of Kocher and Clegg at 166°, 160° and 154°.[10] In Fig. 3 is shown the cross section enhancement vs energy in the lab system, i.e., ratio of the elastic scattering cross section in the lab system to the Rutherford cross section in the lab system. The enhancement increases linearly as the energy of the incident beam increases. The error associated with this measurement is ± 2% for the cross section and less than ± 0.5% for the energy.

IV. DEMONSTRATION OF DEPTH PROFILING TECHNIQUE ON SPECIALLY-PREPARED FILMS

Multilayered deuteride films were prepared to demonstrate deuterium concentration vs depth measurements and also determination of depth resolution. A typical sample was prepared by depositing an erbium-deuterium film on a molybdenum substrate of $\sim$ 1000 Å thickness. This was followed by a chromium deposition of $\sim$ 1800 Å which in turn was followed by erbium deposition of $\sim$ 800 Å. The spectrum for

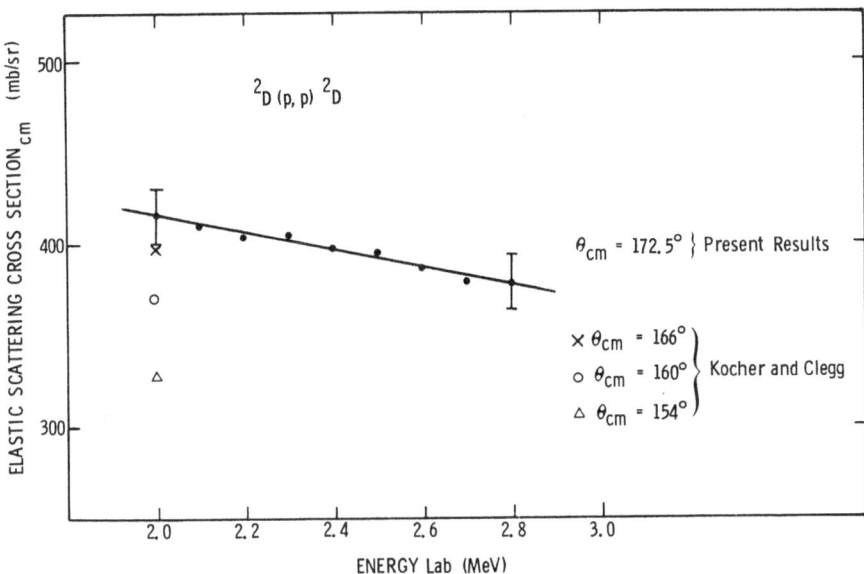

Fig. 2. Experimental elastic scattering cross section for p on D for $\theta_{cm} = 172.5°$. Also included are the results of Kocher and Clegg at other lower scattering angles.

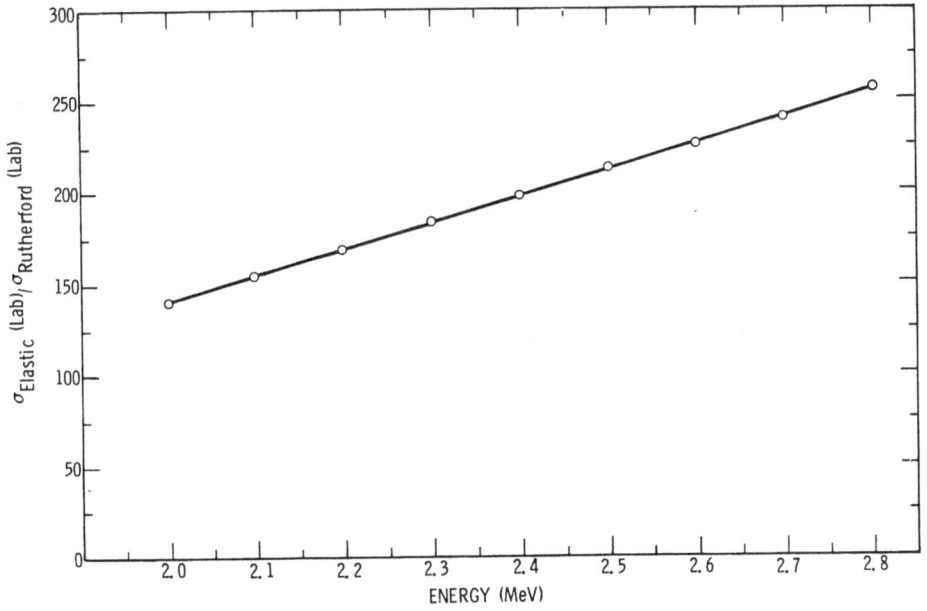

Fig. 3. Elastic cross section enhancement over Rutherford for p on D ($\theta_{cm} = 172.5°$).

2.8 MeV proton backscattered from this multilayered film is shown
in Fig. 4. At energies around 2.6 MeV are shown the two erbium
peaks not completely resolved. At 2.55 MeV is shown the Mo substrate edge with the Cr edge at 2.5 MeV and around 0.3 MeV is shown
the well-separated deuterium peaks. The depth resolution for this
particular spectrum was about 600 Å. If one cools the detector this
could be decreased to about 400 Å. It is noticed from this spectrum
that the depth resolution for the light mass component D of the film
is improved over that for the heavy mass film component Er because
of differences in the stopping cross sections for protons scattered
from the two film components. This is because the stopping cross
section is increasing with decreasing energy in this energy range
and those protons backscattered from deuterium have energies much
lower than those backscattered from erbium.

In addition, an implantation of 2×10^{17} He atoms/cm^2 was made
in an erbium deuteride film. The energy of the implant was 50 keV.
This was made into an erbium dideuteride film on a kovar substrate.
One can easily distinguish the helium which is implanted into the
film as shown in Fig. 5. Its mean range is 1500 Å. From spectra of
this type the enhancement factor is estimated. Since it is difficult
to accurately measure the amount of helium implanted and its unform-
ity in the substrate, only an estimate of the elastic scattering
cross section enhancement over Rutherford can be made. For 2.8 MeV
this enhancement is ~ 300.

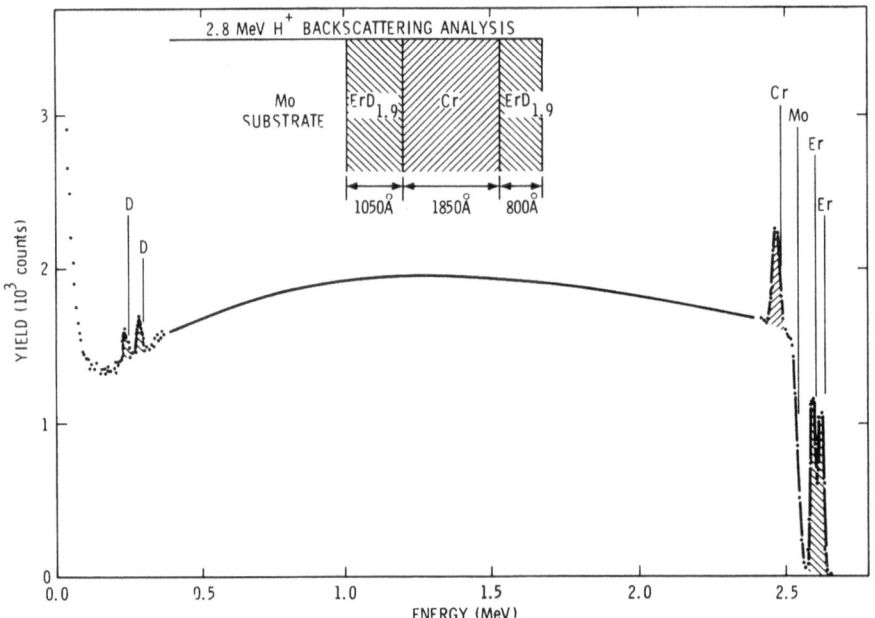

Fig. 4. Energy spectrum for 2.8 MeV protons on a multilayered
deuterided film. The energy edges are identified.

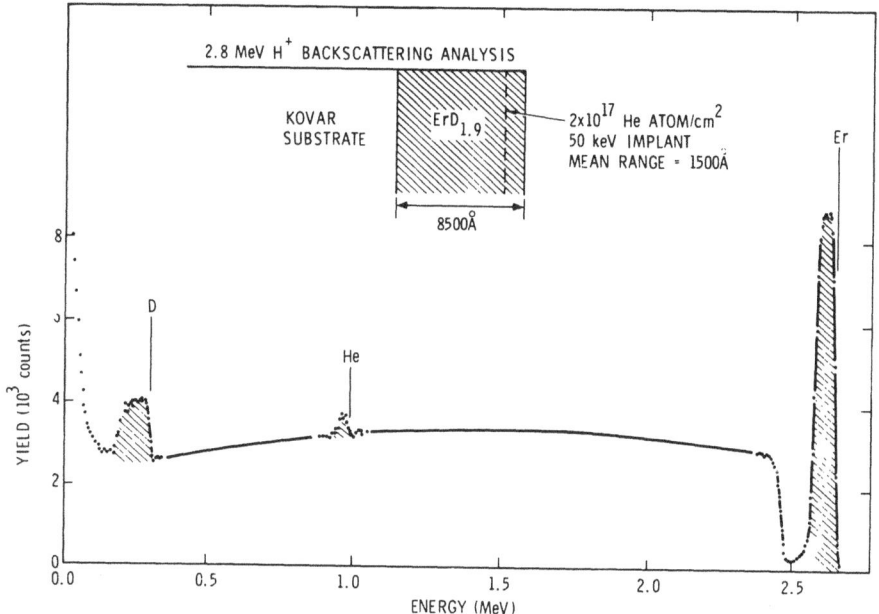

Fig. 5. Energy spectrum for 2.8 MeV protons on an [4]He implanted deuteride film.

V. ANALYSIS TECHNIQUES

In order to reduce yield vs energy spectra to impurity concentration vs depth it is necessary to go through some analytical technique. Such a technique has been developed.[11] Previous techniques for data reduction have consisted of calculating an expected spectra from an assumed distribution and comparing that with an observed spectra[12] or by a direct mathematical approach in which a reference spectrum and the spectrum to be analyzed are compared to give a concentration vs depth.[13] This latter calculational method is only applicable for elastic collisions. This is not the case for the present technique which is applicable to both inelastic and elastic collisions and uses only the spectrum to be analyzed.

The only experimental parameter which remains the same throughout the spectra is the energy per channel. As the stopping cross sections for both the incoming and outgoing particles are a function of energy, the thickness layer associated with each energy channel will change accordingly and is defined as a depth channel. An analytical procedure was used to derive a general formula for the nth depth channel. Since the stopping cross section for low energy protons varies quite rapidly with energy below 300 keV it was found necessary to be able to subdivide each channel. This procedure made it possible to determine thickness and composition independent of

the calculational procedure. The concentration vs depth of deuterium atoms calculated by this technique is given in Fig. 6 for the
spectra shown in Fig. 1. Use of this computer program is not limited to elastic scattering but may be used for nuclear microanalysis
techniques involving inelastic scattering and nuclear reactions.
The analysis technique does not take into effect straggling effects
or multiple scattering effects which depend on energy, mass and
atomic number of the incident ion as well as on the masses and
atomic numbers of all the constituent atoms of the target.

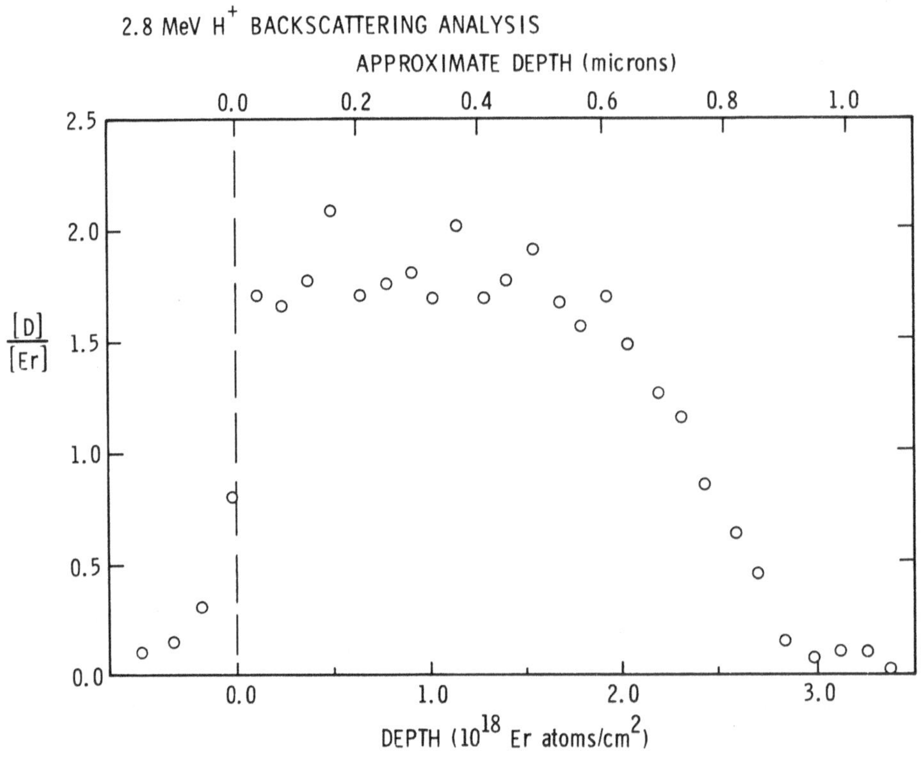

Fig. 6. D concentration vs depth for the spectrum of Fig. 1. These
data are obtained by the calculational method explained in the text.

VI.   CONCLUSION

It has been demonstrated that deuterium and helium profiles
can be measured in films on thick substrates, that the cross section
for the elastic cross sections on deuterium is enhanced over Rutherford, and that this enhancement factor increases linearly with increasing energy. Depth resolutions of ~ 600 Å were obtained. In
addition estimates of the scattering cross section for protons on
^{4}He were made.

ACKNOWLEDGMENT

I would like to thank Len Provo for the preparation of the erbium deuteride films used in the cross section measurements and the analytical measurements made on them.

REFERENCES

1. See, for example, G. Amsel, J. P. Nadai, E. D'Artemare, D. David, E. Girard and J. Moulin, Nucl. Inst. Meth. 92, 481 (1971).

2. R. A. Langley, Bull. Am. Phys. Soc. 20, 499 (1975).

3. L. R. Mervine, R. C. Der, R. J. Fortner, T. M. Kavanagh, and J. M. Khan, Lawrence Livermore Laboratories Report No. UCRL-73087, 1971.

4. R. S. Blewer, Appl. Phys. Lett. 23, 593 (1973).

5. J. B. Marion, Rev. Mod. Phys. 33, 139 (1961).

6. R. A. Langley and R. S. Blewer, Thin Solid Films 19, 187 (1973).

7. A general review of the methods of ion beam analysis has recently been given by W. K. Chu, J. W. Mayer, M.-A. Nicolet, T. M. Buck, G. Amsel and F. Eisen, Thin Solid Films 17, 1 (1973).

8. B. B. McInteer, Los Alamos Scientific Laboratories Report No. LA-2056 (1957).

9. See for example V. G. Mosotti and V. A. Fassel, Spectrochim. Acta 20, 1117 (1964) or J. C. Van Loon, J. H. Galbraith and H. M. Aarden, Analyst 96, 47 (1971).

10. D. C. Kocher and T. B. Clegg, Nucl. Phys. A 132, 455 (1969).

11. R. A. Langley, Sandia Laboratories Report No. SAND75-0331 (1975).

12. See for example A. Turos, L. Wielunski and A. Barcz, Nucl. Instr. and Meth. 111, 605 (1973).

13. D. K. Brice, Sandia Laboratories Report SLA 73-0843 (1973).

DISCUSSION

Q: (E. Wolicki) Do you have any data which show that the proton scattering cross section from Erbium is Rutherford at these large angles?

A: (R. Langley) No. We assumed the cross section to be Rutherford.

Comment: (J. L'Ecuyer) As far as sensitivity is concerned, I believe that the detection of the forward angle recoiling light ions is the best with a sensitivity of 10^{14} atoms/cm^2 when heavy ions are used for bombardment.

Q: (W. Möller) It should be possible to calibrate your method by the nuclear reaction profiling technique $|D(^3He,p)\alpha|$. Did you try that?

A: (R. Langley) The samples studied for this work were also used in the $D(^3He,p)d$ and correlated quite well.

Q: (R. Behrisch) Have you also measured the increase in elastic cross section for backscattering of protons on tritium and what enhancement would you expect?

A: (R. Langley) No, but Claassen et al. have measured it to scattering angles of $\sim 160^\circ$ and is referenced in Phys. Rev. about 1955.

NEAR-SURFACE INVESTIGATION BY BACKSCATTERING OF N$^+$IONS AND GRAZING ANGLE BEAM INCIDENCE*

W. Pabst

Institut für Festkörpertechnologie der Fraun-

hofer-Gesellschaft, Munich, Germany

ABSTRACT

An improvement of depth resolution in backscattering measurements by the use of heavy ions and by increasing the incidence and exit angles of the analyzing ions is investigated. As detector resolution, stopping power, energy straggling and multiple scattering effects are involved in resolution, these data were measured for He$^+$ and N$^+$ ions of energy 1,3 MeV on thin W layers. The geometrical effect is found to be by far more efficient than heavy ions. Actual resolution is demonstrated and discussed by means of the spectrum of a multilayer structure.

I. INTRODUCTION

Ion backscattering in the energy region of some MeV is widely used in the surface layer analysis of solids /1/. He$^+$ ions are commonly used because of their good resolution for medium masses and their depth resolution down to 100 Å. However, few publications reported on measurements with heavy ions, such as C^{12}, N^{14} and O^{18} /2-5/. They allow a better mass resolution and a higher sensivity; about improved depth resolution there are only few data. This work compares the depth resolution of He$^+$ and

*   Work sponsored by Bundesministerium für Forschung und Technologie, Bonn

N^+ ions and discusses an improvement by grazing angle
beam incidence. This latter technique has already been
applied by Williams to measure implanted distributions
/6-9/.

II) DEPTH RESOLUTION

From general theory of backscattering (e.g. /1/,
Appendix I), the energy interval ΔE of backscattered
ions that reflects a depth interval Δt is given by

$$\Delta E = \Delta t \left(\frac{k\, S_{in}}{\cos\Theta_{in}} + \frac{S_{out}}{\cos\Theta_{out}} \right). \qquad (1)$$

S is the stopping power in the material and Θ are the
angles relative to the surface normal for the ingoing
and outgoing particles; $k = k(\Theta_{in} - \Theta_{out})$ is the energy
loss factor in nuclear collision.
From eq. (1) it can be seen that depth resolution is improved either by increasing the stopping power (using
heavier particles or suitable particle energies) or by
increasing the angles Θ_{in} and Θ_{out}. In order to maintain
a constant sensivity, only Θ_{in} may be increased (the number of counts per energy interval is then constant, as
the way in the layer is prolonged). For light target
atoms instead, the small value of k favours a large angle
Θ_{out}. The angles Θ are limited at about 85° because of
scattering effects and too large a surface probed by the
beam.
Consider the ions backscattered by an extremely thin
layer at a depth t. Depth resolution Δt is given by
eq. (1) when ΔE is the measured width of their energy
distribution. The various effects causing this broadening are: detector resolution Δ, straggling χ, and multiple scattering ψ. These parameters can be added quadratically for near-Gaussian distributions; hence resolution
is

$$\Delta t = \frac{\sqrt{\Delta^2 + \chi^2 + \psi^2}}{k\, S_{in}\cos^{-1}\Theta_{in} + S_{out}\cos^{-1}\Theta_{out}} \qquad . \quad (2)$$

Δ, χ, ψ, and S were measured for 1.3 MeV N^+ ions incident
on W layers of various thicknesses (200 ... 2000 Å). The
results are compared to the data for 1.3 MeV He^+ ions,
measured at the same specimens.

III) EXPERIMENTAL DATA

1. Stopping Power

From the spectra (Figs. 1 a and b), the relative
stopping power values for N^+ and He^+ ions were calculated using eq. (1). The result is a mean value instead of
S_{in} and S_{out}. It is associated with a mean energy : in
a simple approximation /12/,

$$\langle E \rangle = E_0 \frac{1+k}{2} - \frac{\delta E}{2}, \qquad (3)$$

where δE is the energy loss in the film (see Figs. 1 a
and b). For the same films, the N^+ stopping cross section
is taken as mean value over a broader energy range than
the He^+ value; but it is sufficient to enter as a mean
value in eq.(2).
For 1400 Å thick W films, the $\frac{N^+}{He^+}$ stopping power ratio
was found to be 2.5 (Tab.1). The N^+ stopping power
values were normalized to the He^+-values measured by
Borders /12/ (which agree with the data of Chu et al.
/13/). The results of Northcliffe and Schilling /14/ for
N^+ are somewhat lower. (Björkquist and Domeij /15/ found
a similar discrepancy for C, N and O ions in Cr, Fe, Co,
Ni, Cu and Zn. Instead, stopping power values for N^+ in
Al, Ni, and Au measured by Porat et al. /16/ agree quite
well with /14/.)

2. Detector Resolution

Detector resolution and energy straggling may be
deduced by differenciating the spectra. The widths of the

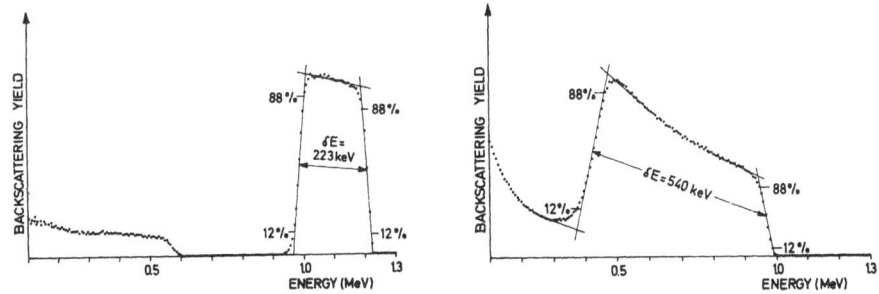

Fig. 1: Backscattering spectra for a 1400 Å thick W layer
        on a Si substrate:left: He^+ ions, right: N^+ ions,
        both of energy 1.3 MeV.

front and rear edges yield detector resolution and detector resolution plus straggling and scattering, respectively. Surface roughness may cause additional broadening. When the derivatives at the front and rear edges are Gaussians, the FWHM can be taken more easily as the energy difference between 12 % and 88 % of full step height /11/ - see. fig. 1. Resolution of a heavy ion surface barrier detector was found to be 33 keV for 1.0 MeV N^+ ions. This result is between the theoretical expectations of Lindhard et al./17/ and Haines and Whitehead /18/ and meets the extrapolation of the curve measured by Petersson /2/. The dependence of detector resolution on energy will be neglected in eq.(2), because Δ contributes only at the surface.

3. Straggling

Straggling data were taken from the rear edges of the spectra, as described in sect. III)2.(For perpendicular incidence, no scattering contributes.) The widths of the front edges (detector resolution and surface roughness) were subtracted quadratically. Tab. 1 shows the parameter χ_o for the best fit of the data to a law

$$\chi = \chi_o \sqrt{t(\cos^{-1}\theta_{in} + \cos^{-1}\theta_{out})} \quad \text{(Bohr's theory /10/).}$$

Measured absolute values are between Bohr theory (upper limit) and experimental data for Pt /11/. For N^+ ions, Bohr theory is difficult to apply because of charge fluctuations of the ions being stripped in the material /19/.

4. Multiple Scattering

When the ion beam is passing through the sample, the ions are not only slowed down, but scattered off the original direction into a cone (increased cross-section for scattering about small angles). For large angles θ_{in} or θ_{out} this affects resolution considerably. The broadening of the energy distribution due to scattering of the ingoing beam is (using eq.(1)):

$$\psi = k \langle S \rangle \, t \left(\frac{1}{\cos(\theta_{in} + \delta\theta_{in})} - \frac{1}{\cos(\theta_{in} - \delta\theta_{in})} \right)$$

$$\approx k \langle S \rangle \, t \, \frac{\sin\theta_{in}}{\cos^2\theta_{in}} \, 2\,\delta\theta_{in} \, . \qquad (4)$$

This broadening becomes asymmetrical for large θ_{in}. Using theory of multiple scattering /20-22/ and thin film approximation /23,24/, θ_{in} can be calculated as the half-angle of the cone containing 76 % of the particles. This corresponds to FWHM of resolution, as in this energy and layer thickness region, multiple scattering distributions are Gaussian to a good approximation /23/. These calculations will be used in the discussion. For He$^+$ and N$^+$ in W, ψ was measured by comparing the spectra from thick ($\sim$ 2000 Å) layers at perpendicular beam incidence with those from thin ($\sim$ 400 Å) layers at large incidence angles. The way of the beam in the layer, and hence straggling, was the same in both cases, and the influence of scattering could be found directly. Then $\delta\theta_{in}$ was calculated using eq.(4). With $\delta\theta_{in} \alpha \sqrt{t}$ (an approximation from theory), ψ (θ_{in},t) was included in eq.(2).

1.3 MeV	K (/)	Δ (keV)	S <S> (keV/μm)	χ_0 (keV$\text{Å}^{-1/2}$)	$\delta\theta_{in}$ (degr.)
He$^+$in W	0.93	13.5	780	0.62	0.54
N$^+$in W	0.74	33	2000	1.38	0.38
He$^+$,Ge in Si	0.81	13.5	325	0.28	0.43

Tab. 1 The experimental values of the parameters used in the calculations for Figs. 2 and 4. $\delta\theta_{in}$ is for a way of 1000 Å in the material.

IV) RESULTS AND DISCUSSION

Fig. 2 shows resolution as a function of depth obtained according to eq.(2). The experimental values used for He$^+$ and N$^+$ ions in W are shown in Tab. 1. $\theta_{in} - \theta_{out}$ was fixed to 15^0; θ_{in} was varied from 0^0 to 85^0. It can be seen that the gain in resolution by using N$^+$ ions instead of He$^+$ ions is not significant. By 85^0 beam incidence, instead, resolution by He$^+$ ions is increased by as much as a factor of 7 at the surface. This is demonstrated in Fig. 3: a multilayer structure of Si and Ge/Si-layers of thickness 50 Å is easily resolved. However, this is possible only for the first 500 Å because of straggling and scattering. The influence of scattering is discussed in Fig. 4. Again, resolution was calculated from straggling and scattering data, measured at various

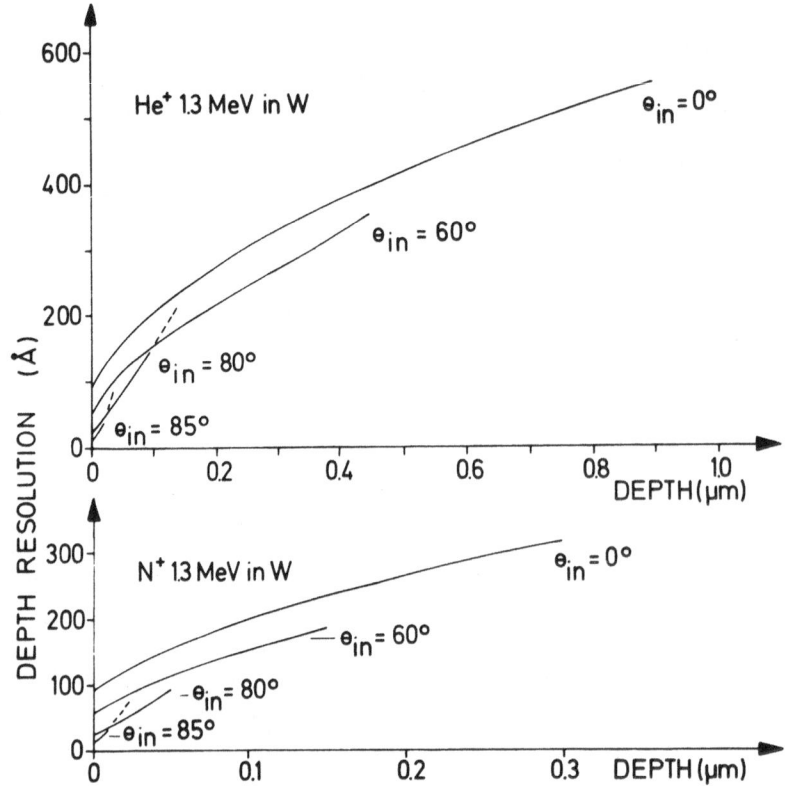

Fig. 2: Depth resolution in W by He$^+$ ions and N$^+$ ions at various incidence angles, as a function of depth. $\Theta_{in} - \Theta_{out} = 15°$.

Ge/Si layer spectra (curve (b)). For comparison, in curve (a) no scattering is included. Actual resolution is much better than estimated by theoretical values for $\delta\Theta_{in}$ /20-24/ (curve (c)). This might give the grazing angle method more importance than previously expected.

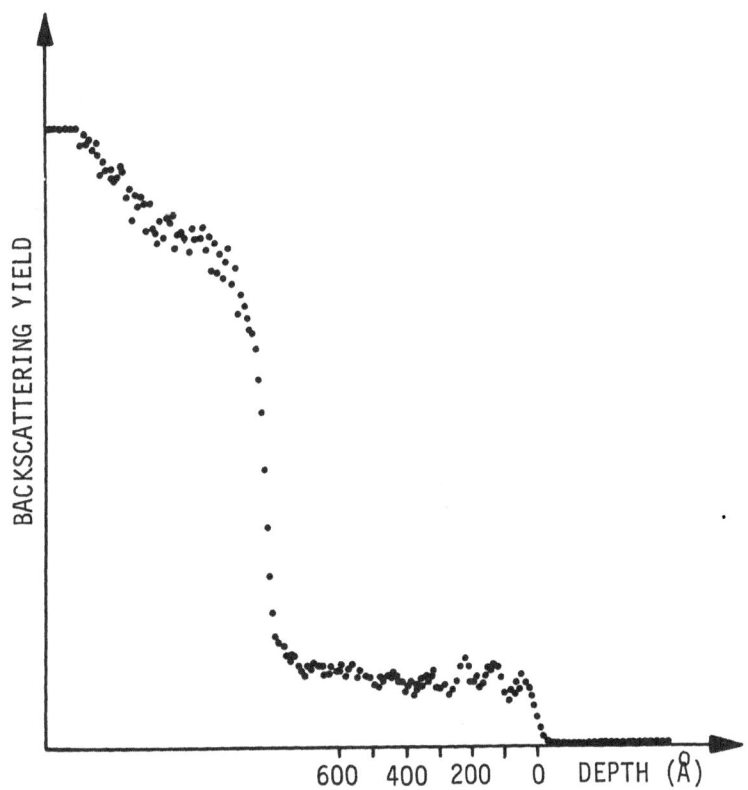

Fig. 3: A 50 Å Si-Si/Ge multilayer structure resolved
by 85° He$^+$ beam incidence.

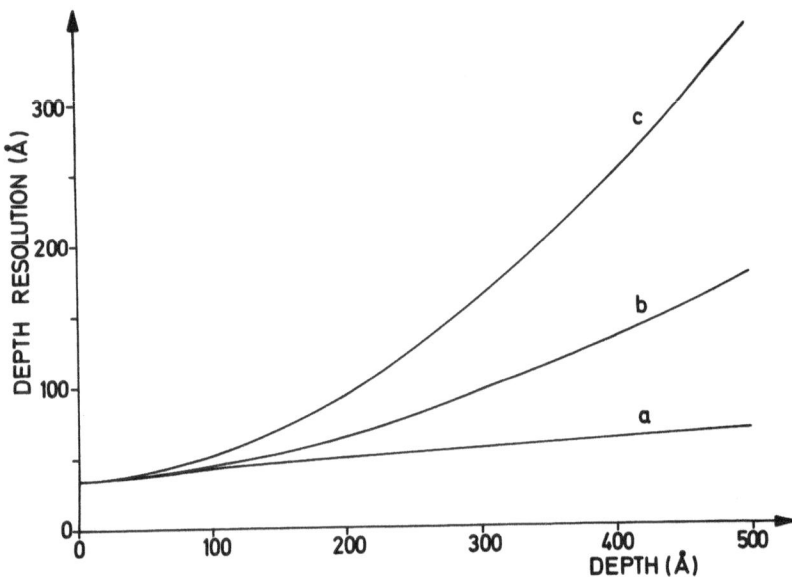

Fig. 4 : Depth resolution for Ge (15 %) in Si, when
a) no multiple scattering, b) experimental
scattering values, c) theoretical values
are included. $\Theta_{in} = 85^{\circ}$, $\Theta_{out} = 70^{\circ}$.

Acknowledgements

The author wishes to thank Dr. P. Eichinger and Dr. H.
Bernt for discussions. Ms. L. Fiesel and Mr. J. Weid-
haas are thanked for technical and programming assistance. We are grateful to Valvo GmbH for the W layers,
and to AEG-Telefunken, Forschungsinstitute Ulm, for
supplying the multilayer structures.

Note: In June 1975, a discussion of the grazing angle
incidence method appeared by J. S. Williams, Nucl.
Instr. and Meth. <u>126</u> (1975) 205.

V. REFERENCES

/1/ See for example Proc. Int. Conf. on Ion Beam Surf. Layer Analysis, edited by J. W. Mayer and J. F. Ziegler, Elsevier Sequoia S.A., Lausanne 1974

/2/ S. Petersson et al., Thin Solid Films $\underline{19}$ (1973) 157, or in /1/

/3/ J. W. Mayer, L. Eriksson and J. A. Davies, Ion Implantation in Semiconductors, Academic Press, New York, 1970

/4/ M. Guttmann et al., Scripta metallurgica $\underline{5}$ (1971) 479

/5/ R. R. Hart et al., Thin Solid Films $\underline{19}$ (1973) 137, or in /1/

/6/ J. S. Williams, Rad. Eff. $\underline{22}$ (1974) 211

/7/ J. S. Williams and W. A. Grant, Rad. Eff. $\underline{25}$ (1975) 55

/8/ J. S. Williams, Physics Letters $\underline{51A}$ (1975) 85

/9/ G. A. Stephens et al., Proc. Int. Conf. on Ion Impl., Osaka 1974

/10/ N. Bohr, Phil Mag. $\underline{30}$ (1915) 581

/11/ J. M. Harris et al., Thin Solid Films $\underline{19}$ (1973) 259

/12/ J. A. Borders, Rad. Eff. $\underline{21}$ (1974) 165

/13/ W. K. Chu et al., Appl. Phys. Lett. $\underline{22}$ (1973) 437

/14/ L. C. Northcliffe and R. F. Schilling, Nuclear Data $\underline{A7}$ (1970) 233

/15/ K. Björkquist and B. Domeij, Rad. Eff. $\underline{13}$ (1972) 191

/16/ D. I. Parat and K. Ramavataram, Proc. Roy. Soc. (London) $\underline{A252}$ (1959) 394

/17/   J. Lindhard et al., Mat. Fys. Medd. Dan. Vid.
       Selsk. 33 (1963) 29

/18/   E. L. Haines and A. B. Whitehead, Rev. Sci.
       Instr. 37 (1966) 190

/19/   V. V. Avdeichikov et al., Nucl. Instr. & Meth.
       118 (1974) 247

/20/   G. Moliere, Z. Naturforschung, 2A (1947) 133

/21/   G. Moliere, Z. Naturforschung, 3A (1948) 78

/22/   B. P. Nigam, Phys. Rev. 115 (1959) 491

/23/   J. B. Marion and B. A. Zimmerman, Nucl. Instr. &
       Meth. 51 (1967) 93

/24/   H. Bichsel, Univ. of Southern California Report,
       Los Angeles 1970

DISCUSSION

Q: (B. Scherzer) How large are deviations from Rutherford scattering cross sections for nitrogen ions around or below 1 MeV? This may considerably complicate valuation of spectra with respect to composition.

A: (W. Pabst)This has not been considered. The measured values for stopping power, straggling (and detector resolution) however were as far as possible compared to values from the literature. The stopping power values are somewhat higher than from Northcliff and Schilling (i.e. not too low), for straggling only few values are available. The applicability of Bohr's theory is very limited due to stripping.   This is discussed already in the paper. What I mean is: the values are not too bad. If the analysis is performed as suggested, the Rutherford correction should   cancel out.

Q:(J. L'Ecuyer) For a nitrogen beam, the maximum of the stopping power curve is somewhere around 10 MeV so that using a 1 MeV beam one is not working under optimum conditions. The comparison between α and nitrogen should be done under optimum conditions for both beams.

A: (W. Pabst) The answer is simply that we are limited to 1.3 MeV (at the moment);so comparison is possible only under conditions available to us. There is no general validity of our results.

Q: (R.F. Lever) If θ_{out} is large, variation in θ_{out}, due to the finite detector angle subtended at the sample, will cause appreciable path length differences. Also, if ($\theta_{in} + \theta_{out}$) is large, the finite detector angle causes changes in recoil coefficient. How do you overcome this ?

A: (W. Pabst) 1.) That's right: aperture of detector is a problem when θ_{out} is large (collimation reduces count rate).
2.) From geometrical limitations, a medium backscattering angle of 165° is possible in our arrangement, where no problems should arise.

THE APPLICATION OF LOW ANGLE RUTHERFORD

BACKSCATTERING TO SURFACE LAYER ANALYSIS

J. S. WILLIAMS

Department of Electrical Engineering

University of Salford, Salford M5 4WT, U.K.

ABSTRACT

By optimization of the RBS geometry, both the system depth resolution and sensitivity can be significantly improved with respect to that obtained from the more usual normal incidence geometry. Typically, a depth resolution of the order of $25\,\text{Å}$ is possible and the sensitivity for surface impurities can be enhanced by about an order of magnitude. The application of the technique to the analysis of thin films, the measurement of ion implantation parameters and channelling studies is examined and illustrated with examples from recent experiments.

INTRODUCTION

Rutherford backscattering (RBS) has received increasing attention over the past decade as a technique for surface analysis and this application has been the subject of a recent, extensive review.[1] RBS has several inherent advantages, most notably that it is the only absolute, non-destructive method of providing both mass and depth analysis of atomic surface constituents, and is therefore a potentially powerful analytical tool for surface studies.

Rather surprisingly and unlike most other 'surface' techniques, very little effort has been directed towards overcoming, or even understanding, the factors which limit both depth resolution and sensitivity of the RBS technique itself. Recently, it has been shown that a simple optimization of the

RBS geometry can considerably improve the near-surface depth resolution[2] and, at the same time, substantially increase the sensitivity for detecting surface impurities. It is the aim of this paper to outline some of the advantages which can be derived from employing an optimized RBS geometry for surface analysis. In particular, the application to i) thin film analysis, ii) the measurement of ion implantation parameters and iii) channelling studies is discussed.   The various techniques are illustrated with examples from recent experiments.

2. OPTIMIZATION OF RBS GEOMETRY

2.1 Depth and Mass Resolution

The depth resolution dz normal to the target surface of an RBS system can be written, with respect to Fig. 1, as

$$dz = dE/\{ k^2 S_1/\cos \phi + S_2/\cos(\phi - \Theta)\}, \quad \text{......(1)}$$

where dE is the system energy resolution, k is a kinematical factor, S_1 and S_2 are the stopping powers of the target for the probe ions before and after large angle scattering through an angle $180° - \Theta$, ϕ is the angle of incidence with respect to the surface normal. For a given probe ion/target combination and a system with a fixed energy resolution, the depth resolution can be improved by a change of geometry such that $\phi \rightarrow 90°$ and $\Theta \rightarrow 0°$. The optimum geometry for maximum mass resolution, for a particular probe ion, is simply that with $\Theta = 0°$ (i.e. 180° scattering). In principle, therefore, both mass and depth resolution can be optimized by employing an RBS geometry such that the target surface is inclined at a glancing angle to both the incident and the scattered beam.

2.2 Limitations

Several factors apart from those contained in Eqn. 1 can influence the depth resolution of an RBS system and include incident beam collimation, detector acceptance angle, energy straggling and multiple scattering of the probe ions within the bulk target, and target surface topography. These contributions are discussed in detail in Ref 2: suffice it to say here that they tend to broaden the system resolution profile and provide a practical limit to the ultimate geometrical improvement in depth resolution.

The final, optimized geometry which requires compromise

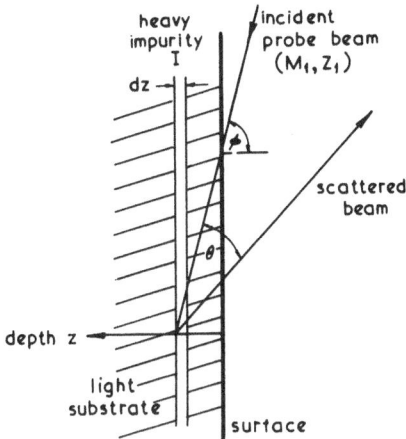

<u>Figure 1</u>   RBS geometry defining relevant parameters
           (see text)

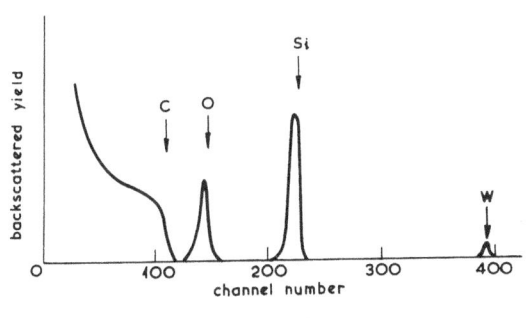

a) Normal incidence

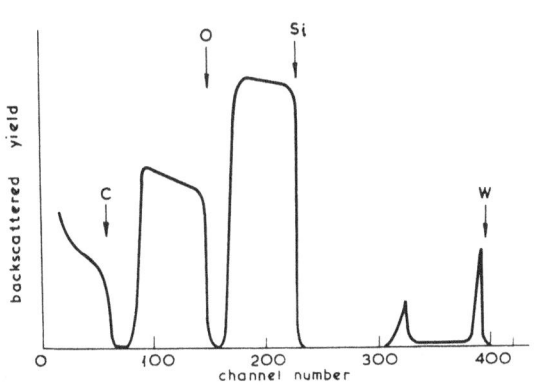

b) Optimised low angle geometry

<u>Figure 2</u> A diagramatic representation of normal incidence and
          optimized RBS spectra from ~ 300 Å $Si O_2$ evaporated
          onto a carbon backing.

between several competing factors will depend ultimately upon the depth and mass resolution requirements. For most general applications, the combination of a 2 MeV He^+ probe beam, a solid state detector (energy resolution 15 keV FwHm and acceptance angle 2-3°) and an RBS geometry of ϕ = 85°, Θ = 12°, optimized in terms of the resolution limiting factors,[2] provides for a simple and direct analysis system with high depth resolution ($\sim$ 25 Å) and reasonable mass resolution. In particular, this arrangement ensures that the analysis time per sample is short ($\sim$ 10-20 min.), the depth resolution remains practically constant up to several hundred Angstroms in low mass targets[2] and the energy to depth conversion of the raw data (backscattered energy spectra) is straightforward (since the depth scale depends linearly on backscattered energy). Alternate RBS arrangements may be employed to further enhance either depth or mass resolution, but always at the expense of simplicity and generality. For example, by either employing a high energy resolution detection system, together with an optimized low angle geometry, or improving beam collimation and reducing the detection acceptance angle to permit 85°$<\phi<$90° and $\Theta <$ 12°, the near-surface depth resolution can be further improved. However, in doing so, the depth resolution may deteriorate rapidly with depth as a result of multiple scattering effects within the bulk target,[2] analyses times will be increased markedly and, if lower energy probe ions are employed to maximize stopping powers (e.g. 0.7 MeV He+ in Si targets), the system mass resolution will suffer.

2.3 Sensitivity for Surface Impurities and Dopants

Low angle RBS can significantly increase the detection sensitivity for surface impurities and dopants and this can be appreciated by expressing the total RBS yield Y_I from an impurity layer, width dz, in terms of the geometry (Fig. 1) as

$$Y_i \propto N_I \sigma_I' \, dz \sec \phi , \quad \text{.........} (2)$$

where N_I is the impurity atomic density and σ_I' is the cross section for scattering from impurity atoms through an angle 180° – Θ. If the scattering angle Θ is kept constant during the change of geometry, then the RBS yield/channel from the substrate constituents will remain almost constant (within a factor of 2 compared with normal incidence geometries[3]) while the total impurity yield, from Eqn. 2, will increase as $\sec \phi$ (typically by a factor of 11.4 for ϕ = 85°). Since the RBS detection sensitivity is limited only by the 'pile up' background from the substrate signal, low angle RBS effectively improves the sensitivity for thin trace impurity layers by increasing the impurity yield without a corresponding increase in background. For example, the geometry ϕ = 85°, θ = 12° provides about an order of magnitude increase in sensitivity.

3. APPLICATIONS

3.1 Thin Film Analysis

A schematic illustrating the advantages of a change in RBS geometry for the measurement of thin film parameters is shown in Fig. 2. In this example a thin film of SiO_2 (evaporated onto a carbon backing) is analysed (a) in normal incidence and (b) in an optimized low angle geometry. The geometry can be chosen so that the Si and O plateaux are just conveniently separated without overlap. It is clear from Fig. 2 that the low angle spectrum facilitates measurements of the film thickness and stoichiometry throughout the film, which could not otherwise have been obtained from the normal incidence spectrum, where the system depth resolution is of the order of the film thickness (i.e. ~ 300 Å for a solid state detector with an energy resolution of 15 keV (FwHm).

The presence and distribution of any heavy impurities which may be introduced during film preparation can also be measured more readily with the low angle geometry. For example, in the analysis of the W impurity in Fig. 2 effects such as the greater concentration of W at the front and back surface of the film are resolved.

Of course, the general principles of optimized RBS geometry, illustrated in Fig. 2, can be employed equally well for other thin film arrangements such as multilayer films of different elemental species (for studying interdiffusion effects) and alloy films of two or more elemental species (for examining stoichiometry). The initial stages of oxidation and corrosive film growth are additional, useful applications: an example of oxide analysis is given in Fig. 3, where RBS spectra are shown for thermal oxidation of Si in air at 700°C for various times. Comparison of the normal incidence spectra in (a) with those in (b) shows the obvious advantages to be gained by employing low angle RBS for the measurement of oxide stoichiometry and growth rate during the first few hundred Angstroms of oxidation.[4] Interesting trends in stoichiometry and growth rate, measured as a function of oxide thickness and impurity concentration, have been made and the results shall be published elsewhere.[5] The limiting oxide thickness which can be measured by low angle RBS is about 25 Å or of the order of the residual (room temperature) oxide thickness on

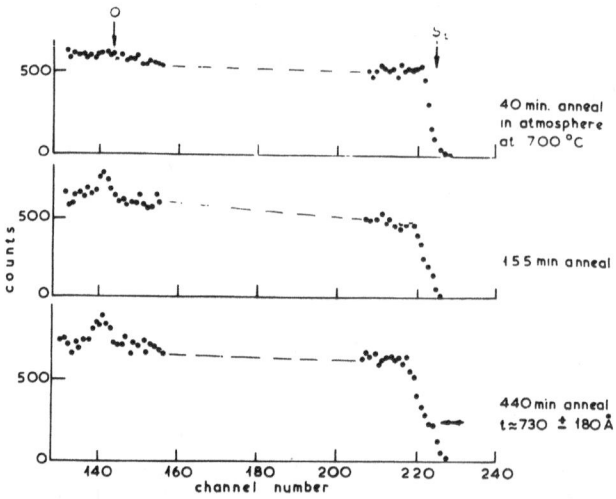

a) Normal Incidence Geometry : detector resolution (16keV)≈3.2ch

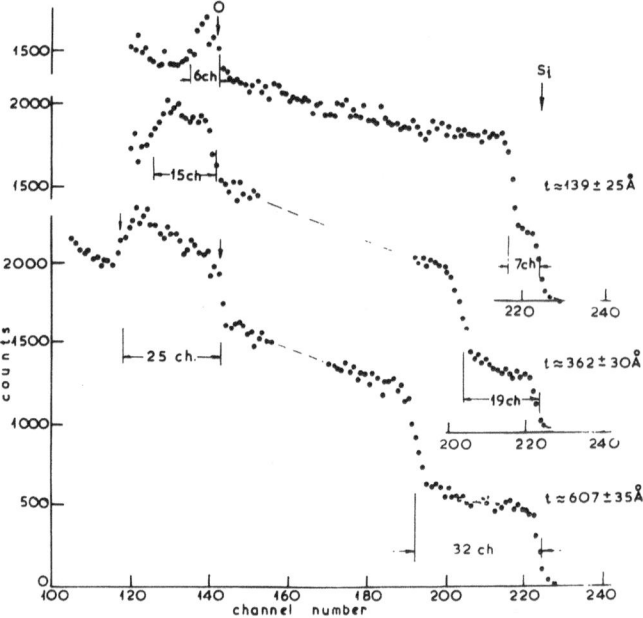

b) Low Angle (∮=85°) Geometry : detector resolution (18keV) = 3.6ch

Figure 3 Normal incidence and low angle RBS spectra showing
the initial stages of thermal oxidation of Si at 700°C
(t is the oxide thickness)

most materials.

3.2 Ion Implanted Layers

The measurement of ion implantation parameters is of fundamental importance as a means of both checking the validity of
accepted theoretical treatments of the slowing down of energetic
ions in matter,[6,7] and enabling a compilation of reliable data
and simple empirical relations for accurately predicting ion range
profiles for device and other applications. The low energy
region ($\epsilon < 0.5$) is interesting both theoretically (here, nuclear
stopping processes should predominate[5]) and practically, in most
device applications, but little experimental data is at present
available in this energy region. Low angle RBS is ideally suited
to speedy accumulation of low energy, heavy-ion range parameters,
where the depth resolution required is of the order of 20-30 Å .
A more comprehensive presentation of this particular application,
together with a compilation of measured range parameters, is
given in a later paper.[8]

The feasibility of employing RBS for measuring low-mass
ion implantation parameters in heavier substrates has recently
been demonstrated[9] - an example of the particular use of low
angle RBS for such measurements is shown in Fig. 5. The
spectrum illustrates the technique for the measurement of
5 keV Cr^+ ions implanted into a thin Ge film evaporated onto a
silicon backing. The geometry can be optimized such that the
RBS yield from the heavy thin film, implanted ion distribution and
light backing material are conveniently separated without overlap
to give maximum depth resolution.

An example of other implantation data obtained from low
angle RBS is shown in Fig. 4. Lead concentration profiles have
been plotted as a function of dose for 20 keV Pb^+ implantations
into Si[10], the change in a 1×10^{15} cm^{-2} profile after a 600°C
(30 min.) anneal is also included. Important modifications to low
energy profiles as a result of sputtering and related effects can
be observed directly and interesting features may be resolved
which could dramatically affect ion collection (such as the
apparent accumulation of Pb at the Si surface at the highest dose).
The marked diffusion of Pb to the Si surface after annealing
(dotted curve) is a further observation which is facilitated by the
improved depth resolution. A more detailed account of the study
of Pb into Si system, by low angle RBS, shall be published
elsewhere.[11]

3.3 Low Angle Channelling

A schematic illustration of low angle channelling is shown

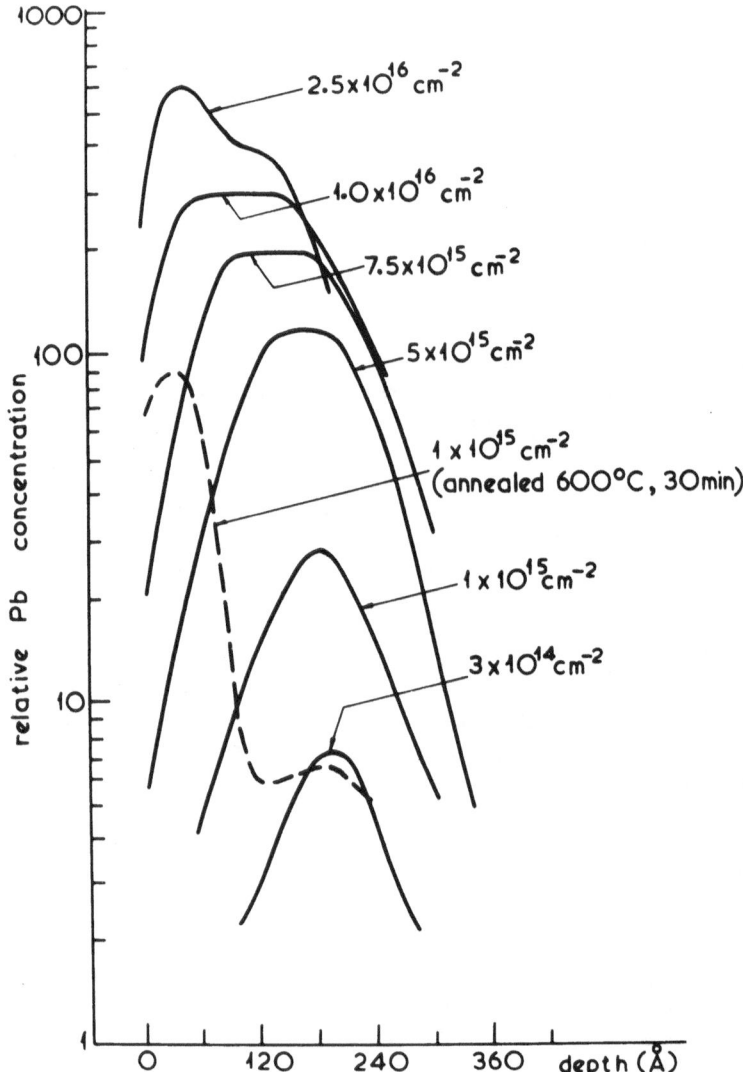

Figure 4 Build up curves for 20 keV Pb implanted into Si.
Dotted curve shows the Pb outdiffusion of 1×10^{15} cm^{-2}
dose annealed for 30 mins. at 600°C.

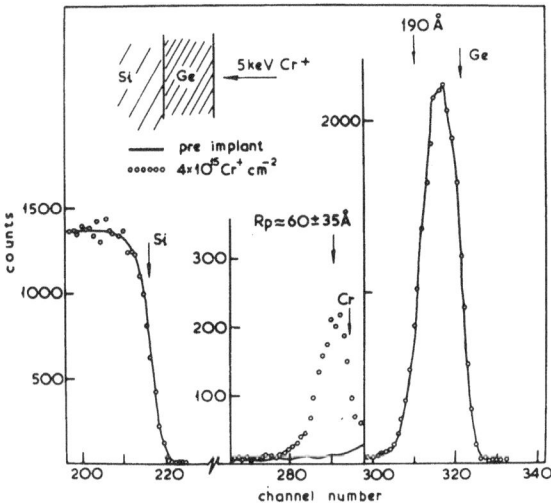

Figure 5 Low angle RBS profile of Cr$^+$ implanted into a thin Ge film evaporated onto a Si backing.

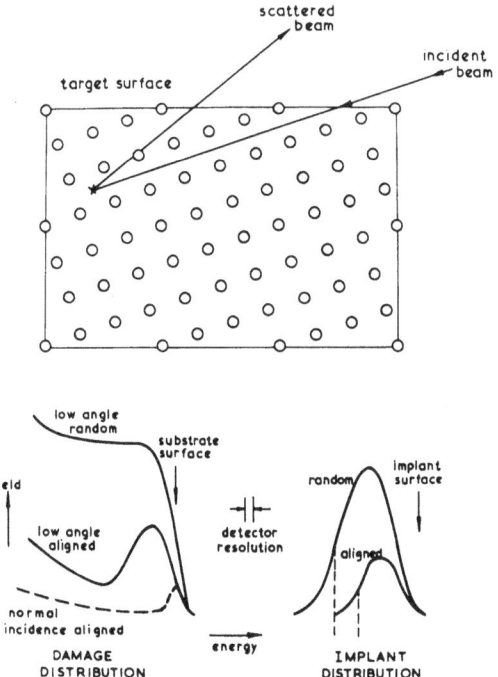

Figure 6 A schematic showing glancing incidence channelling technique with typical damage and implant distributions which can be obtained.

in Fig. 6. The analysis beam is directed along a low index
channelling direction at a small angle with respect to the target
surface and the detector is positioned to collect a fraction of the
beam backscattered along a random crystal direction at a less
inclined angle with respect to the target surface. This geometry
results in an effective increase in thickness for any surface –
damaged layer, allowing channelling analysis with both increased
depth resolution and sensitivity as shown (lower left). The
technique can be applied to the measurement of strain and related
disordering effects of amorphous films (e.g. oxides) on single
crystal substrates, surface preparation damage and implantation
disorder profiles. This last application has recently been
demonstrated for measuring 20 keV Dy^+ implantation disorder in
Ni single crystals which were oriented with the ⟨110⟩ axis
inclined at about 10° with respect to the target surface.[12]

By comparing random and aligned spectra collected under
the same low angle geometry conditions, the improved resolution
facilitates atom location measurements as a function of concentration throughout the implanted distribution as shown in Fig. 6
(lower right). It is worth noting that, as the path length along the
channelling direction in low angle RBS is much greater (in general)
than the scattered path length along a random direction, effects
due to stopping power differences between the aligned and random
spectra will be accentuated. As a result, the lower stopping
power for channelling trajectories will effectively 'compress' the
depth scale (Eqn. 1) of the aligned spectra (see dotted profile
edges of implant distribution in Fig. 6). In fact, the change in
profile width between random and aligned low angle spectra
(particularly for truly interstitial impurities) can be used as a
convenient measure of the stopping power in a channel.

4. CONCLUSION

An optimized low angle RBS geometry affords two main
advantages for surface layer analysis: an improved depth
resolution and increased sensitivity for surface impurities and
dopants. The employment of a solid state detector, of 15 keV
(FwHm) resolution, in a low angle scattering arrangement
preserves the simplicity and directness of RBS analysis and
provides a depth resolution of typically 25 Å which is comparable
to that of more complex, high energy resolution RBS systems.[13,14]

A degree of caution must be exercised when employing RBS
for surface layer analysis since, in averaging the backscattered
signal from the irradiated area, the presence of localized,
structural features in a composite target, such as precipitation
effects, may go unresolved necessitating some complimentary
means of analysis. Combined RBS and TEM studies are already

proving successful, particularly for the analysis of high dose implantations.[11,15,16]

The author is pleased to acknowledge both the SRC and NATO for financial support.

1. W. K. Chu, J. W. Mayer, M. A. Nicolet, G. Amsel, T. Buck and F. H. Eisen. Thin Solid Films 17, 1 (1973)
2. J. S. Williams. Nucl. Instr. Meth. 126, 205 (1975)
3. The yield per channel for substrate components is proportional to the fraction of the energy/channel lost during the incident path; thus, the substrate yield can only be enhanced by upto a factor of 2 over normal incidence.
4. Channelling can be used to improve normal incidence sensitivity. See for example W. K. Chu, E. Luggujjo, J. W. Mayer and T. W. Sigmon. Thin Solid Films 19, 329 (1973)
5. J. S. Williams, C. E. Christodoulides and W. A. Grant, to be published.
6. J. Lindhard, M. Scharff and H. Schiott. Mat. Fys. Medd. Dan. Vid. Selsk. 13, 14 (1963)
7. K. B. Winterbon, P. Sigmund and J. Sanders, Mat. Fys. Medd. Dan. Vid. Selsk. 37, 14 (1972)
8. D. Dodds, W. A. Grant and J. S. Williams, this conference.
9. R. S. Blewer, Appl. Phys. Lett.
10. J. S. Williams, Phys. Letters 51A, 85 (1975)
11. C. E. Christodoulides et al, to be published.
12. G. A. Stephens, E. Robinson and J. S. Williams, "Ion Implantation in Semiconductors and Other Materials" Osaka (1974). Publ. Plenum, New York (1975)
13. E. Bogh, Rad. Effects 12, 13 (1972)
14. A. Feuerstein, S. Kalbitzer and H. Oetzmann, Phys. Lett. 51A, 165 (1975)
15. G. J. Thomas and S. T. Picraux, "Application of Ion Beams to Metals", Albuquerque (1973) Publ. Plenum, New York (1974)
16. L. T. Chadderton and J. L. Whitton, Rad. Effects 23, 63 (1975)

DISCUSSION

Q: (G. Foti) How change the low angle RBS energy spectra, when the topography of surface is modified like valley structure ?

A: (J.S. Williams) Surface topography is a big problem. Surface flatness variations (long range effects) can be investigated optically but short range roughness on an atomic scale (10's of Å's) are more difficult. It would appear from our initial tests that, at 85° geometry, carefully mechanically polished and etched speci-

mens are adequate such that no significant deterioration in resolution is observed.

Q: (L.C. Feldman) Have you experienced difficulty with beam integration at the grazing angles?

A: (J.S. Williams) Yes, but for the most part we can obtain consistent current measurements with changing angle down to 85°C incidence
with respect to the surface normal. To go this we have had to increase
the voltage in our secondary suppression ring to 1000 volts (negative).

Q: (R. Behrisch) How does your depth resolution depend on depth
1) Due to geometry (opening angles)
2) Due to multiple scattering.

A: (J.S. Williams) Angular deviations in the analysis beam due to geometry and multiple scattering increase with depth so that the depth
resolution decreases with increasing depth. The effect is more severe
for high mass targets.

Comment: (J. Davies) Another big advantage of the low-angle technique
in channeling applications is that it can reduce the dependence of
depth scale on the channeled ion stopping power. Of course, you
would have to interchange the positions of incident beam and detector, so as to have the glancing angle on the emitted trajectors.

A: (J.S. Williams) Yes, we have investigated the use of similar a-
ligned directions in different geometries and there are certainly
advantages to be gained.

Q: (W. Gibson) Do depth resolution estimates contain energy straggling effects ? How do such effects limit the application of this
technique ?

A: (J.S. Williams) Yes, both straggling and multiple scattering (more
correctly lateral dispersion) are important limitations. At low
angle geometries multiple scattering is the more detrimental factor
as our initial tests have indicated.

MEASUREMENT OF PROJECTED AND LATERAL RANGE PARAMETERS FOR LOW ENERGY HEAVY IONS IN SILICON BY RUTHERFORD BACKSCATTERING

W A GRANT, J S WILLIAMS AND D DODDS

DEPARTMENT OF ELECTRICAL ENGINEERING, UNIVERSITY

OF SALFORD, SALFORD M5 4WT, LANCASHIRE, U.K.

ABSTRACT

Range parameters for heavy ions implanted into silicon have been measured using Rutherford backscattering. The parameters measured were the projected range R_p and corresponding standard deviation ΔR_p, and the standard deviation ΔR_L in lateral range. Projectile masses and energies were chosen to cover the range $0.005 < \epsilon < 0.5$. Measured values of R_p, ΔR_p and ΔR_L are greater than theoretical predictions based on LSS. For R_p and ΔR_L the discrepancy is least ($\sim 15\%$) as $\epsilon \rightarrow 0.5$ but rises to $\sim 40 - 70\%$ at low ϵ. The relative straggle, $\Delta R_p/R_p$, is in good agreement with theory at low energies and is 30% larger at high energies. The relationship $\rho_p = 1.75 \epsilon^{2/3}$ is found to be a good approximation for the projected range for $0.02 < \epsilon < 0.5$.

INTRODUCTION

For any practical application of ion implantation it is essential to have detailed knowledge of basic range parameters such as the projected range R_p and its standard deviation ΔR_p. The lateral range R_L and its standard deviation ΔR_L are also of importance, particularly where implantation involves defining masks. The theory of Lindhard, Scharff and Schiott[1] (LSS) enables predictions of these various range parameters to be made with good accuracy over a wide range of energies and ion/target combinations. Recent compilations[2-5] based on LSS have provided these predictions in a readily accessible form. The theory is, however, based on a number of approximations and simplifications and is not expected

235

to be exact.   For this reason, experimental measurement of range
distributions is still essential.   In addition, accurate
experimental data provide insight into the validity of theoretical
assumptions.

Rutherford backscattering provides a simple, non-destructive
method for measuring ion ranges when the implant is heavier than
the target.   Chu et al[6] used the technique to measure the range
of ions, with atomic numbers Z_1 = 30 to 52, in amorphous oxides.
The results cover the energy range $0.4 < \epsilon < 2.0$ (where ϵ is the
dimensionless energy parameter used in LSS) and indicates that
experimental ranges are 20 - 40% greater than theoretical
predictions.   Neilson et al[7] measured ion ranges for
Z_1 = 55 to 79 at 100 keV in aluminium and found values differing
from LSS by up to a factor of 2 together with an oscillatory
dependence on Z_1.   Andersen et al[8] report range values $\sim$ 30%
higher than LSS for various ions in Al and Al_2O_3 at energies up
to 100 keV and Feuerstein et al[9] measured similar discrepancies
at 10 to 35 keV in silicon.   This last paper[9] summarises much
recent data.

In the work reported here, projectile masses and energies
were chosen to cover the range $0.005 < \epsilon < 0.5$ for implantations
into silicon: the mass ratio $\mu = M_2/M_1$ is from $\sim$ 0.14 to 0.9.
This choice places our measurements in the low energy region
specified by Schiott[10] where the ranges are small and the more
accurate Thomas-Fermi potential may be approximated by a power law
potential, $V(r) \propto r^{-3}$.   Schiott suggests that a useful
approximation for predicting the range in this energy region is
$\rho = 1.5 \, \epsilon^{2/3}$.

EXPERIMENTAL

Single crystal silicon samples were implanted using the
University of Salford isotope separator.   The ions used were
Ar, Cu, Kr, Cd, Xe, Cs, Dy, W, Au, Pb and Bi at energies up to
100 keV.   Bi was also implanted using an isotope separator at
AERE, Harwell at energies up to 400 keV.   Implant fluences were
kept as low as practical, compatible with obtaining good counting
statistics in subsequent Rutherford analysis.   Measurements at
our own and other laboratories[11] show that range broadening occurs
as the implant fluence is increased.   Typically, for Xe at 40 keV,
a dose of 5×10^{14} ions cm^{-2} was implanted since significant
broadening occurs at 2×10^{15} ions cm^{-2}.   Implant fluxes were also
kept low to minimise target heating.   Separate samples were
implanted perpendicular to the target surface and at an angle θ
to the surface normal (Figure 1).   The normal implant provides
data for R_p and ΔR_p and together with the implant at θ gives ΔR_L

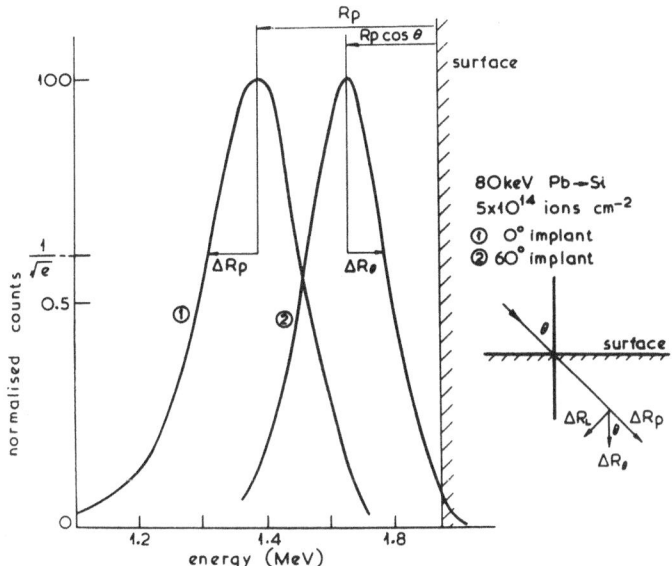

FIGURE 1.   Normalized backscattered 2 MeV He$^+$ profiles for 80 keV
Pb implanted into Si at normal incidence and θ = 60°.   RBS
analysis at 85° to the surface normal.

since

$$\Delta R_\theta^2 = \Delta R_p^2 \cos^2\theta + \Delta R_L^2 \sin^2\theta.$$

This experimental technique for obtaining ΔR_L has been
suggested by Schiott[12] and applied by Furukawa et al[13].

Rutherford analysis was carried out using 2 MeV He$^+$ ions and
a solid state detector, 16 KeV FwHm.   The depth resolution was
optimised, typically to $\sim$ 30A°, by employing an analysing beam that
entered and left at small angles ($\sim$ 5° and 17°, respectively) to
the target surface.   The experimental technique has been discussed
earlier in this conference[14] and presented in detail elsewhere[15].
The improved resolution obtained in this way enables the shallow
ion ranges ($\leqslant$ 1000A°) to be measured with good accuracy.   At low
ϵ values the detector resolution becomes of increasing relevance
in measuring profile widths and was corrected for in quadrature[16].
In converting from scattering energy to depth, the stopping power
of the analysing beam is an important parameter.   In the present
work we have used stopping power data compiled by Northcliffe and
Schilling[17] but if helium stopping powers given by Ziegler and
Chu[18] are used, our range parameters should be reduced by $\sim$ 5%.

TABLE I. Tabulation of experimental and theoretical range parameters for several ions at various energies in Si. Range values (A°) were calculated using a Si density of 2.33 gms cm^{-3}. Theoretical values were taken from Furukawa et al[4] and Schiott[10]

Ion	energy (keV)	R_P (Å)	ΔR_P (Å)	ΔR_L (Å)	$\dfrac{R_{P(EXP)}}{R_{P(THEORY)}}$	$\dfrac{\Delta R_P}{R_P}$ / EXP	$\dfrac{\Delta R_P}{R_P}$ / THEORY	$\dfrac{\Delta R_L(EXP)}{\Delta R_L(THEORY)}$
Pb	40	170	26	33	2.26	0.15	0.16	1.94
	20	209	44	44	1.69	0.21	0.23	1.83
	40	274	70	62	1.35	0.26	0.24	1.72
	80	456	108	92	1.38	0.24	0.24	1.59
Bi	50	371	97	67	1.58	0.26	0.24	1.59
	150	603	178	141	1.14	0.30	0.24	1.56
	200	803	255	137	1.22	0.32	0.24	1.23
	400	1310	404	198	1.15	0.31	0.24	1.06
Ar	40	458	–	–	1.13	–	0.52*	–
Cu	40	354	194	128	1.20	0.56	0.44*	1.02
Kr	40	322	144	75	1.28	0.45	0.35	1.17
Cd	40	301	95	75	1.29	0.32	0.30	1.44
Al	40	254	105	66	1.16	0.40	0.30	1.40
Dy	40	274	104	68	1.32	0.37	0.27	1.66
W	40	278	73	68	1.46	0.26	0.24	1.79

RESULTS AND DISCUSSION

The principal results are given in Table 1 where we list measured values of R_p, ΔR_p and ΔR_L. Other columns in the table compare these range parameters to theoretical predictions of Furukawa et al[4] and Schiott[10] based on LSS. The measured values can be seen to be greater than theory by varying percentages, although an average discrepancy is ~ 30%. The data is shown in graphical form in Figures 2-5, together with other measurements made earlier[19].

Figure 2 illustrates the energy variation of R_p, ΔR_p and ΔR_L for (207)Pb and (209)Bi, using axes of $R_p(A^o)$ and E(KeV)

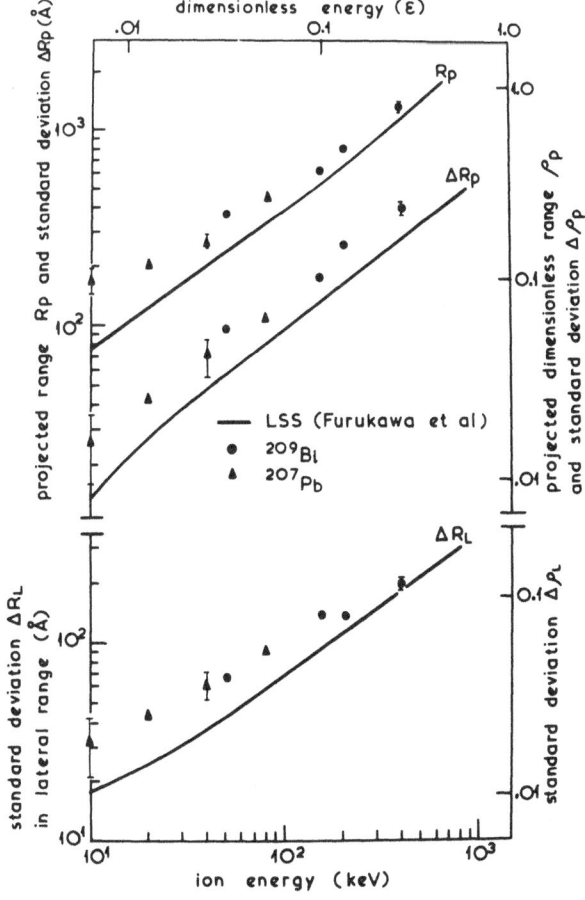

FIGURE 2. Range parameters R_p, ΔR_p and ΔR_L as a function of energy for Bi(●) and Pb(▲) implanted into Si. Theoretical (LSS) curves were calculated from tables supplied by Furukawa et al[4].

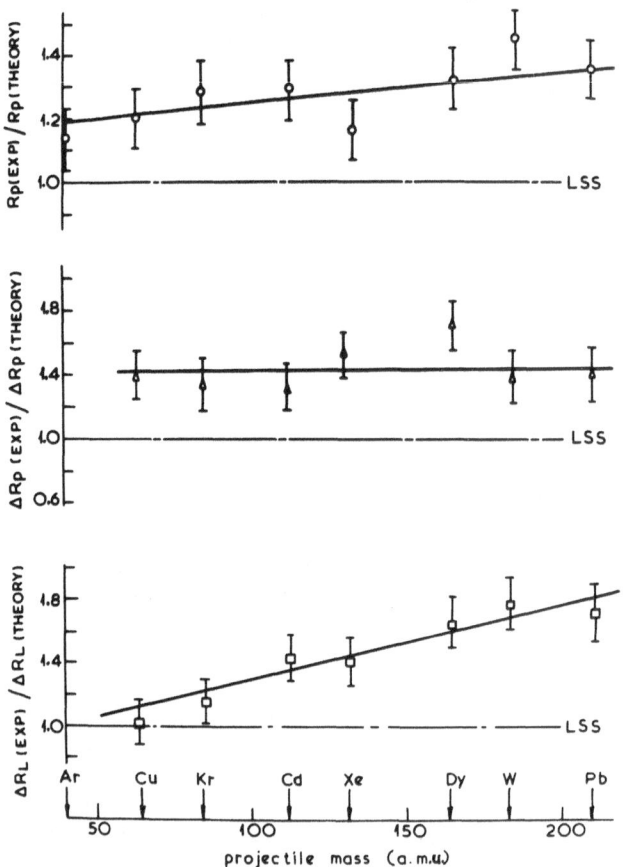

FIGURE 3.   Range parameters R_p, ΔR_p and ΔR_L as a function of projectile mass M_1 for 40 KeV implants into Si compared to theoretical (LSS) values.

together with the dimensionless ρ_p, ϵ parameters of LSS.   The theoretical tabulations of Furukawa et al[4] are shown as a solid line and in the case of R_p and ΔR_L, the agreement between theory and experiment improves as $\epsilon \to 0.5$.   The discrepancies for ΔR_p remain essentially uniform over the whole energy range.   The effect of varying the projectile mass (at a fixed energy of 40 KeV) on the three range parameters is shown in Figure 3.   Again both R_p and ΔR_L depart to a larger extent from theory for higher masses (and hence lower ϵ), a trend not shown by ΔR_p.

The trends in energy and mass dependence of the projected range are combined in Figure 4 by plotting all the results

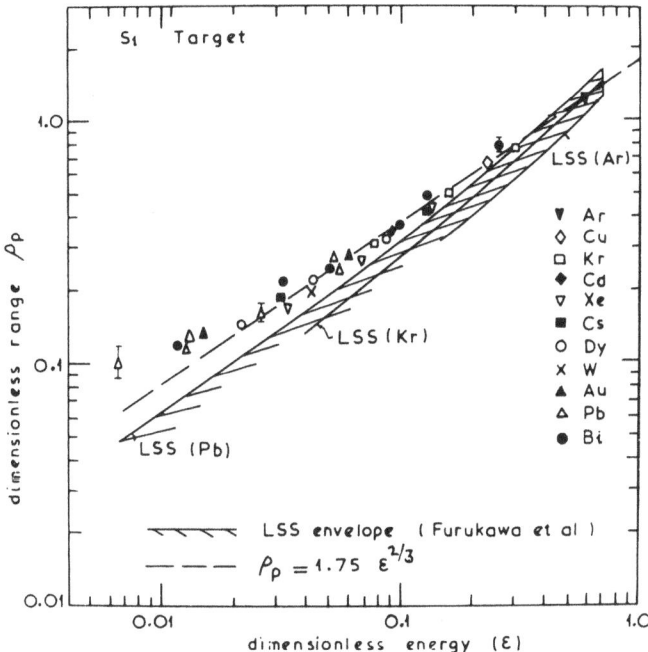

FIGURE 4.   Range ρ_p versus ϵ for several ions at various energies compared to theoretical (LSS) envelope.

(including those previously measured in silicon[19]) in terms of dimensionless ρ_p, ϵ units.   Although theory predicts a unique curve for all target/projectile combinations on a ρ, ϵ plot, the conversion from range to projected range (i.e. $\rho \rightarrow \rho_p$) involves mass dependence[10] so that an envelope of theoretical curves for different masses is the result of a ρ_p, ϵ plot.   Theoretical curves are shown for typical ions (Ar, Kr, Pb) which form a hatched envelope.   The experimental results form a similar envelope which exhibits a greater departure from theory at low ϵ and approaches theory as $\epsilon \rightarrow 0.5$.   Schiott[10] suggests that the relationship $\rho = 1.5\epsilon^{2/3}$ is a useful approximation to LSS in the energy region $0.002 < \epsilon < 0.1$.   In terms of projected range, mass dependence is involved so that $\rho_p = 1.5C(\mu)\epsilon^{2/3}$, where $C(\mu)$ equals R_p/R for mass ratios $\mu < 3$ (R is the total range).   Using an average value for $C(\mu)$ for the mass ratios involved in the present work gives[10] $\rho_p = 1.3\epsilon^{2/3}$.   In Figure 4, the relationship $\rho_p = 1.75\epsilon^{2/3}$ provides a reasonable fit to our data in the energy region $0.02 < \epsilon < 0.5$, but large deviations for $\epsilon \leq 0.02$ suggest that a more screened potential than $V(r) \propto r^{-3}$ is needed at these low energies.

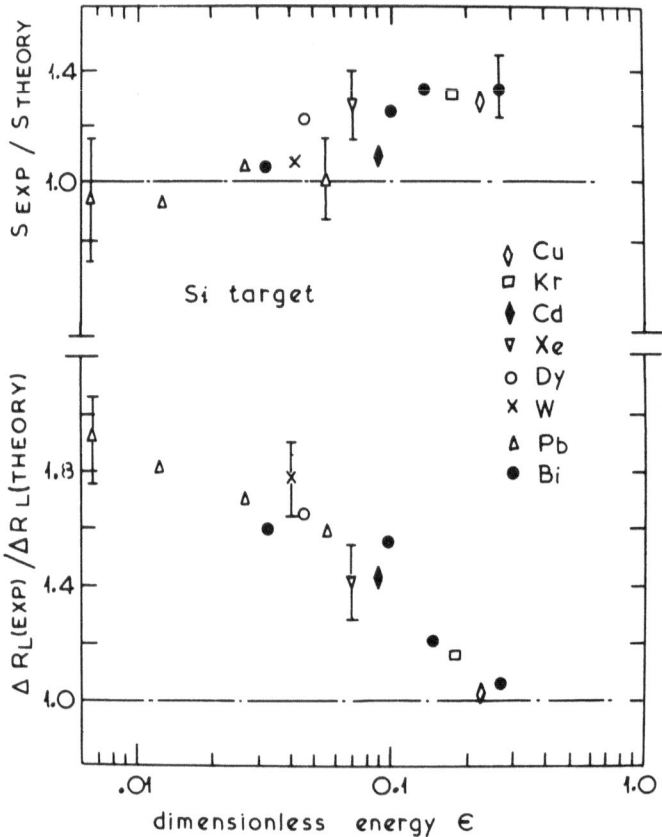

FIGURE 5.   Straggle in projected range, $S(= \Delta R_p/R_p)$, and standard deviation, ΔR_L, in lateral range as a function of dimensionless energy ϵ for various ions compared to theoretical (LSS) values.

The final graphs, in Figure 5, summarise all the results for lateral standard deviation, ΔR_L, and relative straggle in projected range, $\Delta R_p/R_p$, as a function of dimensionless energy ϵ. The ratio of experimental to theoretical data has been plotted to facilitate direct comparison with theory. For ΔR_L, the same general trend as for the projected range is evident: large deviations at low ϵ giving way to a closer fit with theory as $\epsilon \rightarrow 0.5$. The straggle in projected range, however, shows the opposite trend. At low ϵ values the straggle is close to theory, in good agreement with other recent data in this energy region[8], whereas, for $\epsilon \rightarrow 0.5$, the measured straggle is about 30% greater than theory, in qualitative agreement with measurements by Chu et al[6] in the intermediate energy region $0.5 < \epsilon < 2.0$.

CONCLUSIONS

Experimental values for the projected range R_p, standard deviation ΔR_p and standard deviation ΔR_L in lateral range measured for various heavy ions in silicon are greater than tabulated values based on LSS.   In the energy range $0.005 < \epsilon < 0.5$ and for mass ratios $0.14 < \mu < 0.9$, the discrepancies between theory and experiment for R_p and ΔR_L are least as $\epsilon \rightarrow 0.5$. The relative straggle in projected range is in good agreement with theory at low ϵ values but is $\sim 30\%$ greater as $\epsilon \rightarrow 0.5$.   As suggested by Schiott[10], we find that ρ_p is proportional to $\epsilon^{2/3}$ for the energy region $0.02 < \epsilon < 0.5$ but a more screened potential than $V(r) \propto r^{-3}$ is needed to fit our data for $\epsilon \leq 0.02$.

ACKNOWLEDGEMENTS

M J Nobes and P Cardwell are thanked for experimental assistance and J H Freeman (Harwell) for providing the high energy Bi implants.   S Furukawa, H Matsumura and H Ishiwara are thanked for providing theoretical tabulations of range parameters. Financial assistance by NATO and SRC is gratefully acknowledged. This work has been supported in part by the Air Force Office of Scientific Research (AFSC), United States Air Force, under grant No AFOSR 71-2115.

REFERENCES

1.   J Lindhard, M Scharff and H E Schiott, Kgl Danske Videnskab Selskab, Mat-Fys Medd, 33, No 14 (1963)

2.   W S Johnson and J F Gibbons, Projected Range Statistics in Semiconductors, Distributed by Stanford University Bookstore (1969)

3.   B J Smith, AERE Report R6660 Harwell (1971)

4.   S Furukawa, H Matsumura and H Ishiwara, private communication

5.   G Dearnaley, J H Freeman, R S Nelson and J Stephen Ion Implantation, North Holland (1973)

6.   W K Chu, B L Crowder, J W Mayer and J F Ziegler, Proc Int Conf on Ion Implantation in Semiconductors and Other Materials, Yorktown Heights, NY (1972), Plenum Press NY (1973) p 225

7.   G W Neilson, B W Farmery and M W Thompson, Physics Letters 46A, 45 (1973)

8.   H H Andersen, J Bottiger and H Wolder Jorgensen, Applied Physics Letters, 26, No 12, 678 (1975)

9.   A Feuerstein, S Kalbitzer and H Oetzmann, Physics Letters
     51A, 165 (1975)

10. H E Schiott, Proc Int Conf on Ion Implantation, Thousand Oaks
     California, USA (1970). Gordon and Breach, London (1971) p 197

11. P Blank and K Wittmaack, to be published

12. H E Schiott, Kgl Danske Videnskab, Selskab, Mat-Fys Medd, 35,
     No 9 (1966)

13. S Furukawa and H Matsumura, Appl Phys Lett, 22, No 3 (1973)

14. J S Williams, This conference

15. J S Williams, Nuclear Inst and Methods, 126, 205 (1975)

16. D Powers, W K Chu and P D Bourland, Phys Rev, 165, 376 (1968)

17. L C Northcliffe and R F Schilling, Nuclear Data Tables, A7,
     233 (1970)

18. J F Ziegler and W K Chu, Thin Solid Films, 19, 281 (1973)

19. J S Williams and W A Grant, Radiation Effects, 25, 55 (1975)

DISCUSSION

Q: (J. Baglin) in answer to Mayer's question: Our observations on C
build-up are consistent with a constant build-up rate in the direction normal to the film surface. You are of course more sensitive
to it when your probing beam "sees" it at grazing incidence.

A: (J.S. Williams) We have measured the rate of carbon build up with
dose and it is a negligibly small correction to our depth measurements for normal analysis doses. We have not investigated the rate of
carbon growth with changing angle.

Q: (J. Baglin) Does grazing incidence cause you extra experimental
difficulty from the enhanced effect of carbon build-up ?

A: (J.S. Williams) Yes.

RANGE PARAMETERS OF HEAVY IONS IN SILICON AND GERMANIUM

WITH REDUCED ENERGIES FROM 0.001 $\leq$ ϵ $\leq$ 10

H. OETZMANN, A. FEUERSTEIN, H. GRAHMANN AND

S. KALBITZER

MAX-PLANCK-INSTITUT FÜR KERNPHYSIK, HEIDELBERG

ABSTRACT

Heavy ions of various elements have been implanted into amorphous silicon, germanium and aluminium layers at energies of 1- 60 keV. High resolution ion backscattering has been used to determine range parameters and nuclear stopping powers at low reduced energies. In the medium reduced energy range thin window silicon detectors have been exposed to ion beams of the lighter heavy ions. Nuclear stopping powers have been derived from ionization deficits.

INTRODUCTION

The slowing down of energetic particles by the nuclear stopping mechanism is of particular importance for ion implantation processes.

Firstly, range parameters of implanted ions depend on nuclear stopping markedly, or even entirely.

Secondly, formation and distribution of radiation defects are due to this energy loss mechanism.

Although LSS-theory was formulated about ten years ago(1), experimental data on nuclear stopping is still meager. This is due to the fact that stopping power measurements are extremely difficult or even impossible at low energies, and the easier, still difficult, range measurements are usually not precise enough as to derive the desired differential quantity. At medium energies, where electronic and nuclear stopping are comparable, the additional difficulty of separating the two competing stopping mechanisms arises.

The present work reports on range measurements of low energy heavy ion implants using a spectrometer of high depth resolution, and results of measurements of nuclear stopping power in the intermediate energy region with silicon detectors.

THEORETICAL

For elastic scattering, described by power law potentials, the following equations hold:

(1) $V(r) \sim r^{-1/m}$, $1/m = 1,2 \ldots$

(2) $d\sigma \sim E^{-m} T^{-1-m} dT$,

(3) $S_n \sim E^{1-2m}$,

(4) $R \sim E^{2m}$,   if $S_n \gg S_e$

where $V(r)$ = interatomic potential, $d\sigma$ = differential scattering cross section, S_n, S_e = nuclear and electronic stopping power, R = range, E = particle energy, T = transferred energy. Eq. (4) is a good approximation for reduced energies of $\varepsilon \le 0.1$.

Thus, by measuring range-energy curves with sufficient precision one obtains detailed information on the scattering potential.

Most theoretical work has been carried out with using the Thomas-Fermi-potential. At low particle energies the elastic stopping takes place in the outer part of the atom where the validity of this potential becomes questionable.

EXPERIMENTAL

In a previous paper (2) we have shown that high resolution backscattering of light ions in the 100 keV range is well suited to analyze low energy implants with respect to range parameters. A few first results, limited to two implantation energies and two ion species,

were given and compared with experimental data by others
and with theoretical predictions based on LSS - theory.

The present work reports on measurements of the
same kind, the set of experimental parameters, however,
having been greatly extended. A total of 25 different
implants was analyzed being characterized by the following parameters:

ion species	:	As, Ge, Sb, Au, Bi;
ion dose	:	10^{14} - 10^{15} ions/cm^2;
ion energy	:	1 - 60 keV;
target	:	Si, Ge, Al; amorphisation by pre-bombardment with 3 x 10^{15} Ne/cm^2 at 40 keV and room temperature;
analyzing beam	:	200 keV ^{2}H$^+$, 200 keV ^{4}He$^+$.

Actually, when using deuterons, a 600 keV $(^2H)_3^+$ -
molecule beam was preferred because of intensity
reasons. In terms of reduced energy values ε the interval between .0006 - .3 has been covered.

Prior to implantation the silicon wafers were subjected to a special deoxidation procedure. Residual
layers of about 15 Å SiO$_2$ were detected with ellipsometric measurements. This was confirmed by the backscattering spectra. The backscattering analysis was carried
out with an electrostatic analyzer system described in
detail elsewhere (3). The system was operated at about
.3 - 1% resolution according to the experimental requirements. Using measured specific energy loss figures
of 11 (25) and -- (22.5) eV/Å for ^{2}H$^+$ (^{4}He$^+$) in Si and Ge,
respectively, at 200 keV energy theoretical apparatus
widths of 74 (35) and -- (32) Å are expected for the
above four cases at a path length factor of 2.5 and a
system resolution of 1%. These figures compare well with
our experimental data when using the surface edge of the
backscattering spectrum of the respective target material as a measure of the system resolution.

Fig.1 shows a typical backscattering spectrum. The
obtained impurity profiles were analyzed in the following way. The peak position of the depth distribution was
taken for the mean projected range. This assumption is
justified since the profiles are very nearly symmetrical, at least down to the points at half maximum. This
is in accord with calculations and other experimental

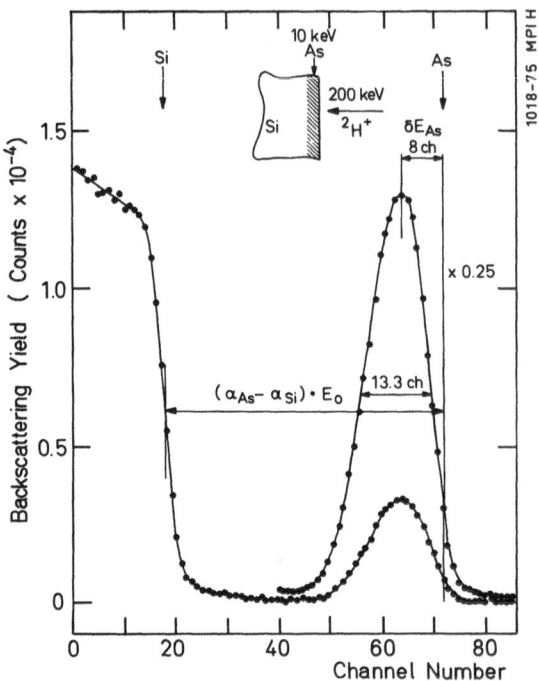

Fig.1   Backscattering spectrum of a 1E15 As/cm^2 silicon
        implant with an electrostatic analyzer.

data on various ion - silicon combinations (4,5).

     The widths of the profiles were corrected by sub-
tracting the system resolution in quadrature. This latter figure was derived from the energy spread of the
substrate surface edge which was nearly gaussian in shape after differentiation. These reduced widths still
contain contributions of straggling effects of the analyzing beam which increase with depth. This additional
broadening may be comparable to the width of the implanted profiles at the lowest implantation energies. Corrections, however, could not be made due to the lack of
sufficiently precise numbers. Accordingly, the straggling parameters may be high by a few 10% for the shal-
lowest ion distributions. In the medium energy region,
where $S_n \sim S_e$, a different experimental technique was
applied. Here a thin window ion implanted silicon particle detector was exposed to ion beams of pA-intensity.
Due to the finite window thickness of about 400 Å sili-

con and the limitation in energy to 65 keV, only the
lighter heavy ions H, He, B, C, N and Ne could be measured. A full description of the experimental details is
given elsewhere (9). Here only the results concerning
nuclear stopping will be given.

RESULTS

The results of our range measurements are shown in
fig.2. In the entire nuclear stopping regime, where the
electronic stopping cross section is negligible, the experimental projected ranges are considerably larger than
the theoretical values. The discrepancies increase from
about 30% at ε = 0.1 to about 100% at ε = 0.001 with
respect to the theoretical figures. The solid curve
through the experimental points is well approximated by
a straight line following the relation ρ_p = 1.0 x $\varepsilon^{1/2}$

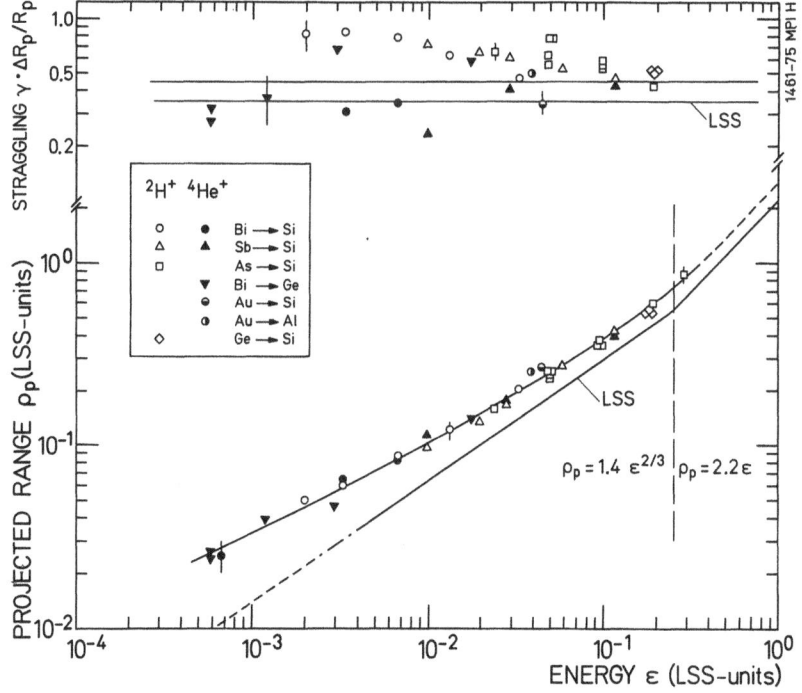

Fig.2   Experimental and theoretical range-energy and
straggling-energy data. Note that the LSS straggling value of 0.35 refers to $\gamma\Delta R/R$.

for values of ε less than 0.02. In the energy range of
$0.02 < \varepsilon < 0.25$ the approximation $\rho_p = 1.8 \times \varepsilon^{2/3}$ gives
a good fit to the experimental data which is about 10%
higher than the previously reported figure of 1.6 obtained by averaging over all published data (2). The statistical errors in the individual range values are estimated to be within $\pm$ 10% at the higher and $\pm$ 20% at the
lower energies. No range oscillations with Z are observed.

Recently published backscattering measurements on
Si-implants (6) in $0.015 < \varepsilon < 0.3$ are in excellent agreement with our range data.

The straggling parameter data are seen to scatter
within $\pm$ 50%. The average value of the ^{4}He - data amounts
to $0.4 \pm$ 10% which is close to the LSS - estimate of
0.35 and also to 0.45, the latter figure being suggested
by previous investigators from data in the ε interval
1 - 10 (7). The figures derived from measurements using
the molecular deuteron beam are systematically high and
perhaps show a little increase towards lower energies.
An additional spread in energy of the analyzing $(^2H)_3^+$-
beam is caused by the break-up of the molecule-ion. It
is of the order of 1 keV which is equivalent to about
50 Å spread in depth. This contribution, however, should
cancel after reduction of the measured total width with
the width of the surface edge. The previous analysis(2)
suggesting a pronounced increase of straggling parameters with lower energies, has obviously suffered from
the small number of available data points in a relatively limited energy interval.

Fig.3 shows nuclear stopping powers derived from
the experimental range - energy curve of fig.1 by numerical differentiation. From the relation used to fit the
range-energy curve, $\rho_p = 1.0 \times \varepsilon^{1/2}$, a total stopping
power of $d\varepsilon/d\rho = 1.8 \times \varepsilon^{1/2}$ is derived using a mean pro-
jection factor of 0.9. Correction for electronic stop-
ping, with a mean value of $k \sim 0.1$, gives the nuclear
stopping power of $(d\varepsilon/d\rho)_n = 1.7 \varepsilon^{1/2}$. This relation may
be used up to about $\varepsilon \sim 0.01$. In the energy decade above
the approximation $(d\varepsilon/d\rho)_n = 0.7 \times \varepsilon^{1/3}$ is less satisfying.

The approximation of the Thomas-Fermi-potential by
$V(r) \sim r^{-3}$ leads to the range-energy relation of $\rho \sim \varepsilon^{2/3}$.
Inversely it follows from $\rho \sim \varepsilon^{1/2}$ that in the low energy range a potential of the form $V(r) \sim r^{-4}$ should be a

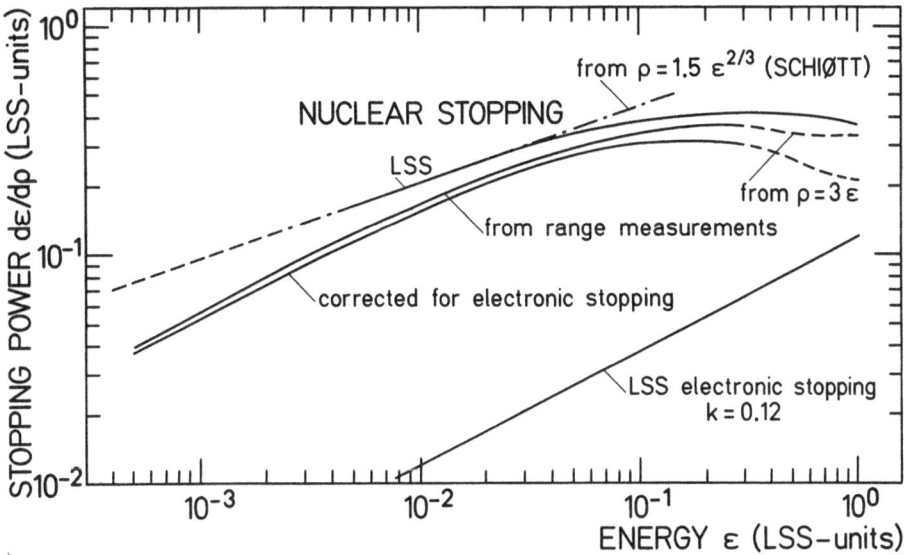

Fig.3   Stopping power as a function of energy. Data
         derived from the curves of fig.2.

better choice for theoretical calculations. This is consistent with recent findings (8), where a Lenz-Jensen
potential gave better agreement with experimental ranges of low energy heavy ions in solids, whereas the
weaker screened Thomas-Fermi potential yielded much too
small values.

Fig.4 compares theoretical and experimental stop-
ping powers using the standard $d\varepsilon/d\rho - \varepsilon^{1/2}$ - plot. The
data of fig.3 are represented here by the half-open
circles. Continuing this curve, the open triangles have
been obtained by taking the empirical relationship
$\rho = 3\varepsilon$ (10) and correcting for electronic stopping with
the same value of k as before. Obviously, at values of
$\varepsilon > 5$ this curve runs low and does not tend to approach
the Rutherford backscattering level.

The data points of Högberg (11) obtained for five
ion species in carbon at one energy are in good agreement with our data, whereas the agreement with Sidenius'(12) data is not so good, especially at the lower
energies in the vicinity of $\varepsilon \sim 1$. On the other hand,
by using higher values for electronic stopping the data

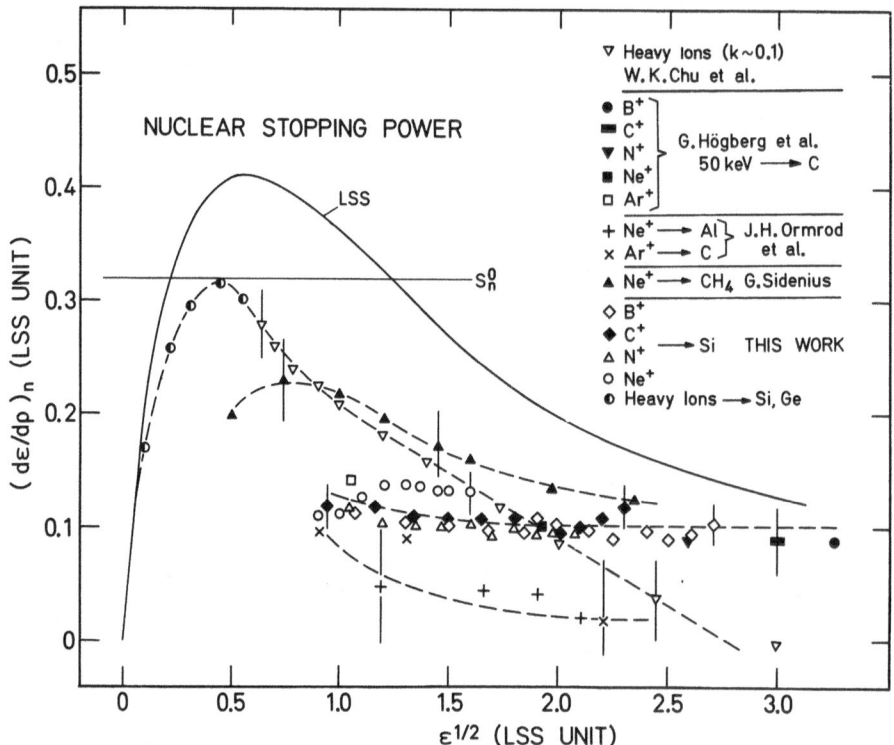

Fig.4 Nuclear stopping vs. energy in LSS-units.

points of Sidenius would come closer. Although the
scatter of our data points which are purely experimental (9) is relatively small we cannot exclude systematic
errors of up to 20%. The figures taken from Ormrod's(13)
measurements mark - within a relatively wide range of
scatter - the lowest experimental limit. All experimental points in the medium energy region are still un-
corrected for projected path effects which may lower
the figures by as much as 30%.

The general conclusion is that LSS-theory overesti-
mates nuclear stopping in the entire energy region of
about $0 < \varepsilon < 10$, at maximum by about a factor of two.

ACKNOWLEDGEMENTS

The preparation of the silicon wafers and the ellipsometric measurements were performed at AEG-Telefunken, Heilbronn. We thank Drs. Goldbach and Kostka for their kind cooperation.

REFERENCES

1. J. Lindhard, M. Scharff and H.E. Schiøtt
    Mat. Fys. Medd. Dan. Vid. Selsk. <u>33</u> (1963) No. 14

2. A. Feuerstein, S. Kalbitzer and H. Oetzmann,
    Phys. Lett. <u>51A</u> (1975) 165

3. A. Feuerstein, H. Grahmann, S. Kalbitzer and
    H. Oetzmann, Proc. Int. Conf. Ion Beam Surface
    Layer Analysis, Karlsruhe, 1975.

4. S. Mylroie and J.F. Gibbons,
    Proc. III. Int. Conf. Ion Implantation in Semicon-
    ductors and Other Materials, 1973, Yorktown Heights,
    Plenum Press, p. 243

5. S. Furukawa and H. Ishiwara, p. 147,
    Y. Ohmura, K. Koike and H. Kobayashi, p. 183
    K. Wittmaack, F. Schulz and B. Hietel, p. 193,
    Proc. IV. Int. Conf. Ion Implantation in Semi-
    conductors, 1974, Plenum Press.

6. J.S. Williams and W.A. Grant, Rad. Effects <u>25</u>
    (1975) 55

7. J.W. Mayer, L. Eriksson and J.A. Davies,
    Ion Implantation in Semiconductors,
    (Academic Press, New York, 1970), p. 34

8. T. Ishitani and R. Shimizu,
    App. Phys. <u>6</u> (1975) 241

9. H. Grahmann and S. Kalbitzer, Int. Conf. Atomic
    Collisions in Solids VI, Amsterdam, The Netherlands,
    1975.

10. W.K. Chu, B.L. Crowder, J.W. Mayer and J.F. Ziegler,
    Proc. III. Int. Conf. Ion Implantation in Semicon-
    ductors and Other Materials, Yorktown Heights, New
    York, 1972 (1973 Plenum Press, New York) p. 225

11. G. Högberg and R. Skoog, Rad. Effects <u>13</u> (1972) 197

12. G. Sidenius, Mat. Fys. Medd. Dan. Vid. Selsk.
    <u>39</u> (1974) No.4

13.   J.H. Ormrod, J.R. McDonald and H.E. Duckworth
      Can. J. Phys. 43 (1965) 275
      J.H. Ormrod, H.E. Duckworth, Can. J. Phys. 41
      (1963) 1424

DISCUSSION

Q:(J. Davies) Did you investigate the dependence of range profile
on dose and, if so, over what dose range ?

A: (S. Kalbitzer) Doses ranged between $10^{14} - 10^{15}$ ions/cm^2. No effects were observed with respect to range parameters.

Comment: (K. Wittmaack) It is not correct to assume that $\gamma \Delta Rp/Rp$
normalizes the relative straggling. $\Delta Rp/Rp$ increases monotonically
with increasing M_2/M_1 whereas γ has a maximum (of 1) at $M_2/M_1 = 1$.
Therefore, $\gamma \Delta Rp/Rp$ is not constant but depends upon M_2/M_1.

A: (S. Kalbitzer) For the present experimental conditions LSS predicts the straggling parameter $\gamma \Delta Rp/Rp$ to be nearly constant.

ON PROBLEMS OF RESOLVING POWER IN RUTHERFORD BACKSCATTERING

J. Schou, S. Steenstrup, A. Johansen and
L.T. Chadderton
Physical Laboratory II
Universitetsparken 5, DK-2100 Copenhagen Ø
Denmark

The Rutherford backscattering technique has found increasing applications in the analysis of surface layers. At high energies - above 1 MeV for example - the resolving power of the detectorsystem is usually small compared to the incident energy. At lower energies, between 50 and 300 keV, however, this need not be the case. Since interpretations of spectra depend on the relative resolving power it is important both to determine the latter for a given system, and to examine in detail the influence on the spectra.

Several important cases are examined and definite methods of extracting data from measured spectra are presented.

Introduction

Rutherford backscattering (RBS) has become a very versatile and precise method of analyzing properties in the solid state[1]. The versatility of the method is based on two important facts: 1) that a beam of particles is scattered off different atomic species with different energy and 2) that a beam of particles is scattered back from different depths with different energy. The energy of the backscattered particles is in both cases the important quantity to measure. In view of the very precise information that can be obtained from the energyspectra of backscattered particles it is of paramount importance to know how these spectra are influenced by the detectorsystem, and to what extent one has to take such an influence into account.

Theory

As is well known[2] the process of measuring a spectrum corresponds to the mathematical operation of a folding (or a convolution):

$$g(x) = \int_{-\infty}^{\infty} S(x-y)f(y)dy = \int_{-\infty}^{\infty} S(y)f(x-y)dy. \quad (1)$$

In this expression $f(y)$ is the 'true' spectrum, $S(y)$ is a detector response function and $g(x)$ is the measured spectrum. It is almost always assumed - implicitly - that the detector response function is a Gaussian density:

$$S(y) = \frac{1}{\sqrt{2\pi}\sigma} \exp(-y^2/2\sigma^2), \quad (2)$$

a function that is thus characterized by one parameter σ, the standard deviation. It will be shown that in the case of a solid state detector for RBS this assumption is indeed well justified. It should be mentioned that experimentally the Gaussian density is most easily characterized by the Full Width at Half Maximum (FWHM) $\lambda_{1/2}$. $\lambda_{1/2}$ is related to σ by

$$\lambda_{1/2} = 2\sqrt{\ln 4}\sigma \simeq 2.35 \cdot \sigma \quad (3)$$

A number of cases of interest will be examined in the following.

Experiments

The experiments to be described were performed on an isotope separator with a maximum energy of 130 keV and a $90°$ analyzing magnet. An Ortec silicon surface-barrier detector cooled around 200 K was employed, and pressure in the target chamber was $\lesssim 10^{-6}$ Torr. For experiments with the detector as target the attenuation of the beam

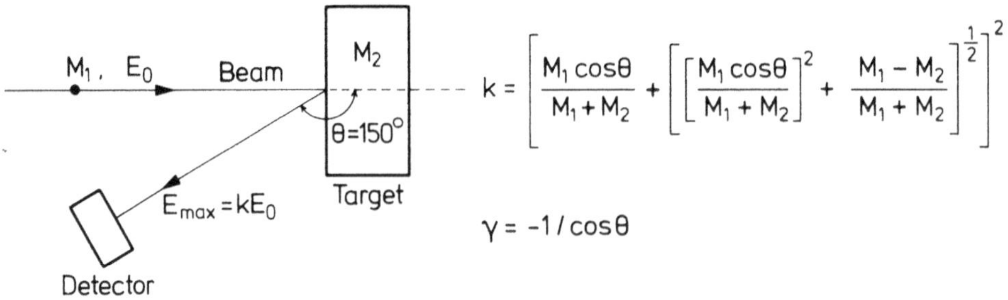

$$k = \left[\frac{M_1 \cos\theta}{M_1 + M_2} + \left[\left[\frac{M_1 \cos\theta}{M_1 + M_2} \right]^2 + \frac{M_1 - M_2}{M_1 + M_2} \right]^{\frac{1}{2}} \right]^2$$

$$\gamma = -1/\cos\theta$$

Fig. 1 Geometry and notation for a typical RBS experiment

was accomplished by placing apertures both before and
after the magnet with diameters of 50 μm and 10 mm respectively. A 1mm aperture was placed directly in front
of the detector. With this arrangement the energy spread
of the beam was estimated in the present case to be about
200 V determined by the analyzing magnet. For RBS measurements the geometry shown in Fig. 1 was used. Our notation
is also shown on Fig. 1.

Results

The simplest case is that of a monoenergetic beam
striking the detector directly. $f(x)$ will then correspond
to a δ-function and $S(x)$ will be equal to $g(x)$. This method
thus is ideally suited for determining the standard deviation of the detector system. However, it is not possible
to produce a δ-function beam form; one will always have
a certain spread. If the beam is assumed to be a Gaussian
density in energy, then $g(x)$ will again be a Gaussian but with a stan-
dard deviation equal to the geometric sum of the standard deviations of the two Gaussians $f(x)$ and $S(x)$. What is then obtained is an upper limit to the detector resolution. If on the other hand the same beam that has been used for measuring the resolution is also used for other analyzing purposes then the total standard deviation is the relevant pa-
rameter, due to the fact that the operation of convolution is both a commu-
tative and an associative operation.

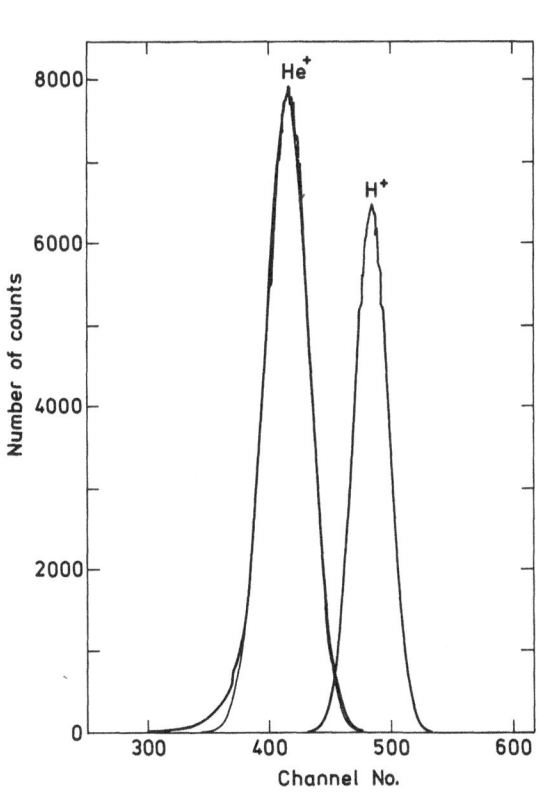

—— Measured spectra of 100 keV He⁺ and H⁺
—— Calculated Gaussian distribution

Fig. 2 Response from the detector when hit directly by the beam.

Fig. 2 shows two examples of such direct spectra. Calculated Gaussians with the same halfwidth are shown. Only in the case of the He$^+$ beam is there any detectable difference between the spectrum and the Gaussian and then merely in the tail.

Consider now the case of backscattering from a random solid medium. The theoretical forms of such spectra have been derived previously to different degrees of approximation[3-5].

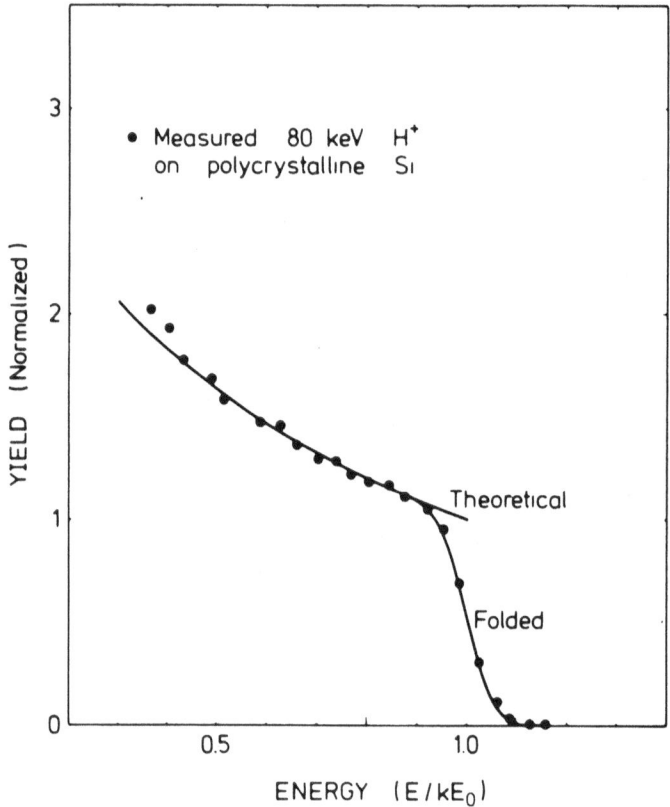

Fig. 3 Typical RBS spectrum. The theoretical was calculated with k = 0,878 and γ = 1,15 (see fig. 1). Energy independent stopping power and Rutherford cross section ($\sigma \alpha E^{-2}$) was assumed. The theoretical was folded with a gaussian with σ/kE_o=0,042.

Fig. 3 shows one example of a calculated spectrum using the formula given by Jack (3). The effect of the detector is also shown, as well as experimental points. It is to be noted that the energy $E_{\frac{1}{2}}$ corresponding to the energy where the folded spectrum has fallen to one half of the extrapolation (linear) of the theoretical is <u>not</u> equal to kE_O. The magnitude of this difference between $E_{\frac{1}{2}}$ and kE_O can be calculated. The result is displayed in Fig. 4.

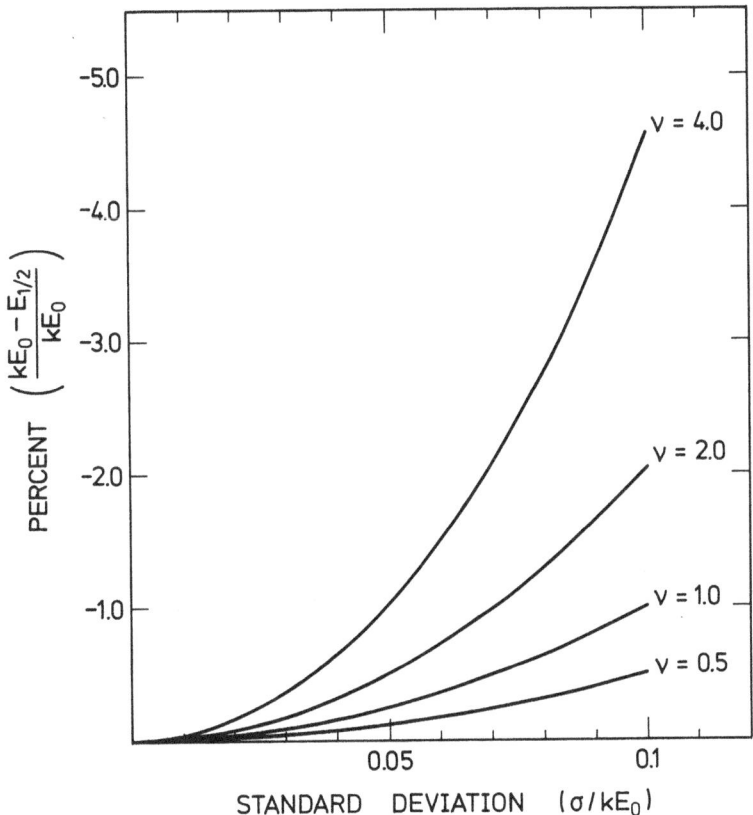

Fig. 4 Percent deviation between $E_{\frac{1}{2}}$ and kE_O as a function of σ/kE_O with $\nu = \eta kE_O/\chi$ as parameter.

For the calculation the theoretical spectrum was assumed linear near the edge with a slope η and a yield χ at kE_O. The percentage difference $(kE_O - E_{\frac{1}{2}})/kE_O$ can be expressed universally as function of σ/kE_O depending on only one parameter $\nu = \eta\ kE_O/\chi$. Similarly from the slope $(\frac{1}{2}\eta+\chi/\sqrt{2\pi}\sigma)$ at $E=kE_O$ the standard deviation can be calculated. Note that if $E_{\frac{1}{2}}$ is confounded with kE_O and if the slope at $E_{\frac{1}{2}}$

is set equal to $\eta/2+\chi/\sqrt{2\pi}\sigma$ a standard deviation σ_1 different from σ will result. When σ_1 and σ are expressed in u-
nits of kE_0 a universal plot depending only on one parame-
ter ν giving the relation between σ_1 and σ can be calculated and is shown in Fig. 5.

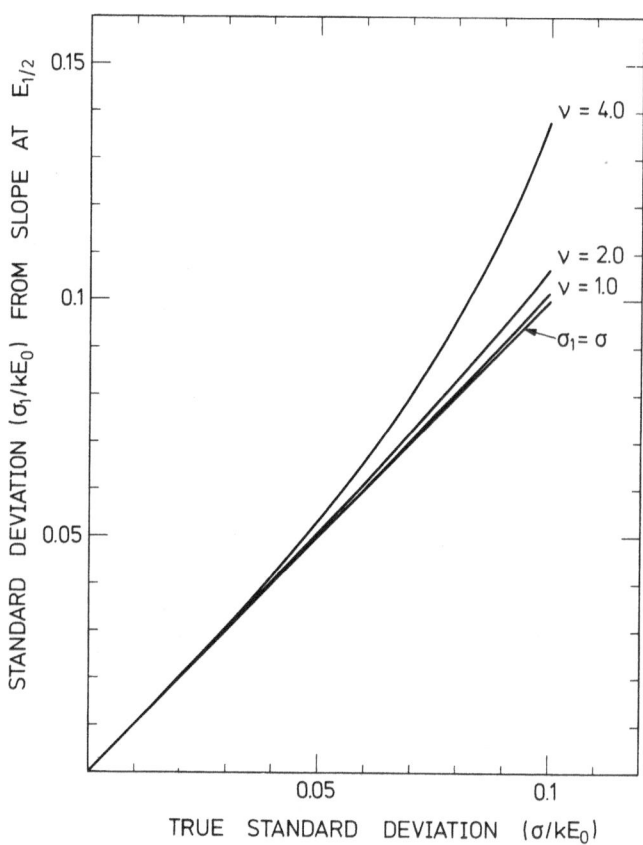

Fig. 5 Relation between σ_1 and σ (both in units of kE_o).

From Figs. 4 and 5 it is seen that the correction to the
standard deviation is negligible in almost all cases (but
not all), while correction to the channel number corresponding to kE_0 in many cases can be considerable. In any case
the corrections are most important for low energies (large
σ/kE_o) and heavy ions (large ν). As an example giving the
magnitude of the corrections take He^+ (80 keV) on Si where
$\nu = 4$ and σ/kE_o was measured to around 0.07. The correction was then 6% to the standard deviation and 2% to channel

number corresponding to kE_0.

Yet another case of fundamental interest is that of an axially 'aligned' spectrum from a single crystal with a surface peak like the one shown in Fig. 6. Again both the 'true' and folded spectra are shown. The problem which often arises in experimental studies is that of determining the area of the surface peak. Normally it is assumed that the spectrum consists of two parts - one , that comes from the scattering from a thin (mono) layer of imperfect crystal, the other part arising from the dechanneling of the beam. The latter contribution has a more or less 'random' shape, only the yield being much smaller. This is illustrated in Fig. 6, where the summing has been performed in the folded spectrum but not in the theoretical. For these assumptions a simple geometrical construction is sufficient to determine this area. Thus the area under the surface peak is found rather directly by subtracting $3/2\eta\sigma^2$ from the crosshatched area in Fig.6, where η is now the slope of the curve, assumed linear, behind the surface peak. The construction in Fig.6 is not dependent on the form of the surface peak. The result would be the same if instead of a 'semi'-Gaussian a δ-function or a rectangular shaped peak had been used. The only important points are that the position of kE_0 can be determined precisely and that the yield behind the surface peak is sufficiently well behaved to be extrapolated to kE_0.

The theoretical spectrum from a surface layer can be expected to be very nearly of rectangular form. The effect of the RBS detector system on such a spectrum is to broaden it out, and the halfwidth $\lambda 1/2$ of such

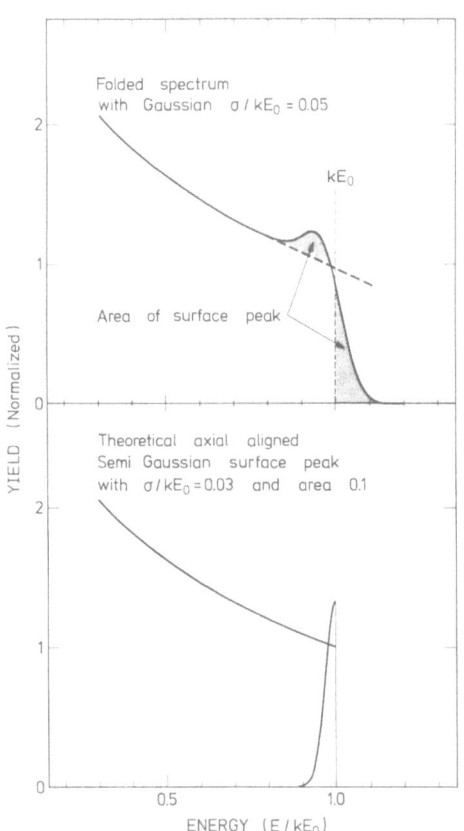

Fig. 6 Construction to find the area of surfacepeak in an axially aligned spectrum. From the area indicated as crosshatched should be subtracted $3/2\eta\sigma^2$ (η being the slope of the yield curve behind the surfacepeak).

a spectrum should be an easily measurable quantity. However, the simple relationship (3) between the halfwidth
and the standard deviation of the detector system is not
valid. None the less this relationship can be calculated
numerically if rectangular shape is assumed. Fig. 7 shows
the relation between λ1/2 and σ as a function of the
thickness of the layer ξ, measured in the same units as
λ1/2 and σ (keV). ξ is accordially the difference in energy between an ion scattered from the front of the layer
and from the back. This plot provides an easy way of
finding the thickness (in energy units) of a thin layer
if of course one is not too near the region corresponding
to the relation (3).

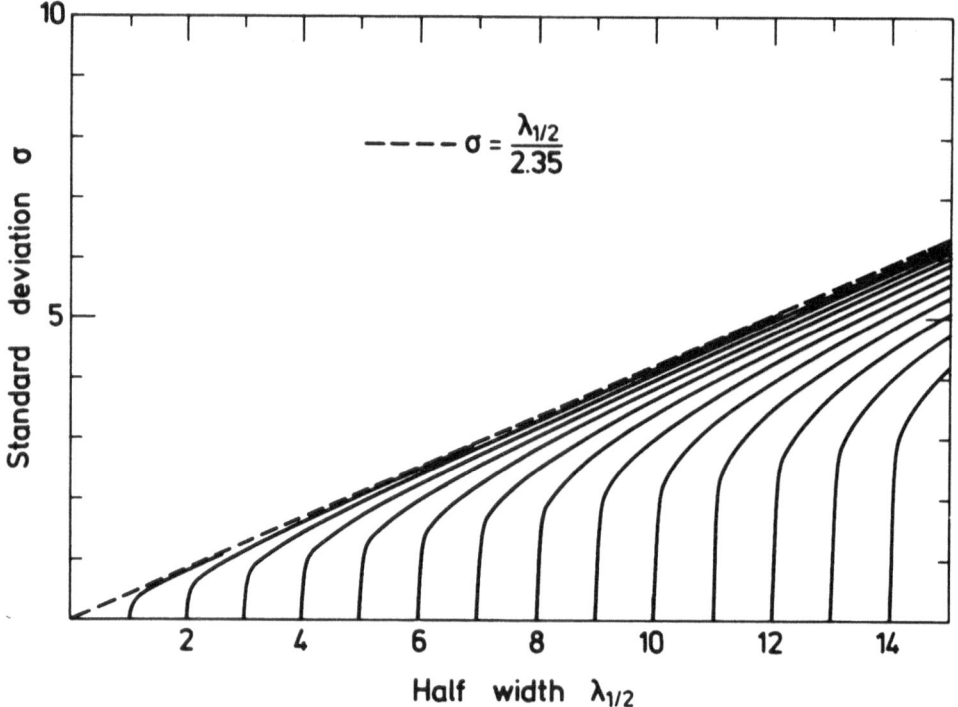

Fig. 7 Relation between a measured halfwidth from the
spectrum of a thin surfacelayer and the detectors standard deviation with the energyloss as parameter.

Conclusion

We have briefly shown how it is that the resolving power of the detector system RBS experiments can influence simple spectra from poly- and single crystalline media (random and axially aligned), and have indicated a method for extracting the resolving power from experimental observations. Not surprisingly the influence of the detection system is most pronounced when the resolving power is non negligible in comparison with the energy of the back scattered ions. In some such cases estimates of the necessary corrections can be made on the basis of a single parameter applicable to all energies and all types of analyzing beam.

A more detailed account of part of this work appears elsewhere[6].

Acknowledgements

The authors are grateful for financial support from the Danish Scientific Research Council (Forskningsråd) and the NATO Research Grants Programme.

References

1. D.S. Gemmel   Rev. Mod. Phys. 46, 129,(1974).

2. A.F. Jones and
   D.L. Misell   J. Phys. A 3, 462, (1970).

3. H.E. Jack, Jr.   Thin Solid Films 19,267,(1973).

4. R. Behrisch and
   B.M.U. Scherzer   Thin Solid Films 19,247,(1973).

5. D.K. Brice   Thin Solid Films 19,121,(1973).

6. J. Schou   Thesis, University of
                        Copenhagen, (1975).

STUDIES OF SURFACE CONTAMINATIONS, COMPOSITION AND FORMATION OF
SUPERCONDUCTING LAYERS OF V, Nb_3Sn AND OF TUNNELING ELEMENTS
USING HIGH ENERGETIC PROTONS COMBINED WITH HEAVY IONS

P. Müller, G. Ischenko and F. Gabler

Physikalisches Institut der Universität

852 Erlangen, West-Germany

ABSTRACT: An introductory short description of the principles of
analytic methods is given for the elastic scattering with protons
of (5-8) MeV - with energies well above the Coulomb barrier - with
large depth information and high sensitivity for the light elements
(^{1}H-^{19}F) and Rutherford backscattering with ^{12}C and ^{16}O ions with
better depth and mass resolution for the heavier elements. As realistic applications, several topics in sample preparation for
superconducting cavities and irradiation experiments on supercon-
ductors are presented.

1. INTRODUCTION

As proposed earlier[1] and now applied to materials analysis[2] beams of high energetic heavy ions can improve in some cases
the quality of obtainable information with respect to the more conventional Rutherford Backscattering (RBS) methods. Furthermore non-
Coulomb scattering with (5-8) MeV protons gives advantages concerning high detection sensitivity for the light elements and the
large analyzing depth. The combination of these two methods permits
analyses ranging from trace contaminations with light isotopes on
surfaces to depth profiles of compounds or alloys of the heavier
elements. This is demonstrated by realistic applications resulting
from investigations of irradiation damage effects in superconduc-
tors and metals made during the last years[3,4].

2. DESCRIPTION OF THE METHOD-ANALYZING PARAMETERS

Principal conceptions like kinematics or backscattering energy
loss parameter are well known from the RBS method and will not be
regarded here. For the energy particle combination 6 MeV p and
20 MeV ^{16}O frequently used in our studies, a short comparison of
analyzing parameters with those of RBS with 2 MeV ^{4}He-Ions is given.

Separation regions of neighbour isotopes and approximate ratios of depth resolutions (defined as the areal density which corresponds via the energy depth relation to the detector resolution) and analyzing depth (Range of scattered particle) are presented in table 1. More detailed information is summarized in ref. 2.

Separation regions for neighbour isotopes	p : $1 < M < 35$ ^4He : $5 < M < 35$ ^{16}O : $20 < M < 90$
depth resolution	p : ^4He : $^{16}O \approx 20$: 2 : 1
analyzing depth	p : ^4He : $^{16}O \approx 25$: 1 : 1

Table 1: Analyzing Parameters for the example described in the text. The ratio mentioned are computed for $^9Be(p)$, $^{19}F(\alpha)$ and $^{48}Ti(^{16}O)$ as scattering centres and are nearly independent from the mass of stopping material (c.f. ref. 2).

3. EXPERIMENTAL LAYOUT

The experiments were performed at the facilities for low temperature irradiations or at the scattering chamber of the heavy ion arm of the Erlangen HVEC Tandem Accelerator. The ions have been detected by surface barrier detectors using standard NIM module electronics. The analyzing time for a "good" spectrum was (20-120) min with an ion current of about 0,1 µA after passing an aperture of 2 mm diameter. Data were fed into a PDP 7 computer as multichannel analyzer and stored on magnetic tapes. Depth profiles are computed by mean values of energy losses (thin targets) or by exact integration using analytical approximation formulas[5].

4. APPLICATIONS

The examples and applications presented here are not the socalled typicals but reach in some special cases the limit of the method. On the other hand they are realistic and due to the large analysing depth required or the specital topology of the sandwich structures, backscattering was found to be the appropriate analysis method.

a) Vanadium Foils

An important preparation step to produce good residual resistance ratios (RRR) which are necessary for irradation experiments, is the degassing of the foils by heating in UHV. The information obtainable from scattering with (5-8) MeV protons are surface contaminations like carbon or oxygen and determination of the hydrogen contents of the foils, as is demonstrated in fig. 1. The foil has been cleaned in a HNO_3-CH_3COOH aqueous solution and heated 40 min at 1500°C in UHV ($< 10^{-8}$ Torr). During this cycle 1 at %

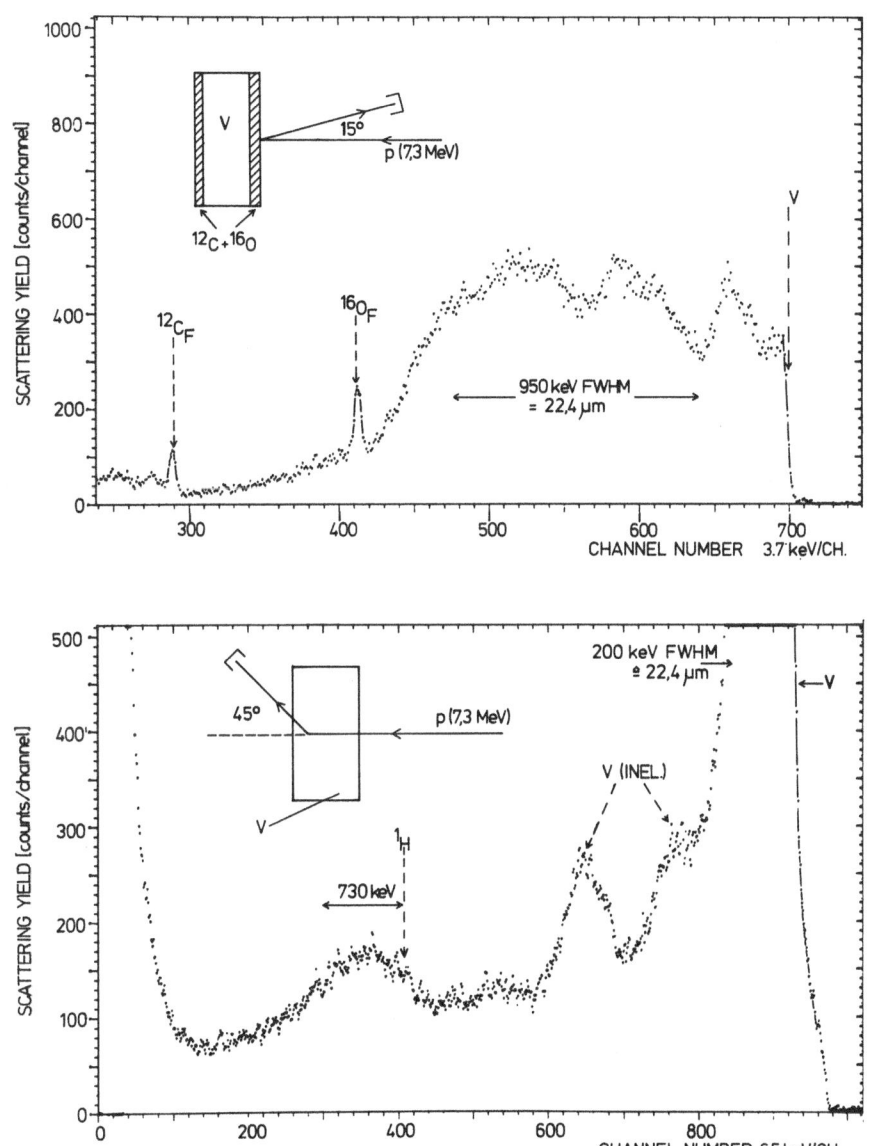

Fig. 1: Spectra obtained simultaneously by scattering of 7,3
MeV protons on a vanadium foil at 165° (top) and 45°(bottom).
The modulation of the vanadium peak at 165° is caused by
variation of the scattering cross section with energy. The
arrow-marked peak widths of 950 keV, 200 keV and 730 keV
correspond according to the energy-depth relation to the
same foil thickness of 22,4 μm.
(Index F: contamination of the front surface).

*H was absorbed by the foil. Determination of the residual gases in
the vacuum chamber by a quadrupole mass spectrometer shows the decomposition of water (major contamination in the chamber) above
about 1200°C by the heated foil. H-absorbing materials like a heated
Zr-wire are therefore needed to protect the foil during the heating
cycle. The raw material shows carbon contaminated oxide layers of
more than a factor 10 thicker but no detectable hydrogen contents
within the detection limit of ≈ 300 ppm.*

b) NbTi Foils

*Problem: Surface contamination due to thinning by electropoli-
shing of a commercially available NbTi sheet. Polishing has been
performed in a mixture of $Mg(ClO_4)_2$ and LiCl soluted in CH_3OH at
about -35°C. A proton backscattering spectrum of such a foil is
shown in fig. 2. The nominal composition has been determined by
RBS with 20 MeV ^{12}C ions to be $NbTi_{2.3}$.*

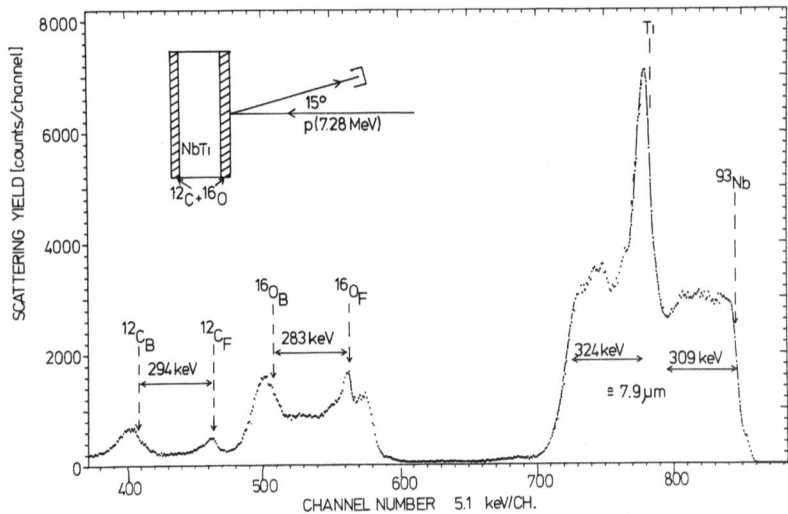

Fig. 2: *Spectrum from a study of surface contaminations of NbTi.
Because of a not optimally chosen polishing voltage the
contamination of the back surface (Index B) is bigger than
that of the front. The shape of the Nb and Ti peak is due
to signals overlap.*

c) Nb₃Sn Layers

*There are three species of interest: i) layers deposited by
Chemical Vapour Deposition (CVD) onto Hastelloy substrates as used
by our group in low temperature irradiation experiments[4];
ii) layers fabricated by diffusion of Sn-vapour into heated Nb-
sheets used as coating of Nb-cavities, which have been developed
for particle separation in elementary particle physics during the
last years and iii) the preparation of superconducting tunneling
elements like Nb_3Sn-oxide-Pb-junctions. The properties of the last*

*two species depend very strongly on the surface condition of the Nb_3Sn layer: the rf properties of cavities are degraded **and** the junctions **show** an undesired point contact behaviour[6]. Fig. 3 shows a spectrum of a Nb_3Sn diffusion layer taken with 20 MeV ^{12}C-ions.*

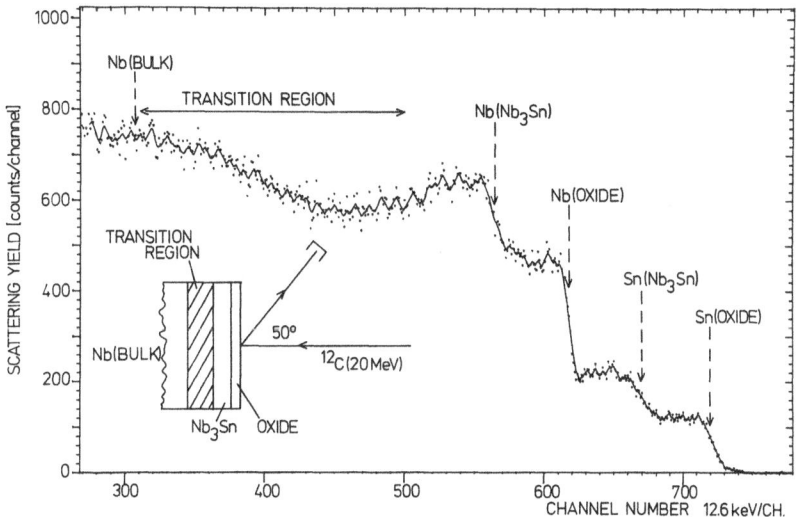

Fig. 3: *Spectrum resulting from RBS with 20 MeV ^{12}C-ions on an anodized Nb_3Sn-diffusion layer. Further explanations see text. The solid line drawn corresponds to a mean value of the scattering yield.*

*This layer was oxidized anodically in a 20 % NH_3 solution. The particle energy E chosen was a compromise between separation of the overlapping Nb and Sn "mountains" (raising with E) and an optimum depth resolution (deteriorating with E). The continuous transition region between $Nb(Nb_3Sn)$ and Nb(bulk) should be noticed. After the phase diagram in the temperature and concentration region regarded, a NbSn mixture segregates into Nb and Nb_3Sn and therefore the Nb_3Sn penetrates **probably** like needles into the Nb. The averaging of the analyzing ^{12}C-beam suggests a continuous transition.*

d) Tunneling Elements
As illustration for the analysis of thin film sandwiches serves a spectrum of a $Al-Al_2O_3$-Sn tunneling element - as used by our group for low temperature experiments - deposited onto a thin transparent carbon foil. The corresponding spectrum obtained with proton backscattering is shown in fig. 4. Assuming bulk Al_2O_3 composition and density of the oxide, the ^{16}O Peak corresponds to a thickness of 80 Å and therefore to detection sensitivities for oxygen in the

submonolayer region obtainable in a reasonable measuring time.

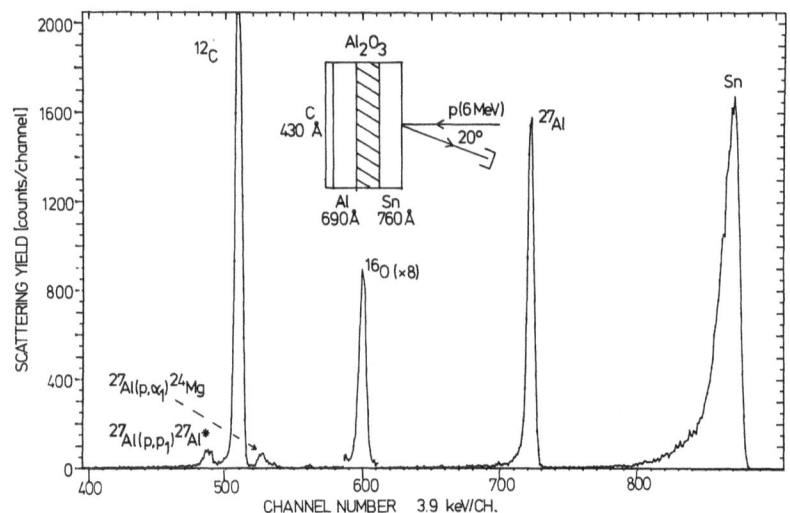

<u>Fig. 4</u>: *Spectrum of a $Al-Al_2O_3$-Sn junction deposited on a thin carbon film. For clarity the ^{16}O-Peak is enlarged by a factor of eight.*

5. CONCLUSIONS

Summarizing the preceding results the *characteristic advantages of the analyses may be stressed:*

-*isotope separation within the mass numbers 1 to 100*
-*detection of light elements with high sensitivity*
-*analyzing depth extendable to thick "metal sheets"*
-*complementary properties like the conventional $^4He^+$ RBS but for the heavier elements combined with the special advantages of elastic non Rutherford backscattering with (5-8) MeV protons.*

ACKNOWLEDGEMENTS

We thank all our colleagues from the low temperature division for their help during irradiation cycles and for preparing the samples. We also thank the Siemens AG, FL, Erlangen for supplying materials and samples.

REFERENCES

1) S. Petersson, P. A. Tove, O. Meyer, B. Sundqvist,
A. Johansson, Thin Solid Films *19*, 157 (1973)

2) P. Müller and G. Ischenko, Paper submitted for publication in
J. Appl. Phys.

3) S. Klaumünzer, G. Ischenko, P. Müller, Z. Physik *268*, 189 (1974)

4) B. Besslein, G. Ischenko, S. Klaumünzer, P. Müller, H. Neumüller,
K. Schmelz, H. Adrian, Phys. Lett. *53A*, 49 (1975)

5) C. S. Zaidins, Nucl. Instr. and Meth. *120*, 125 (1974)

6) S. I. Vedeneev, A. I. Golovaskin, G. P. Motulerich,
Sov. Phys. Solid State *16*, 439 (1974)

DISCUSSION

Q: (J.F. Ziegler) One of the difficulties of using high energy
(4-8 MeV) protons on complex samples is the lack of published
proton cross sections over this continuous energy range for many
elements. Where are such cross-section surveys available?

A: (P. Müller) A data index is published in Nuclear Data Tables:
F.K. McGowan, W.T. Milner, H.J. Kim, W. Hyatt, Nucl. Data A6,353
(1969); A8, 199 (1970); A9, 469 (1971), continued in Nucl. Data
Sheets.

This covers investigations from the years 1950-1971 and contains
excitation functions of the elastic (p,p) cross sections in the interesting energy region (5-8 MeV) at large angles (150°-170°) for
example: ^{12}C, ^{13}C, ^{14}N, ^{15}N, ^{16}O, ^{17}O, ^{19}F, Mg-Isotopes Al, etc.

DETERMINATION OF IMPLANTED CARBON PROFILES IN NbC SINGLE CRYSTALS FROM RANDOM BACKSCATTERING SPECTRA

K.G. Langguth, G. Linker, and J. Geerk

Institut für Angewandte Kernphysik

Kernforschungszentrum Karlsruhe

ABSTRACT

The depth profile of carbon atoms implanted into NbC_x single crystals has been determined from random spectra of α-particles backscattered from the Nb atoms. Ziegler and Crowder reported a decrease of the random height of an amorphous surface region of a silicon single crystal. Such effects could give rise to errors in the determination of concentration profiles and were therefore additionally studied in amorphousized silicon and germanium single crystals. An increase of the random height of the amorphous regions as compared to undisturbed crystals has been detected for our scattering geometry.

INTRODUCTION

The superconducting transition temperature T_c of the refractory carbides NbC and TaC depends strongly on the carbon to metal atom ratio[1]. The highest T_c is expected in compounds having the ratio 1.0. As it is impossible to produce compounds with this ratio using conventional techniques like diffusion we implanted carbon ions into NbC single crystals having a concentration ratio of carbon to niobium of 0.89 (corresponding to a T_c of 4 K) in order to increase this ratio and hence T_c. After a homogeneous implantation with $1.5 \cdot 10^{17}$ C^+/cm^2, 200 keV and $5 \cdot 10^{16}$ C^+/cm^2, 80 keV the transition temperature rose to about 7 K, and a maximum T_c of 11.7 K was reached after subsequent annealing at 1100°C in an UHV-oven for 3 min. This value is 0.5 K higher than the maximum T_c reported so far in literature. In Fig.1 T_c is shown as a function of the annealing temperature. In order to determine the atomic concentration ratio in the implanted crystal the question

273

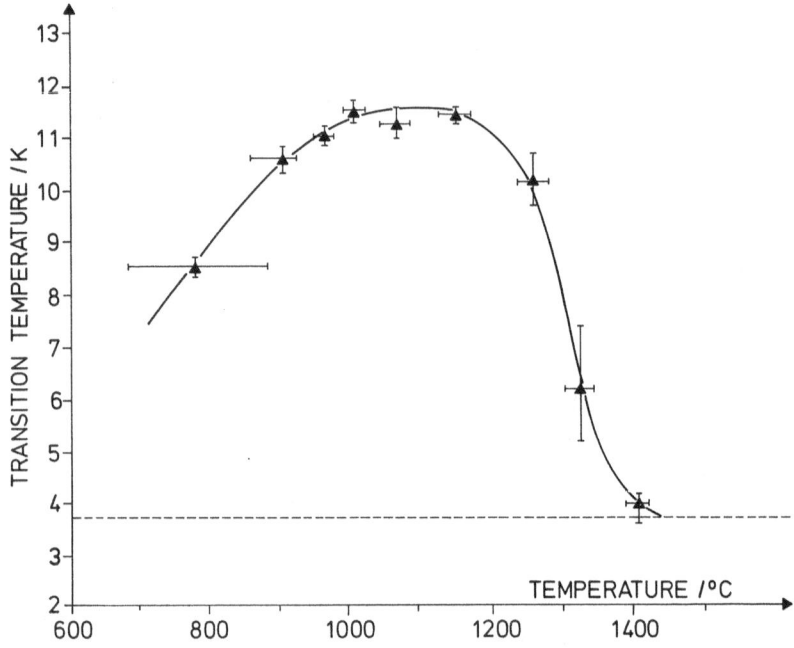

Fig. 1 Superconducting transition temperature of a carbon implanted NbC single crystal as a function of annealing temperature.

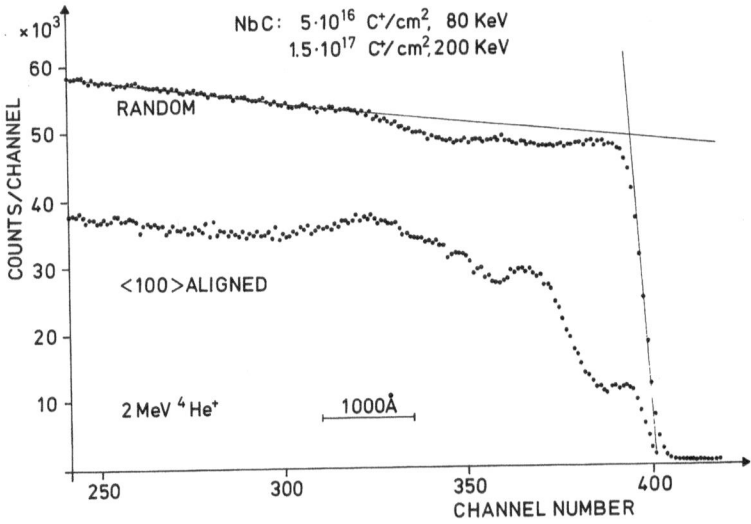

Fig. 2 Random and <100> aligned backscattering spectra of a NbC
single crystal after carbon implantation.

arose whether the backscattering technique could be applied for
this problem with reasonable accuracy.

The carbon concentration profile in the implanted crystal has
been determined from random backscattering spectra in the energy
range of particles scattered from the Nb atoms in the niobium car-
bide. The random spectra were taken by rotating the crystal around
an axis 10 degrees off the <100> channelling direction. As an example such a spectrum is shown in Fig. 2. The carbon profile can
be calculated from the difference between the measured spectrum and
a curve which is fitted to the undisturbed part of the spectrum and
extrapolated over the implanted region.

Ziegler and Crowder[2] investigated the conditions for determining reproducible random backscattering spectra. Using the crystal
spinning procedure they found, that there is a decrease of the random height in the amorphousized surface region of a silicon single
crystal as compared to the random height of an undamaged crystal.
As our implanted NbC single crystals are appreciably damaged in the
implanted surface region even after the annealing procedure the
effect reported by Ziegler and Crowder could possibly lead to serious errors in the determination of the carbon content. Therefore
this effect was further studied in silicon and germanium single
crystals.

EXPERIMENTAL AND ANALYSIS

Amorphous layers of about 3000 $\overset{\circ}{A}$ thickness on Si and Ge single
crystals were produced by bombardment with 175 keV Si^+ ions and
320 keV Ge+ ions, respectively. As implantation of $^{28}Si^+$ ions always bears the risk of simultaneously implanting $^{28}N_2^+$ ions due to
the same charge to mass ratio for these ions, additional bombard-
ments were performed with $^{29}Si^+$ ions which easily can be separated from $^{28}N^+$, and for comparison $^{28}N_2^+$ implantations were also
performed. Fig. 3 shows random and <111> aligned backscattering
spectra of 2.0 MeV $^4He^+$ particles from an amorphousized silicon
single crystal, the aligned spectrum indicating the amorphous region on top of the crystal. The random spectra were taken by rotating the crystals around an axis 12 degrees off the <111> channelling direction with statistics of about 24000 counts/channel
and a channel to energy conversion of 3.9 keV/channel.

A method similar to that of Ziegler and Crowder[2] was used to
compare the heights of the backscattering spectra from crystalline
and amorphous silicon. Figure 4 illustrates this analysis procedure. A curve of the type $Y(E) = A + B/E$ is fitted through that
part of the spectrum which corresponds to the undisturbed crystal
and extended over the damaged region; E is the actual energy of the
particles on their path through the crystal and A and B are fitting parameters. The difference D between the values $Y(E)$ of the

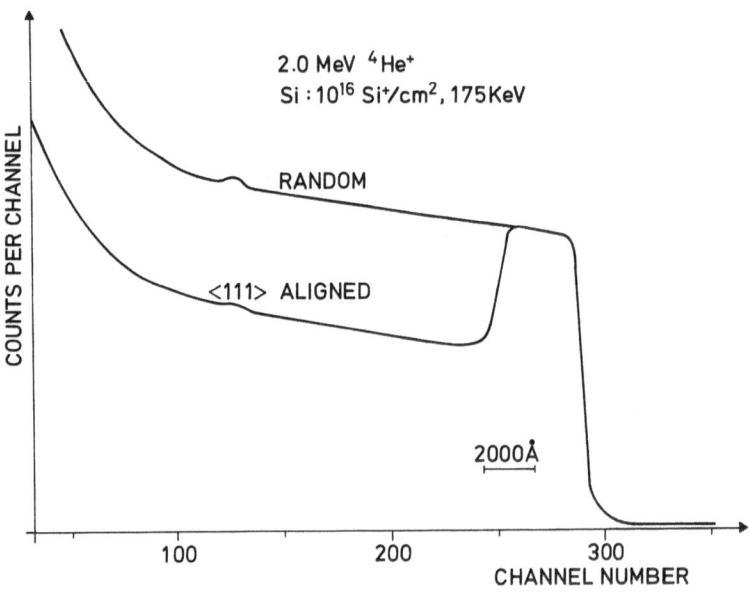

Fig. 3 Random and <111> aligned backscattering spectra of a silicon single crystal bombarded with Si⁺ ions showing the amorphousized
region on top of the crystal.

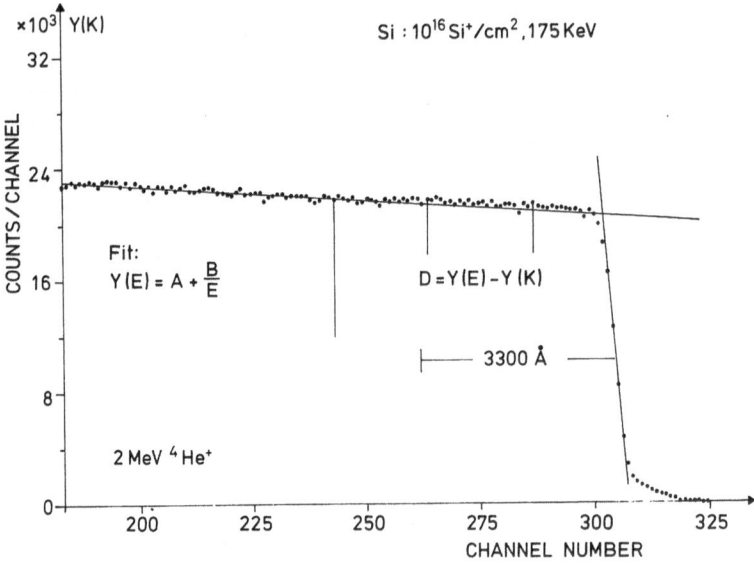

Fig. 4 Random backscattering spectrum of a bombarded silicon single
crystal with the fitted curve extended over the amorphousized region.

fitted curve and the heights Y(K) (K = channel)of the corresponding values from the measured spectra points is then calculated.

RESULTS AND DISCUSSION

The results of the difference calculations for a Si single
crystal bombarded with Si^+ ions are demonstrated in Fig. 5. Here
the differences between the fitted curve and the spectra points
are plotted as a function of depth into the crystal. In order to
test if the extended part of the fitted curve really matches the
random spectrum this difference has also been calculated for random spectra of unbombarded single crystals and is included in the
figure. For comparison also the result of a difference calculation
for a silicon crystal implanted with $10^{16}N_2^+/cm^2$, 175 keV is
shown.

One finds that the fitted curve matches the spectrum of the unbombarded crystal reasonably well. For the single crystal bombarded with Si^+ ions we get a marked negative difference. This result is in contrast to that of Ziegler and Crowder[2] who observed
a decrease of the random height of an amorphousized crystal region as compared to a spectrum from a single crystal. For the N_2^+
implantations the expected positive difference is observed as
the implanted nitrogen atoms cause a decrease of the random height
of the silicon spectrum. As we obtained the same results for
$^{28}Si^+$ and $^{29}Si^+$ implantations we conclude that the beam of $^{28}Si^+$
ions was free from $^{28}N_2^+$ ions.

Similar results were obtained for Ge single crystals bombarded with Ge^+ ions and support the effect observed in silicon.

To compare the results more quantitatively a single magnitude,
which characterizes the damage effect in the random spectrum was
defined by averaging the depth dependent difference over 22 channels
of the central part of the damaged region and by normalizing it
to the total spectrum height. Table 1 shows the so defined mean
relative deviations found for Si and Ge single crystals.

The different backscattering heights from amorphous and crystalline material are thought to be due to channelling effects occu-
ring during rotation, as the number of backscattered particles from
the undisturbed region of the crystal is reduced. The deviation in
germanium is a factor of 2 higher than in silicon. This cannot be
completely explained by the greater critical angles in germanium
as compared to silicon.

As at the present stage of our work the physical reasons for
the deviations found in the random spectra are still unclear, it is
impossible to extrapolate the results of the Si and Ge single crystals to NbC single crystals. From the comparison of the implanted

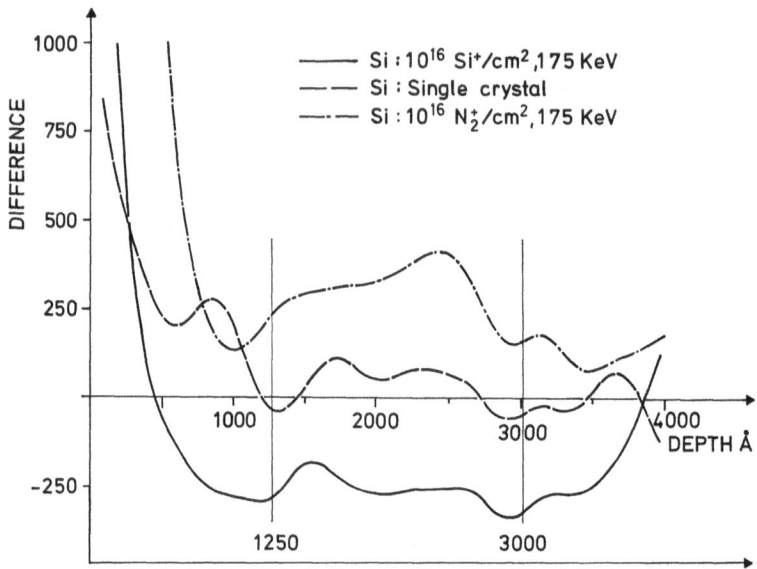

Fig. 5 Differences between the fitted curves and the measured
backscattering spectra for   unbombarded, and Si^+ and N_2^+ implanted
silicon single crystals as a function of depth.

Crystal	Ion	Fluence cm^{-2}	Energy (Impl.) keV	Channel	Energy (Backsc.) MeV	Mean relative deviation
Si	Si$^+$	10^{16}	175	⟨111⟩	2	$- 0.90 \pm 0.14$
Si	N$_2^+$	10^{16}	175	⟨111⟩	2	$+ 1.20$
Si	not bombarded			⟨111⟩	2	$+ 0.27 \pm 0.04$
Ge	Ge$^+$	10^{16}	175	⟨111⟩	2	$- 2.48 \pm 0.1$
Ge	not bombarded			⟨111⟩	2	$- 0.15 \pm 0.25$

Table 1 Mean relative deviations of the fit curves from the backscattering spectra in the amorphousized region of different Si and
Ge crystals.

amount of carbon as determined from the random spectrum with the
fluence we determined the mean deviation in the random spectrum of
NbC caused by amorphisation to be smaller than 0.6 %, which is of
the same order like the deviation found in Si. As the error in the
carbon content determined from the random spectrum is roughly a
factor of 4 higher than the error in the random height itself,
an error of about 2.5 % in the determination of the carbon content
from the random spectrum can still not be excluded. Further work is
in progress to clarify the physical reasons for deviations in the
random spectra of amorphousized single crystals.

ACKNOWLEDGEMENTS

We thank Mr. M. Kraatz for performing the implantations and
Dr. O. Meyer for valuable discussions.

REFERENCES

1. A.L. Giorgi, E.G. Szklarz, E.K. Storms, A.L. Bowman and
    B.T. Matthias, Phys. Rev. 125, 837 (1962).

2. J.F. Ziegler, B.L. Crowder, IBM-Report RC 3551.

PORE SIZE FROM RESONANT CHARGED PARTICLE BACKSCATTERING

C. D. Mackenzie* and B. H. Armitage

U.K.A.E.A., Harwell

Didcot, Oxon, OX11 ORA, United Kingdom

1.   INTRODUCTION

Since porous materials are in extensive use throughout a number of industries a knowledge of pore size, together with related quantities such as specific surface area, porosity and density is often required. The need is not even confined to basically porous materials since many non-porous materials develop porous surface layers when exposed to environmental conditions. A significant example in the electrical power industry is concerned with boiler tubes which can develop a "scale" on the inside metal surface. This scale is often overlaid with a more porous "deposit" layer. Here there is a need to characterize physical properties such as open and closed porosity and pore size. In the United Kingdom atomic energy industry graphite is in extensive use and problems arise from chemical attack by radiolyzed CO_2 and He. Consequently there is a need to have a knowledge of the pore structure accessible from the graphite surface.

Another important field is the study of powder compaction mechanisms. When a powder is placed in a mould and compressed the particles adhere and a strong porous material can result. The application of powder compaction covers a wide range of materials through pharmaceutical tablet making, fuel briquetting to metallurgy and ceramics.[1]

*On leave from University of Melbourne, Australia.

In the above fields there are a few well established techniques used in physical characterization. One of these is gas adsorption from which the specific surface area is obtained. From a knowledge of the surface area and the porosity of a given material it is possible with the aid of a suitable model to obtain pore size information. Information on the size distribution of pores access-- ible from the sample surface is obtainable by the penetration of a non-wetting liquid such as mercury. In the mercury porosimeter pressure is applied to the mercury such that as the pressure is increased open pores are penetrated of progressively decreasing entrance diameter. The scanning electron microscope is also valuable as a means of obtaining a visual picture of the shape and size of pores.

The above well established techniques are of course comple- mentary in that they give, respectively, information on surface area, pore size and visual characteristics of the samples under investigation. The present method in making use of charged particle beams gives pore size information and is thus super- ficially more akin to the mercury penetration method than either gas adsorption or electron microscopy. In practice, however, the charged particle beam method is so different as to be more complementary than competitive with mercury porosimetry.

This report begins with a description of a method whereby by measuring the average energy loss and energy spread of proton beams after passage through thin uniform samples of porous materials it is possible to determine the porosity and also the average pore size. The fact that the width of the distribution of transmitted protons is enhanced for a specimen containing pores in comparison with a non-porous sample of the same material forms the basis for measuring mean pore size. The major part of this report is concerned with an account of the application of the same concept to charged particle backscattering in materials containing oxygen. In this latter case enhancement in width of particular resonances observed in proton and alpha-backscattering is measured.

2.   PROTON TRANSMISSION METHOD: EXPERIMENTAL AND CALCULATIONAL PROCEDURE

Although a brief account of this technique has been given elsewhere[2], it is appropriate to describe it here as it helps in the understanding of the resonant charged particle backscattering approach. Furthermore the latter method represents a development of the proton transmission technique which has been successfully applied over a period of years.

These measurements have been made with protons of up to 12 MeV from the Harwell tandem accelerator. The layout of the experimental equipment is shown in Fig. 1. A low intensity collimated beam of

protons is transmitted through a prepared uniform sample such
that no more than 80% of the energy is lost. The energy distribution of the emergent protons is then measured in a Si semiconductor detector. The width of the distribution (a) obtained
with a porous specimen (Fig. 1) is much greater than that (b)
obtained with a non-porous specimen. The width of the latter
distribution is due mainly to energy straggling of the beam with
small contributions due to finite detector resolution, non-
uniformity of the specimen and energy spread of the incident beam.
The difference between the widths of the two peaks is due to
differences in areal density over the dimensions of the incident
beam. These differences in areal density are in turn due to
variations in the number or size of pores encountered by individual
protons.

If we take the simple case in which all protons intercept equal
lengths of pore (a) we can see how the proton distributions are
widened. Suppose N pores are traversed on average by the proton
beam in passing through a total thickness t_o and an effective
thickness t_o of solid material, then

$$Na = t_o - t_e$$

If the pores are randomly distributed, N is large and the area
covered by the proton beam is large compared with the pore size,
then the distribution of distances travelled by the protons through
solid material will be Gaussian with standard deviation $N^{\frac{1}{2}}a$ and
FWHM of $2.34N^{\frac{1}{2}}a$. The corresponding FWHM of the measured energy

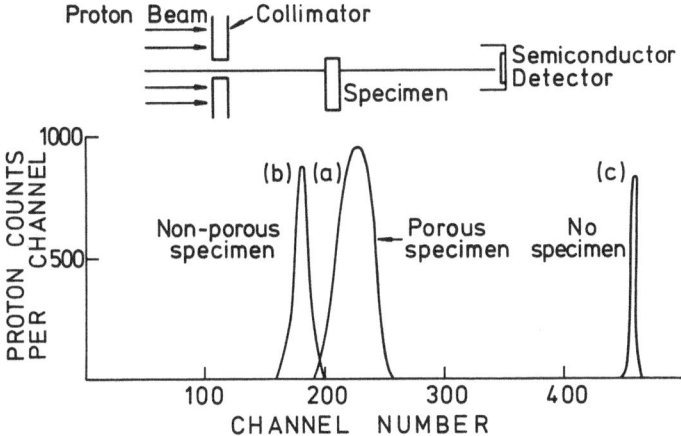

Fig. 1 Proton spectra obtained (a) with a porous specimen, (b) a
non-porous specimen and (c) with no specimen. The layout of the
experimental system for the proton transmission method is also shown.

distribution ΔE will be

$$\Delta E = 2.34 N^{\frac{1}{2}} \, a \, \frac{dE}{dR}$$

where dE/dR is the differential energy loss which can be obtained
from the known range-energy relations such as those of Northcliffe
and Schilling[3]. Here ΔE refers to the FWHM of the porous
specimen after the FWHM of the equivalent non-porous specimen has
been subtracted in quadrature.

From the above we obtain

$$a = \frac{1}{5.48(t_o - t_e)} \quad \left(\Delta E \, / \, \frac{dE}{dR} \right)^2$$

The effective thickness (t_e) can be obtained from the generally
known theoretical density and the measurable superficial area of
the specimen. Since t_o and ΔE can also be measured the above
relation gives an approximate value for the mean pore size. Mean
pore sizes in the range 0.1 μm to 100 μm can usually be measured by
this method which has been extensively applied to graphite. Here
the lower limit is based on the assumption that adequately uniform
specimens are thicker than 0.1 gcm^{-2} and also that the presence of
the pores produces a measurable enhancement of the proton distribution. On the other hand, the upper limit is dictated by the need
for the proton beam to intercept a statistically significant number
of pores during transmission through the sample.

The above calculational method is very approximate in that it
assumes that all interception lengths are equal. Even in the
idealized case of uniform spherical pores this does not apply, and
in real cases there will be a finite distribution of pore sizes in
the specimen. A thorough mathematical analysis of the distribution
of path lengths through pores has been given by Clement[4]. From a
knowledge of the first few moments of the distribution the means and
variances of the path lengths through and between pores may be
calculated.

3.   RESONANT CHARGED PARTICLE BACKSCATTERING METHOD

A method has been developed for measuring pore size in materials
containing oxygen using a charged particle backscattering technique.
The approach is to make use of the strong resonances that exist at
backward angles in the cross-sections for $^{16}O(p,p)^{16}O$ and $^{16}O(\alpha,\alpha)^{16}O$.
The resonances so far used in proton and alpha-particle elastic
scattering are at 2.66 MeV and 3.05 MeV respectively, and can thus
be conveniently reached with the Harwell IBIS 3.5 MV accelerator.

A backscattering spectrum obtained with 2.8 MeV protons on a

porous sintered iron oxide specimen is compared in Fig. 2 with a
similar spectrum obtained with non-porous iron oxide crystals. The
laboratory angle of observation is 170° and a Si semi-conductor
detector of 20 Kev energy resolution was used. The beam was
collimated to 20 μm x 20 μm. In the spectra the Fe and O edges are
clearly visible as is a dip in the spectrum which is associated
with the 2.66 MeV resonance. The enhanced width of the resonance
from the porous material over that obtained with the non-porous
material is clearly observable. The explanation for the enhanced
width of the resonance follows from that given for the proton
transmission case. Due to statistical variations in the number of
pores encountered by the incident charged particle beam in losing
energy down to that associated with the resonance (2.66 MeV),
resonant backscattering will take place from a spread of distances
below the sample surface. This together with a statistical
variation in the number of pores encountered after scattering,
leads to the spread in the width of the observed resonance.

We can adapt the simple calculational method used for proton
transmission to cover the present case if we suppose again that all
charged particles intercept equal lengths of pore, and that N pores
are traversed before resonant backscattering occurs. We also assume
that the natural width of the resonance is very small compared with
the energy lost by the incident beam before resonant backscattering.
In this case the distribution of total distances penetrated by the
beam will be Gaussian with standard deviation $N^{\frac{1}{2}}a$. It should be
remembered here that, straggling and multiple scattering apart, the
total distance travelled in <u>solid</u> material will be the same for all
charged particles.

After scattering the final distribution of distances travelled
by the charged particles through solid material will be Gaussian

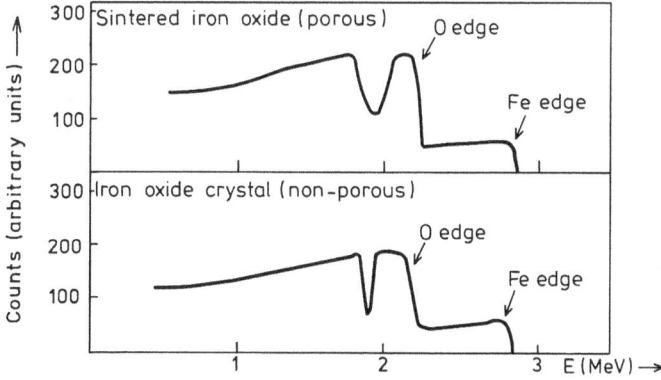

Fig. 2 Proton backscattering showing the enhanced width of the
resonance observed with a porous specimen

with standard deviation $(2N)^{\frac{1}{2}}a$. This arises by adding together
contributions from the spread of distances below the surface at
which resonant scattering takes place, and a contribution due to
variations in the number of pores which would be encountered after
scattering if the _total_ distance through pores and solid were the
same for all charged particles. The above result only applies if
each charged particle does not encounter the same pores before and
after scattering. This means that for a given angle of observation
there is a maximum pore size which can be conveniently measured.
In practice this condition does not appear to be very restrictive,
since in many cases it is only necessary to increase the distance
penetrated before resonance, i.e. increase the incident energy, to
make the effect negligible.

As for the transmission case the contribution to the width of
the resonance due to pores will be

$$\Delta E = 2.34(2N)^{\frac{1}{2}}a \frac{dE}{dR}$$

except that N is replaced by 2N.

Again $Na = t_o - t_e$

where t refers to the distance penetrated before resonant scattering.

It is convenient at this point to introduce the porosity P where

$$P = \frac{t_o - t_e}{t_o}$$

$$a = \frac{(1-P)}{10.96P \; t_e} \; \left(\Delta E / \frac{dE}{dR}\right)^2$$

The thickness of solid material traversed after scattering is
approximately equal to

$$t_e = (E_r - E_d)/ \left(\frac{dE}{dR}\right)_1$$

where E_r and E_d **are the** energies at resonance and at departure
from the specimen respectively. The energy loss $\left(\frac{dE}{dR}\right)_1$ is evaluated
at the mean energy $(E_r + E_d)/2$.

Finally

$$a = \frac{(1 - P)}{10.96P \; (E_r - E_d)} \; \left(\frac{dE}{dR}\right)_1 \; \left(\Delta E / \frac{dE}{dR}\right)^2$$

As before dE/dR and $(dE/dR)_1$ can be obtained from range energy
relations. It should be noted that dE/dR is to be evaluated for
the energy at which the charged particle emerges from the sample.

Here again ΔE refers to the FWHM of the porous specimen after the
FWHM of the equivalent non-porous specimen has been subtracted in
quadrature.

The method enables pore sizes in the range 1 nm to 1000 nm to
be measured, for which alpha-particle and proton scattering respectively cover the lower and higher parts of the range. The lower limit
of 1 nm is set by the need for a discernable enhancement of the
resonance width above that due to straggling, and on an adequate
depth of penetration prior to resonant backscattering so that the

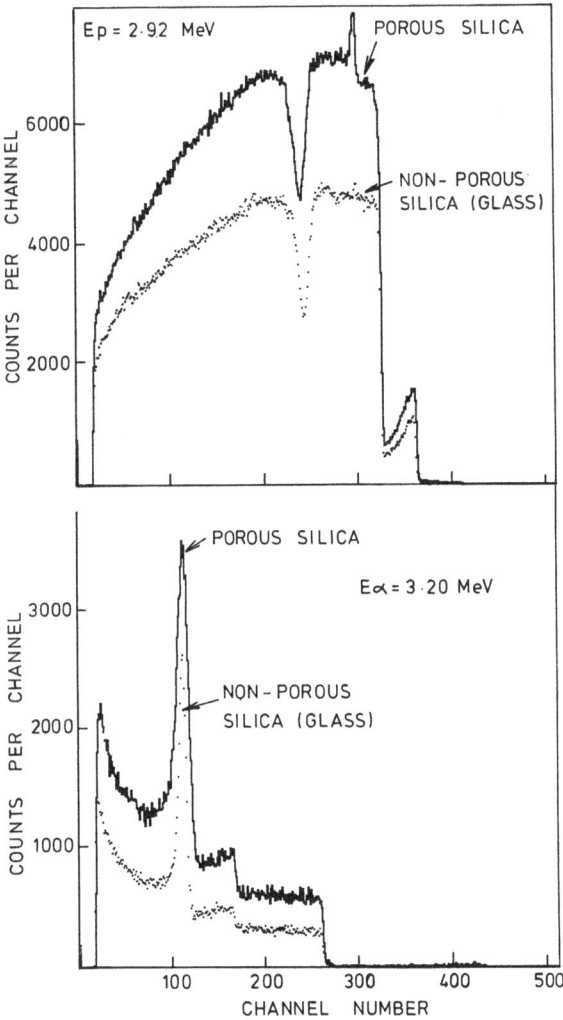

Fig. 3 Proton backscattering spectra (upper) and alpha-
        backscattering spectra (lower) obtained both with
        porous silica and non-porous silica specimens

resonance is superimposed on a flat backscattering background. Again,
as in the proton transmission case the calculational method is very
approximate, but it is supposed that a useful parameterization of
pore size is obtainable from this approach.

   It should be noted that the pore size range 1 nm to 100 μm is
accessible with a combination of the backscattering and transmission
methods. An advantage of the backscattering method is that the
specimen does not have to be prepared in thin uniform slices.

4. BACKSCATTERING MEASUREMENTS ON POROUS SILICA

   Pore size measurements have been made on compacted powders of
silica prepared at Harwell by Avery and Ramsay[5]. Briefly this
material was made by electron beam vapourisation of vitreosil in a
chamber containing oxygen at 0.2 Torr. A powder consisting of
spherical particles about 4 nm diameter was collected from the walls
of the chamber. This powder was then compacted in a hydraulic press
at a pressure of 150 M Pa $(10$ tons in$^{-2})$, and charged particle beam
measurements were made on a sample of area about 10 mm^2 and with a
thickness of about 0.1 mm.

   Pore size measurements were made by proton-backscattering and
also alpha-backscattering. The proton-backscattering measurements
were made at 2.82 MeV and 2.92 MeV. The spectra obtained at the
higher energy is shown in fig. 3 together with a measurement made on

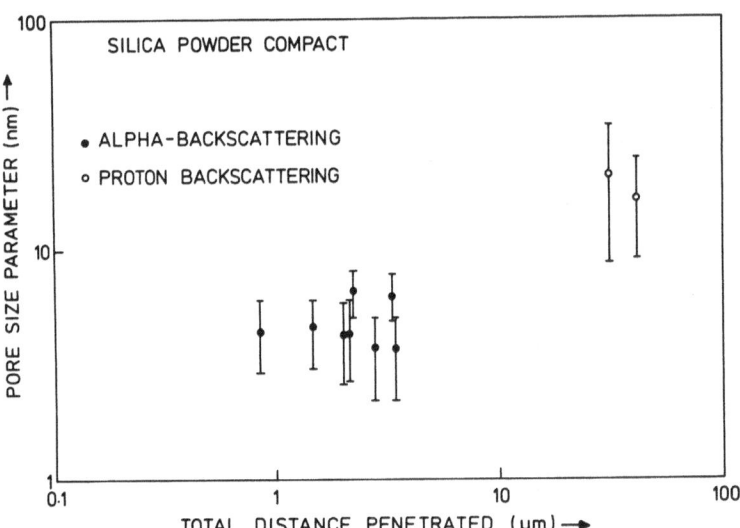

Fig. 4 Pore size parameter obtained as a function of depth of
        penetration by proton and alpha-backscattering on
        porous silica in the form of a silica powder compact.
        The porosity or void fraction is 0.51.

a non-porous silica glass specimen. The figure also shows alpha-
backscattering measurements on the same specimens. It should be
noted that whereas the 3.05 MeV resonance in alpha-backscattering
appears as a peak in the spectrum, the 2.66 MeV resonance in the
proton backscattering case appears as a dip. In fig. 4 the pore
size parameter is plotted as a function of distance penetrated by
the charged particle beam prior to resonant backscattering. This
means that the individual values for the pore size parameter are

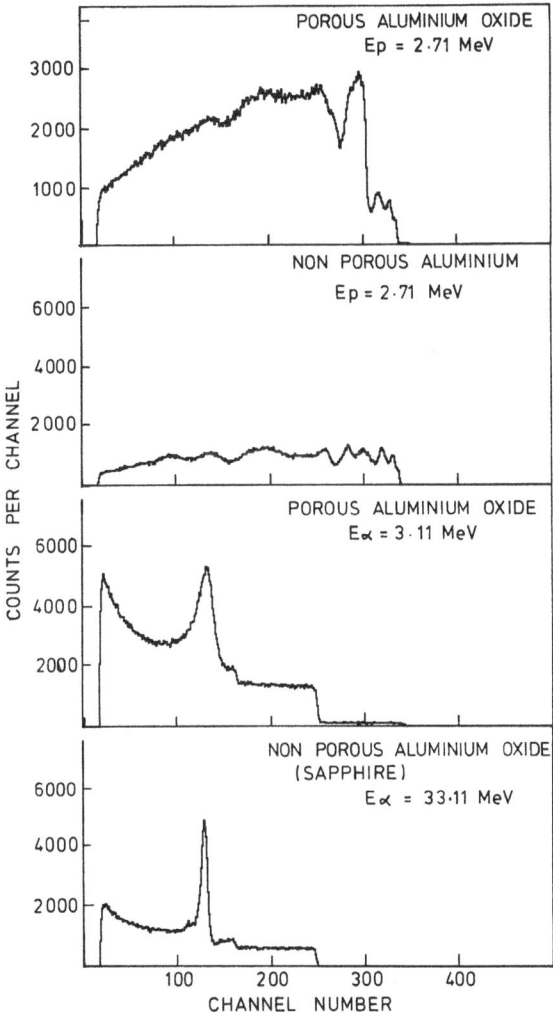

Fig. 5 Proton backscattering spectra (upper) and alpha-
backscattering spectra (lower) obtained with porous alumina and
non-porous alumina specimens. The resonant nature of proton
backscattering spectra from aluminium is also shown.

related to the distances travelled by the charged particle beam from the surface to the depth at which resonant backscattering occurs. Owing to the shorter range of alpha-particles it is advantageous to make such measurements within several µm of the surface and to use proton backscattering for measurements in the range 10 to 100 µm from the surface. The error bars in fig. 5 are of statistical origin and make no attempt to include systematic effects.

The average value for the mean pore size parameter of 5 nm agrees well with the value of 4.6 nm obtained by Avery and Ramsay[5] for the mean pore diameter using gas adsorption techniques. The agreement may to a certain extent be fortuitous since very different calculational procedures and assumptions are used in the two measurements.

6.   ADDITIONAL BACKSCATTERING MEASUREMENTS

Apart from silicon compacts a variety of other porous materials including alumina compacts, corroded steel and sintered iron oxide have been examined. In fig. 5 spectra obtained by proton and alpha-backscattering on alumina compacts are shown. Proton backscattering from alumina is complicated by the existence of discrete resonances in aluminium. The problem of analysing the 2.66 MeV oxygen resonance is partially overcome by subtracting from the alumina spectra a spectrum obtained by proton backscattering from aluminium.

In conclusion it should be stated that further work needs to be done to fully establish the viability of the backscattering technique. It is also proposed to extend the work to materials other than those containing oxygen. Thanks are due to R. G. Avery and J. D. F. Ramsay of Harwell and Dr. H. D. Jonker of Philips, Eindhoven for supplying valuable specimens.

REFERENCES

(1)   J. S. Hardman and B. A. Lilley, Contemporary Physics 15 (1974) 517.

(2)   J. A. Cookson, B. H. Armitage and A. T. G. Ferguson, Non-Destructive Testing 5 (1972) 225.

(3)   L. C. Northcliffe and R. F. Schilling, Nuclear Data Tables A7, 3-4 (1970) 233.

(4)   C. F. Clement, Journal of Physics D, Appl. Phys. 5 (1972) 793.

(5)   R. G. Avery and J. D. F. Ramsay, Journal of Colloid and Interface Science, 42 (1973) 597.

DISCUSSION

Q: (E. Wolicki) Did you say you could measure pores 1 nanometer in diameter?

A: (B.H. Armitage) Yes, we think we should be able to see pores down to about this size.

Q: (J.F. Ziegler) How do you separate out the pores within the target grains (in the powder samples) from the gaps between each of the grains?

A: (B.H. Armitage) The silica compacts are prepared from spheres of silica which are then compacted together under pressure. Hence the spaces between the spheres represent the pores which are being measured. The spheres are assumed to be themselves non-porous.

MEASUREMENT OF THERMAL DIFFUSION PROFILES OF GOLD ELECTRODES ON AMORPHOUS SEMICONDUCTOR DEVICES BY DECONVOLUTION OF ION BACKSCATTERING SPECTRA

J.P.THOMAS S.MARSAUD M.FALLAVIER

Institut de Physique Nucléaire

Université Cl. Bernard LYON 1 43 Bd du 11 nov.1918
69621 VILLEURBANNE FRANCE

ABSTRACT

Using 3.5MeV α-particles backscattering, thermal diffusion of gold electrode into amorphous semiconductor thin films has been pointed out. A computer program for the deconvolution of the experimental spectrum and a simplified numerical calculation allows depth profiling to be easily performed upon about 1600 Å. Preliminary results are given showing different components in the diffusion process below the glass transition temperature; above this point a compound formation between gold and tellurium is discussed.

INTRODUCTION

The study of the thermal diffusion of metallic electrodes into semiconductor devices is of growing interest in electronics techno-logy. For systems with limited number of elements with sufficiently different masses the non destructive character of backscattering and the resolution achieved in mass and depth makes this technique very powerful. As significant applications one can report the diffusion of gold in silicon(1) silicon oxide(2) and cadmium telluride(3) for which bismuth(3) and copper(4) diffusion has also been investigated. More recently developed and amorphous semiconductors have shown very promising characteristics when used in ultra-fast electronics and optical mass memory devices. The thermal behavior of such devices and especially the evolution of the electrode contact are currently of fundamental interest for the viability assessment and the under-standing of the degradation of electrical performances. As a pre-

293

liminary investigation 3.5 MeV α-particle backscattering has been performed for the simpler chalcogenide system As-Te (0.365% . 635% by
weight) and for thin films (2000 to 10 000 Å) covered with gold electrode (∿300 Å). The diffusion profile determination is discussed for
the most general case where the continuous variation of the medium
nature has to be accounted. A simplified procedure is described, for
our particular study, allowing, from numerical graphs to obtain
$C_{Au} = f(x)$. Although no interpretation of the complex diffusion
process is done at this early stage, experimental features can be
noticed. Among them a deep diffusion occurs along the total thickness
range at moderate heating times and temperatures and above the glass
transition temperature (Tg∿129°C) a compound formation is observed,
tentatively attributed to Au Te$_2$.

EXPERIMENTAL

The α - particles beam is delivered by the 4 MeV VAN DE GRAAFF
accelerator of the Institut de physique nucléaire de LYON. Inside
the target chamber a 10^{-6} torr pressure is maintained by a 600ℓ/s
turbomolecular pump, several liquid nitrogen traps placed inside the
chamber and on the beam line insuring a clean vacuum. The 25 mm^2 and
100 μm thick silicon barrier surface detector has a total experimental
resolution of 20 KeV for α - particles. From conventional Tennelec
electronics, the signals are transmitted to a A.D. converter, input
of a 2116-C Hewlett-Packard computer with which numerical treatments

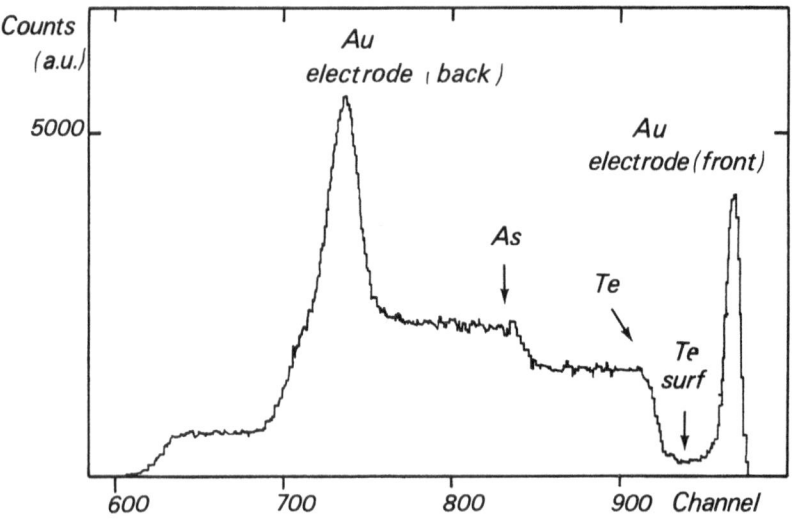

Fig.1. Backscattering spectrum (3.5 MeV , θ = 160°) of the target
       Au/AsTe/Au film.

are performed off-line. The analyses are done from a glass slide
upon which rectangular electrodes are deposited on the back or
upon the AsTe film. The crossing of the back and front electrodes
allows four configurations to be obtained: Backing / AsTe / Au,
B./ AsTe, B./Au / AsTe and B./ Au / AsTe / Au (from which electrical
measurements can be performed). The layers are deposited from H.F.
sputtering in argon atmosphere (5) for As Te and from thermal evaporation for Au. Thicknesses determined by interferometry are typically 10000 Å for AsTe and 300 Å for Au. For such layers, at 3.5 MeV
and θ = 160° the analyzing depth for gold, as represented in
figure 1, is limited to $\sim$ 1700 Å (surface leading edge of the tellurium). The depth resolution for Au is then 180 Å in the Au layer but
390 Å in the AsTe medium (assuming BRAGG's rule). These performances are sufficient in the scope of this study but can be improved
ved using ^{6}Li ions (6) allowing at 2.6 MeV depth resolutions of
respectively $\sim$100 Å in Au and $\sim$230 Å in AsTe. The analyzing depth
is yet limited to 650 Å.

ANALYSIS OF DATA

     Depth profiling requires, as a first step, the determination
of the F(E) function from the energy spectrum A(E) and the instrument function I(E) for :

$$A(E) = F(E) \ast I(E)$$

     If no preliminary conditions are imposed to the shape of F(E)
one has to deal with the old problem of deconvolution mainly
investigated for charged particles spectra by ZIEGLER et al (7).
Severe restrictions are pointed out when the Fourier transforms or
the inversion method are used due to the statistical fluctuations of
the data (8). Better results are obtained from the iterative method
based on the algorithm of BURGER and VAN CITTERT (9) and with several
modifications (10) non physical solutions (like GIBBS phenomena) can
be avoided , but the zero depth discontinuity of the ideal spectrum
is hardly worked out. An intermediate procedure is thus proposed:
- determination of a trial analytical function (continuous) the most
simple shape of which being linear segments, horizontal (stepped),
or oblique.
- use of this function as initial solution for the iterative method.
Efficiency and reliability of the computer program performing these
operations requires visualisation and conversational units to be
connected in order to select the optimum initial solutions. The possibility of numerous tests avoids artefacts production.
     From our particular study the best choice for the initial solu-
tion of the iterative method appears to be the experimental spectrum
itself but after a smoothing treatment described by RITOUT (11) and
based on the SPENCER formula. An example of the results given by
the computer program is shown on figure 2 together with the

reconvoluted spectrum from which the quality of the treatment can be
appreciated. The "pure" spectrum values correspond to channel numbers
i.e. to   given backscattered energy widths.A step concentration profile of about 60 Å (see later on) is so allowed but this value cannot
be considered as the ultimate precision of the method limited in any
case by the depth resolution (8). The **straggling** effect is neglected
for the depth range investigated, the instrument function being experimentally obtained from backscattering on a very thin gold layer
($\sim$50 Å).

   The second step of profile **obtainment requires** the use of backscattered energy - emission depth relationships. For the most general case in a complex matrix upon a ΔE channel width corresponding
to a $\Delta_2 X$ thickness at the depth x_2 (energy E_2), the relation between
number of counts N_i and concentration C_i of the element of interest
is given by :

$$N_i = \frac{\Delta E}{[S]_{E_2}} \times \frac{C_i}{\sum\limits_{y} C_y M_y} \times \mathcal{N} \times \frac{(dE/dx)K^2 E_2}{(dE/dx)E_f} \times \sigma_{E_2} \times I$$

with $[S]_{E_2}$ backscattering energy loss parameter at the energy E_2
where the RUTHERFORD cross section is σ_{E_2}
C_y and M_y atomic concentration and atomic mass of the y element in
the matrix.

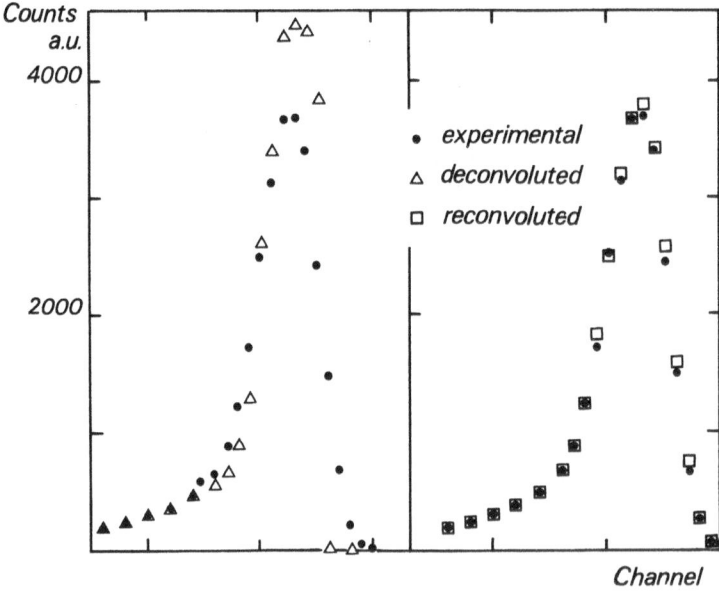

Fig.2. Computer program results for the gold peak (experimental-
       deconvoluted - reconvoluted).

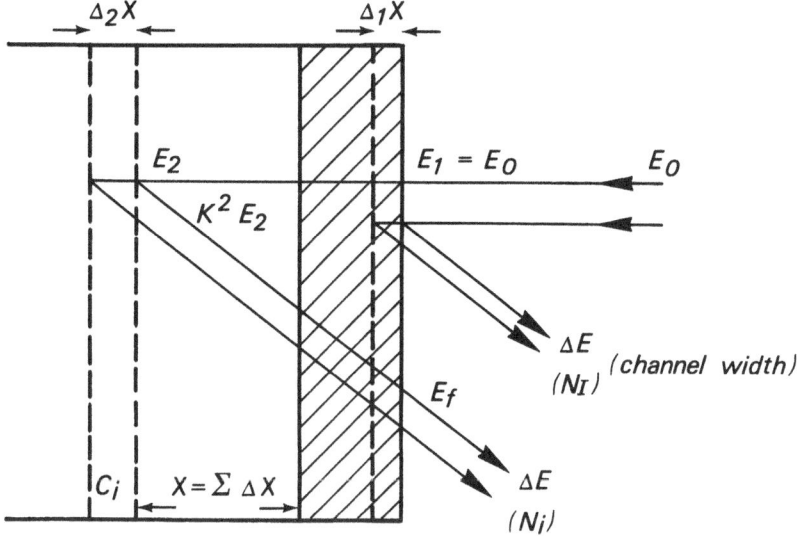

Fig.3 Principles of backscattering analysis using the front layer
as internal standard.

$\mathcal{N}$ Avogadro's number
(dE/dx) stopping power at the energy K^2E_2 or E_f(see fig.3)
I detection parameter (beam flux, counting time, geometry...)

The same relation applies if the matrix is the element of interest itself (atomic mass M),giving N_I for the same E corresponding
then to a Δ_1x thickness at a x_1 depth and E_1 energy. As illustrated
in fig. 3 for our study the surface layer can be considered as
an internal standard ($E_1 = E_0$), the concentration below this layer
being obtained from :

$$\frac{N_i}{N_I} = \frac{[S]_{E_2}}{[S]_{E_2}} E_0 \times \frac{C_i M}{\sum_y C_y M_y} \times \frac{E_0^2}{E_2^2} \times \frac{(dE/dx)_{K^2E_2}}{(dE/dx)_{E_f}}^*$$

* this term, taking account of the variation of the energy width
along the backscattered particle range, applies for the simple case of
a homogeneous target. One has generally to consider a succession of
different layers for the numerical application.

Such an expression requires [S], C_y and E_2 to be known in order
to obtain the different C_i. As these terms are dependent on the
concentration itself, only an iterative procedure would give the solutions. Like in any specific case approximations can be used in our
study. As a test to estimate the resulting precision, the $\frac{N_i}{N_I}$ ratio
has been numerically determined for given Au concentra-
tions at different depths of a given medium (Au or AsTe) (in order

Composition of the thickness of interest	(1) Ni / N_I (2) Δx g/cm^2			
		surface	>2000 Å AsTe	>2000 Å Au
As$_{0.45}$Te$_{0.45}$Au$_{0.1}$	(1) 0.122 (2) 4.26		0.124 4.24	0.128 4.21
As$_{0.35}$Te$_{0.35}$Au$_{0.3}$	(1) 0.349 (2) 4.7 7		0.356 4.75	0,371 4.72
As$_{0.25}$Te$_{0.25}$Au$_{0.5}$	(1) 0.556 (2) 5.24		0.568 5.22	0.584 5.17

Table 1

to maximize its influence). According to the channel width, corresponding Δx thicknesses are also calculated. From the set of values given in ·table 1   and taken from these calculated curves, it must be noticed :

- $\frac{Ni}{N_I}$ and Δx can be considered as independent on the depth from which they are obtained in the limits of the Au concentration range observed (< 5 % deviation)

- The variation of these parameters versus C_{Au} allows to determine it (upon a Δx interval) for each depth x, x resulting from the summation of the Δx successively obtained.

A mean density determination from the bulk material constituants allows to express more classically the depth in Å but it must be kept in mind that systematic deviations may occur from this conversion.

RESULTS

Heating times ranging from one to ten hours and temperatures between 60 and 120°C have been selected as operating conditions for semiconductor devices. The diffusion profiles obtained under these conditions are represented in fig. 4   for 1 and 10 h. It can be noticed that at low temperature (60°C) no marked differences appear in this time range. For higher temperatures the same phenomenon occurs up to about 500 Å, then, between 80 and 100°C no differences are observed after 5 h while for 120°C the differences occur after these 5h. As expected, the temperature effect becomes important near the glass transition temperature (Tg ∿ 130°C).

At this preliminary stage of the diffusion process study it appears difficult to interpret the phenomenon in terms of FICK's laws. Nevertheless, the flatness of the curves observed after about 1000 Å could be attributed to a deep diffusion.

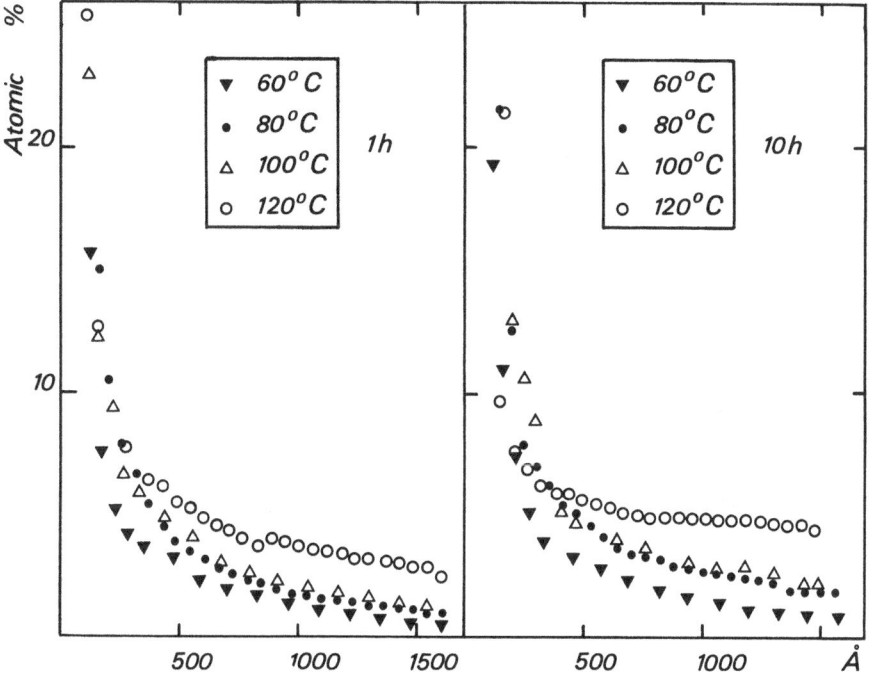

Fig.4. Diffusion profiles of Au into AsTe for 1 and 10h heating time.

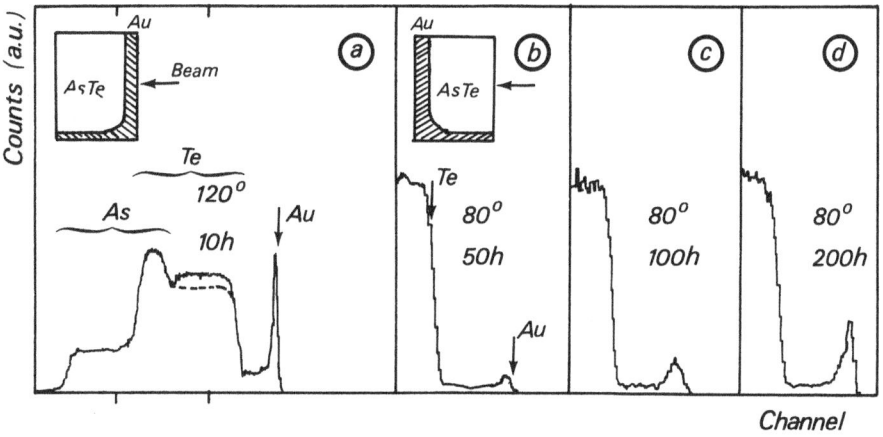

Fig.5. Illustration of the deep diffusion a/ from the front
electrode b/ c/ d/ from the back electrode.

This feature is illustrated in fig. 5 a   the hump appearing
on the Te plateau resulting from the total gold diffusion of the
front electrode (to be compared with fig. 1). The hypothesis of
a gold agglomeration into islands on the semiconductor layer which
can be proposed in such a case is hardly substantiated by   the
results obtained from thinner films : the maximum diffusion depth
depends on the layer thickness. Besides, this temperature is well
below the fusion temperature of the gold (comparing with the lead
agglomeration shown by CAMPISANO et al (12)on a silicon substrate).
At last,the diffusion from the back electrode alone,. showed in fig.5
b c d, rules out this possibility, no superficial diffusion along
the edges of the film having been detected. For fairly high heating
times a strong gold accumulation at the surface is also appearing.

For temperatures higher than Tg a compound formation seems to
take place between Au and Te, as shown in fig. 6, from 5 to 200 h
heating times at 150°C. After 200 h due to the quasi-total evaporation of As, a simple calculation from the plateau heights of Au and
Te leads to $C_{Au}/C_{Te} \simeq 0.52$.   This value suggests a mean stoichiometry
$AuTe_2$ in agreement with the previously published data (13). Due to
the limitation of the backscattering informations (depth resolution,
and structure informations) this interpretation must be substantiated
by other techniques like S.E.M (in connection with the already mentioned problems of gold agglomeration) and ESCA, despite the difficulties appearing in the chemical shifts determination for such
heavy elements.

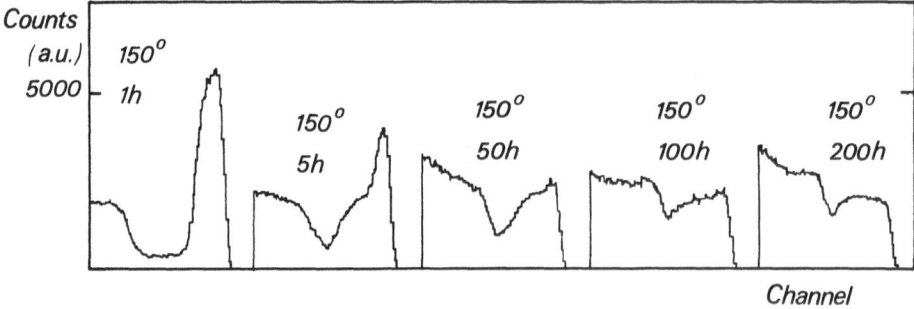

Fig.6. Evolution of the Au-Te interface at 150°C for increasing
       heating times.

CONCLUSION

Using 3.5 MeV α - particles backscattering, thermal diffusion of gold electrode into AsTe chalcogenide glass films has been pointed out far below the glass-transition temperature. Concerning the technique, numerical treatment of the spectra allows the diffusion profile to be obtained the uncertainty of the calculations being on the order of 10 % (neglecting straggling effect). From recent experiments two improvements are proposed : better depth resolution ($\sim$ x 0.7) with 2.6 MeV ^{6}Li ions, better analyzing depth ($\sim$ x 2) with 7 MeV $^\alpha$ - particles (from 3.5 MeV α^{++})(14). From the experimental features pointed out in this study, as well as for other chalcogenide systems (quaternary compounds), different phenomena have to be considered for the diffusion process interpretation. A comparison between electrical measurements (resistivity - activation energy) and diffusion profiles, apart from the evidence of the correlation of diffusion and degradation rate, cannot explain the evolution of electrical parameters. With respect to these measurements no classical diffusion mechanism can be proposed (15).

ACKNOWLEDGEMENTS

This work was supported by the CNRS (A.T.P. n° AA33). Thanks are due to J.M. MACKOWSKI and his staff for targets preparing and to J.A. ENGERRAN for the relialibity of his programs.

REFERENCES

(1) A.HIRAKI M.A.NICOLET J.W.MAYER Appl. Phys. Lett. 18.5(1971)178

(2) C.J.MADAMS D.V.MORGAN M.J.HOWES J. of Appl.Phys. 45,11(1974)5088

(3) M. HAGE-ALI I.V.MITCHELL J.J.GROB P.SIFFERT ·Thin solid films 19 (1973) 409

(4) H.MANN G.LINKER O.MEYER Solid state com. 11 (1972) 475

(5) J.M.MACKOWSKI P.KUMURDJIAN Proceedings of 5th int.conf. on amorphous and liquid semiconductors GARMISCH 1974 (p.692)

(6) J.TOUSSET J.P.THOMAS A.CACHARD International nuclear and atomic activation analysis conference october 14-16 1975 GATLINBURG TENNESSEE U.S.A.

(7) J.F.ZIEGLER J.E.E.BAGLIN J. Appl. Phys. 42 (1971) 2031

(8) A.F. JONES D.L.MISELL J. of Phys. A-3 (1970) 462

(9) H.C.BURGER Ph. VAN CITTERT Z. Phys. 79 (1932) 722

(10) J.A.ENGERRAN Thesis LYON (1975)

(11) M.RITOUT Thesis TOULOUSE (1963)

12) S.U.CAMPISANO G.FOTI F.GRASSO E.RIMINI Thin solid films 25
     (1975) 431

(13) ANDON R.J.L. MARTIN J.F. MILLS K.C. J. of Chem. Soc.
     (1971) 1788.

(14) S.MARSAUD et al. to be published

(15) J.M. MACKOWSKI et al.to appear in the 6th int. conf. on amor-
     phous and liquid semiconductors - Leningrad 1975.

DISCUSSION

Q: (J.A. Borders) How can you be sure that there is an Au-compound
formed when you have made no microstructure sensitive measurements.
There may for example be microscopic lateral variations throughout
the film. There is no evidence here that the Te and Au atoms have
chemically combined to form a compound.

A: (J.P. Thomas) Of course there is no evidence of the chemical
bond between Au and Te from backscattering. The spectrum shows only
a saturation level for gold and the stoichiometry obtained from the
plateau heights agrees with a $AuTe_2$ formation. Such a compound re-
puted in literature has to be confirmed in the lab with ESCA experiments.

Q: (S.U. Campisano) Can you explain the presence and the growth of
the Au revealed to the sample surface ?

A: (J.P. Thomas) At this time and for the chalcogenide systems
very little is known about the diffusion mechanism. Electrical measurements like degradation rate cannot be accounted to clarify the
problem. The deep diffusion, as well as the gold accumulation at the
surface is currently under investigation.

ENHANCED SENSITIVITY OF OXYGEN DETECTION BY THE

3.05 MeV (α,α) ELASTIC SCATTERING

G. Mezey, J. Gyulai, T. Nagy, E. Kotai,
A. Manuaba

Central Research Institute for Physics
1525. Bpest, P.O.B. 49, Hungary

A simple theoretical background for the use of the ^{16}O (α,α) ^{16}O elastic scattering was given and the technique was demonstrated through experimental examples $Si-SiO_2$ and quartz-CoGd layers. The method has advantages over other techniques in the analysis of layer structures with thin (native) oxide coverage, as it gives composition of the layer (backscattering) and the amount of the oxygen.

Introduction

Analysis of oxygen content and oxidation properties are important topics of solid state physics and chemistry. Nuclear techniques are useful candidates to lower detection limits of this element.

For medium energy projectiles four methods were proposed till now. The nuclear reaction techniques,(d,p) and (p,α)[1-2], the conventional backscattering [3-4], the resonant proton scattering at 4 MeV (p,p) reaction [5-6] and the (α,α) compound reaction at 3.05 MeV. Based on Cameron's work [7], this latter method was demonstrated in early measurements of our group (L.Keszthelyi, I.Demeter, G.Mezey, Z.Szőkefalvy-Nagy, L.Varga, 1972)[8] and was also proposed in a review paper by W.K. Chu, J.W. Mayer, M-A. Nicolet, T.M.Buck, G.Amsel and F.Eisen [9].

In this work, we'll present some applications of this resonance showing examples, were the technique has definite advantages.

Method and experimental

The detailed study of the $^{16}O(\alpha,\alpha)^{16}$ reaction in gaseous target, from the point of view of nuclear physics, was published in 1953 [7]. The resonance is centered around E_R = 3.048 MeV, with a cross-section maximum of σ_R = 0.95 barn, a factor 25 over an extrapolated regular Rutherford cross section. The half-width of resonance Γ=30 keV was found. Sharp dependence on detection angle was established, the reaction takes place only for backscattering geometries.

The cross section, unfortunately, varies even for a moderately thin target (comparable with the half-width, i.e. approximately 1200 Å SiO_2). This variation, however, obeys the Breit-Wigner dispersion relation, except some tail regions. Therefore, a simple analytical calculation helps in calibration of scale factors (see App.). For the thin target approximation, the formula reduces to the normal backscattering relation with an increased cross-section (σ_R).

Experiments were made on the 5 MeV Van de Graaff of our laboratory using a goniometer and standard backscattering detector and electronics (Ortec). The energy stability of the generator, an important factor in these investigations, was better than 250 eV. The vacuum at the the goniometer was $\leq 10^{-6}$ torr.

Experimental results

a) SiO$_2$ studies. Dry thermal oxides on silicon were used to measure the resonance parameters. Fig.1 shows the excitation function for different projectile energies and oxide thicknesses. The measured points represent integrated number of counts within the resonance peak (<111> aligned backgrounds subtracted). The position of maximum yield shifted with increasing oxide thickness according to

$$E = E_R + \frac{\beta}{2} = E_R + nt\varepsilon/2 , \qquad\qquad (A7)$$

where β is the thickness of the oxide in eV-s, n the atomic density of the oxygen (cm^{-3}), t the target thickness (cm) and ε the stopping cross section of the incident beam at E_R in eV cm^2/atom$\cdot 10^{15}$. In the region of the thin target approximation, the maximum values

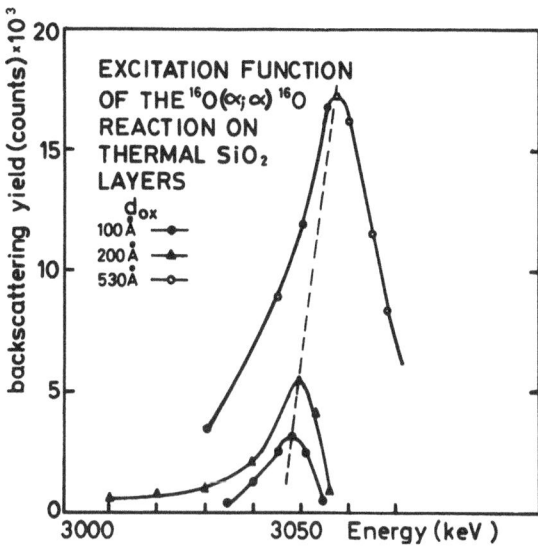

Fig.1. Excitation function vs. energy for different
        oxide thicknesses

increase linearly with target thickness (Eq. A 10).

     For thicker SiO_2 layers, the relations are somewhat
complicated and precautions are necessary to evaluate
the results. Fig.2 shows a limiting case, where the
Breit-Wigner approximation still can be used. The curve
on Fig.3, where $\Gamma \approx 3\beta$, looses its symmetrical shape and
quantitative evaluation becomes complicated. For these
thicknesses other methods, say, simple backscattering
experiment yields results much easier.

     As a consequence, the straightforward approximation
of the method is the analysis of thin (e.g. native)
oxide layers. As an example, $4.3 \pm 0.2 \times 10^{15} cm^{-2}$ oxygen
atoms were measured on a <111> oriented silicon surface.

     On Fig.4 we present an enhanced oxidation measure-
ment on Sb-implanted silicon, as part of the results of
the Caltech-Central Res.Inst.Phys. co-operation. For
the conditions given on the figure, the antimony increases the oxide growth rate by a factor 7.5.

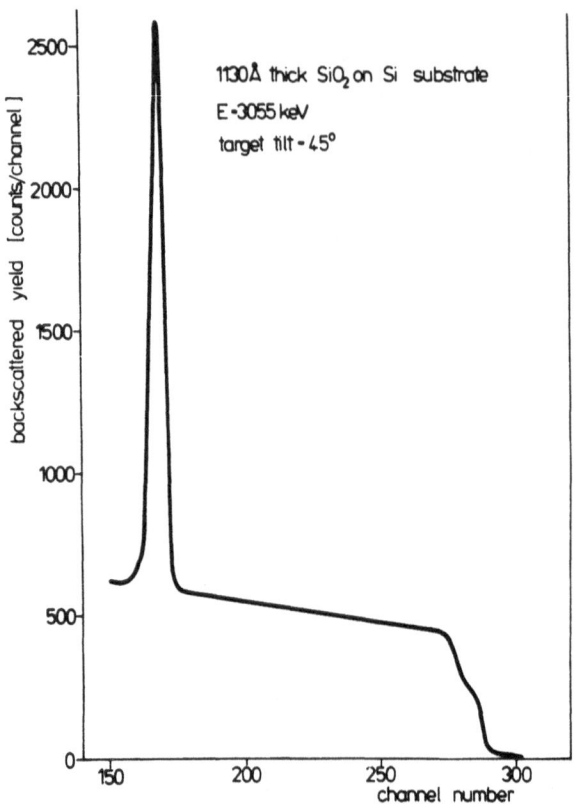

Fig.2. Backscattering spectrum for a silicon-silicon
dioxide structure (Breit-Wigner approximation
still valid)

b)Oxygen on thin metal films. There are some important
metal sandwiches, where this technique yields a very
complex analysis of the structure. The spectra give both
the enhanced sensitivity for adsorbed or bound oxygen
and the composition analysis achievable by normal backscattering. Figs. 5-7 show spectra on preparation of
amorphous magnetic layers. In these Co-Gd layers the
oxygen severely influences the magnetic properties. On
comparison of Figs. 5 and 6, the difference in oxygen
adsorption of cobalt and gadolinum is clearly visible.
The substrate was quartz in all cases.

    Fig.7 illustrates the power of the method, where
both the amount of adsorbed oxygen and composition of
the sandwich were analyzed.

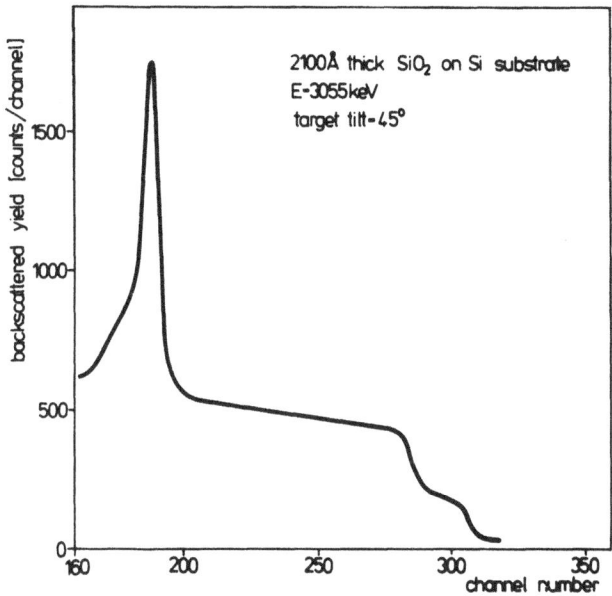

Fig.3. Backscattering spectrum for a thick silicon-
       -silicon dioxide structure (Breit-Wigner approxi-
       mation is not valid)

Discussion

     On comparison of the method with other oxygen detec-
tion procedures, there are cases, where the 3.05 MeV
reaction turns out most favorable.

     Background sensitivity is similar to that of normal
backscattering or (p,p) reaction. An alignment of the
substrate helps. For the (p,p) reaction the resonances
from background may also disturb. Detection limits are
not extremely low, an order of magnitude worse than
Auger spectroscopy, but the depth scale may compensate
for this inconvenience. Nuclear reactions are also
somewhat more sensitive than the (α,α) reaction, but,
for natural mixture of isotopes, the (α,α) case for thin
films is more favorable because of better energy resolution and depth scale.

     The method is capable to give normal backscattering
spectra (with same electronics) together with enhanced
sensitivity of oxygen detection, therefore the analysis
of a film covered, say, by a native oxide is probably

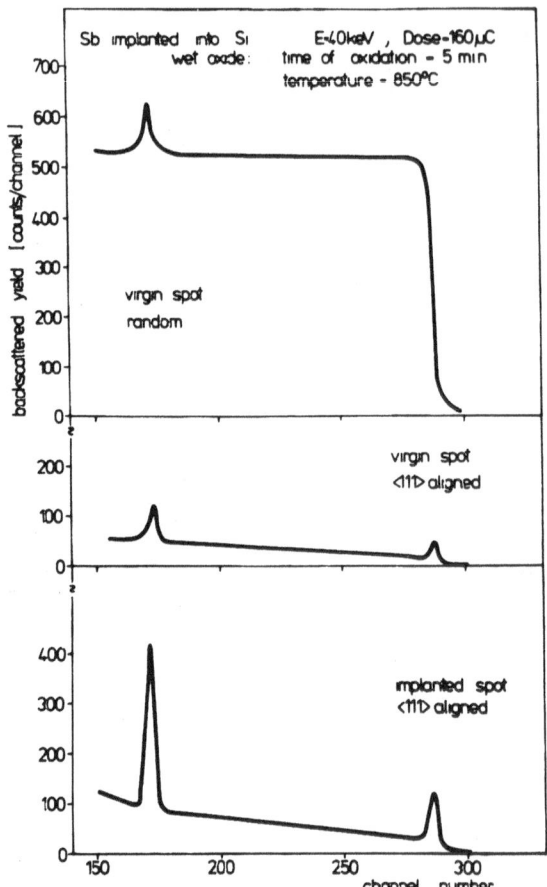

Fig.4. Enhanced oxidation caused by implanted antimony
into silicon

the mostly fitted problem. We are also re-investigating
the enhanced oxidation of implanted layers by this technique. In this case the higher yields shorten measuring
times or increase accuracy.

Acknowledgements

     Thanks are due to Prof. J.W.Mayer (Caltech) for
discussions. Implanted samples are acknowledged to the
group of E. Pasztor (Implantation Program, Centreal Res.
Inst. Phys.) and the magnetic films to Dr.I.Nagy (Magn.
Labs., Central Res. Inst. Phys.).

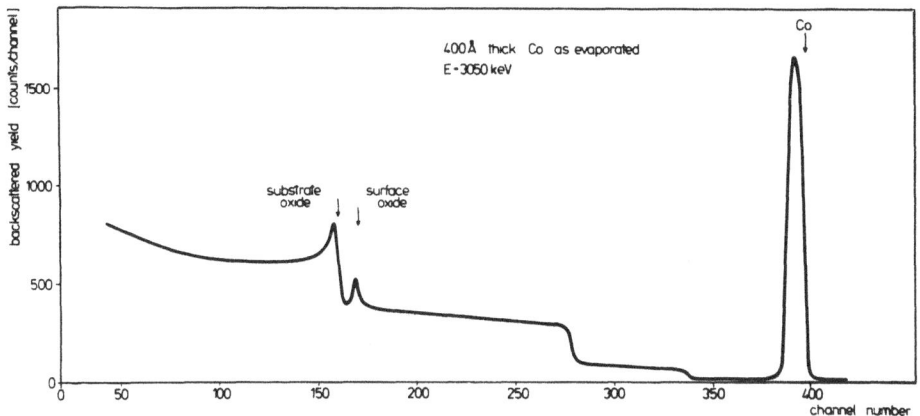

Fig.5. Backscattering spectrum of a cobalt layer on quartz

Appendix

Evaluation of surface oxygen content from a resonance with a half-width Γ

We assume that the cross section $\sigma(E)$ obeys the Breit-Wigner dispersion formula:

$$\sigma = \sigma_R \frac{\Gamma^2/4}{\left(E - E_R\right)^2 + \Gamma^2/4} \quad , \qquad (A1)$$

where σ_R is the cross section at the resonance energy E_R. The area of the oxygen peak can be written as

$$A = \Omega Q \int_{E-\beta}^{E} \left(\frac{\sigma}{\varepsilon}\right) \, dE \quad , \qquad (A2)$$

(Ω solid angle, Q the particle flux). Assuming $\varepsilon =$ const. in the oxide layer and using Eq. A1,

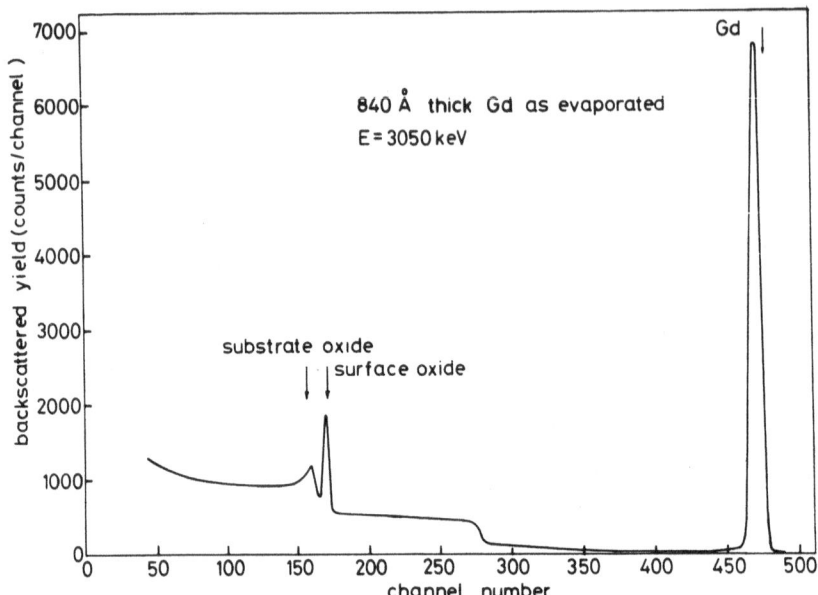

Fig.6. Backscattering spectrum of a gadolinium layer
on quartz

$$A = \Omega Q \; \frac{\sigma_R \Gamma}{2\varepsilon} \left[\tan^{-1} \frac{E-E_R}{\Gamma/2} - \tan^{-1} \frac{E-E_R-\beta}{\Gamma/2} \right]. \quad (A3)$$

At the resonance energy Eq.A3 has a form

$$A = \Omega Q \; \frac{\sigma_R \Gamma}{2\varepsilon} \tan^{-1} \frac{\beta}{\Gamma/2} . \quad (A4)$$

From Eq.A4, one can get the surface density of oxygen
atoms:

$$(nt) = \frac{\Gamma}{2\varepsilon} \tan (\Omega Q)^{-1} \frac{2\varepsilon A}{\sigma_R \Gamma} \quad (A5)$$

To use Eq.A5, the accelerator must be calibrated
in energy, say, by using a thin oxide layer. This procedure can be avoided, if one looks for the maximum
yield for a given target, because the maximum of Eq.A3
is

$$A_0 = (\Omega Q) \frac{\sigma_R \Gamma}{\varepsilon} \tan^{-1} \frac{\beta}{\Gamma} , \quad (A6)$$

at the energy $E = E_R + \beta/2 ,$ (A7)

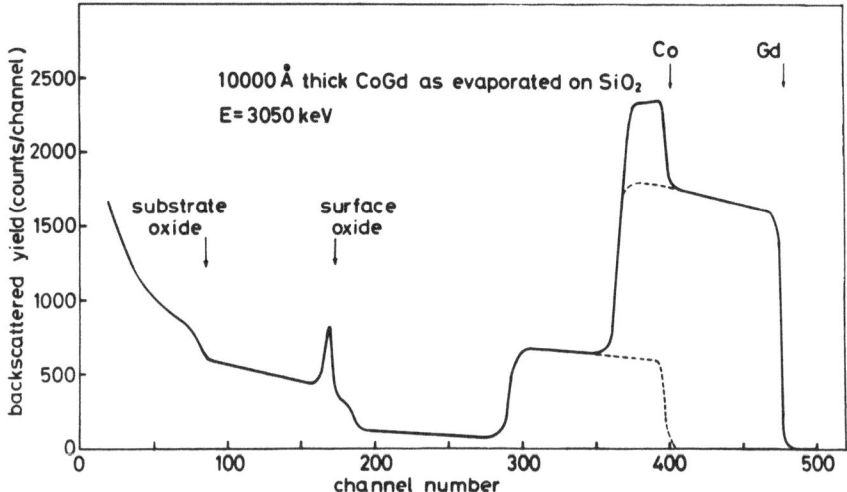

Fig.7. Backscattering spectrum of a simultaneously
       evaporated cobalt-gadolinium layer

Eliminating ΩQ using the height of the silicon
edge, H_{Si}, introducing normals backscattering parameters,
the resolution δE_1 (keV/channel) and the backscattering
energy loss parameter $[\epsilon]_{Si}^{Si}$ and cross section σ_{Si},

$$Q = \frac{H_{Si} \; \delta E_1}{\left[\epsilon\right]_{Si}^{Si} \sigma_{Si}} \qquad (A8)$$

the density per unit area is

$$(nt)_o = \frac{\Gamma}{2\epsilon} \tan \frac{2\epsilon A_o}{\sigma_R \Gamma} \cdot \frac{\sigma_{Si} \delta E_1}{H_{Si}\left[\epsilon\right]_{Si}^{Si}} \qquad (A9)$$

For thin targets, where

$$(nt)_o = A_o \frac{\sigma_{Si}}{\sigma_R H_{Si}\left[\epsilon\right]_{Si}^{Si}} \frac{\delta E_1}{} \qquad (A10)$$

as in the case of constant cross section[4].

REFERENCES

1. G. Amsel, J.P. Nadai, E.D Artemare, D. David,
   E. Girard and J. Moulin: Nucl. Instr. Meth. 92,
   481 /1971/.

2. S.T. Picraux, G. Amsel, L.C. Feldman, Catania Working Data, Selected Low Energy Nuclear Reaction Data,
   to be published.

3. O. Meyer, J. Gyulai and J.W. Mayer: Surf. Science
   22, 263 /1970/.

4. W.K. Chu, E. Lugujjo, J.W. Mayer and T.W. Sigmon:
   Thin Solid Films 19, 329 /1973/.

5. G. Dearnaley, P.D. Goode, W.S. Miller and J.F.
   Turner: Ion Implantation in Semiconductors and
   Other Materials E.d. B.L. Crowder, Plenum Press,
   New York, 1973, p. 405

6. G. Dearnaley, J.M. Freeman, R.S. Nelson and
   J. Stephen: Ion Implantation, North-Holland,
   Amsterdam-London-New York, 1973, p. 734.

7. J. Cameron: Phys. Rev. 90, 839 /1953/.

8. L. Keszthelyi, I. Demeter, G. Mezey, Z. Szőkefalvi-
   -Nagy, L. Varga: Proc. Int. Meeting on Ion Implantation in Semiconductors, Rossendorf, GDR, ZKF-236,
   p. 111, 1972.

9. W.K. Chu, J.W. Mayer, M-A. Nicolet, G. Amsel,
   T. Buck, F. Eisen: Thin Solid Films 17, 1 /1973/.

DISCUSSION

Q: (J.F. Ziegler) What is the width of the resonance? How do you center it over your oxide layer?

A: (J. Gyulai) The width of the resonance was found considerably smaller than the Cameron value, 17-20 keV compared with 30 keV. We can only speculate on the reasons. The peak is centered by an energy scan done by changing the energy of the incoming particles. Our Van-de-Graaff has an energy stability of 250 eV at 3 MeV.

PROGRESS REPORT ON THE BACKSCATTERING STANDARDS PROJECT

J.E.E. Baglin

IBM Thomas J. Watson Research Center

Yorktown Heights, N.Y.

ABSTRACT

An international group of laboratories have collaborated in a standards project aimed at examining the principal (and sometimes elusive) factors which can influence the absolute quantitative interpretation of backscattering analysis spectra. Identically prepared samples of Bi implanted[+] in Si, and others of 2000 Å Pt on oxidized silicon substrates were distributed to the participants together with a detailed prescription for comparative runs which all labs would then follow in nominally identical fashion.

Spectra were taken for the Bi implants, the Pt films and thin SiO_2/Si substrates scattering $^1H^+$ and $^4He^+$ ions at 1.9 and 1.0 MeV and using Si detectors at 165° to the incident beam.

The intercomparison of results obtained so far will be made, with special reference to energy straggling effects, absolute mass (atoms/cm^2) and energy calibrations, and system linearity and resolution.

We will also discuss "reference" spectra from the same Pt films, obtained with the high resolution magnetic spectrometer at Bell Laboratories, with a system resolution of 650 eV.

[+] G. Dearnaley and H. Freeman, AERE, Harwell

Applications of Backscattering and Combined Techniques

ION BEAM STUDIES OF THIN FILMS AND INTERFACIAL REACTIONS

J. M. Poate

Bell Laboratories

Murray Hill, New Jersey 07974

Ion beam techniques, principally Rutherford backscattering, are now contributing significantly to the understanding of thin film and interfacial phenomena. We review three such areas: 1) Superconducting Thin Films, 2) Metal-Metal Interdiffusion and 3) Metal-Semiconductor Reactions. It is emphasized that complementary analytical techniques are necessary for detailed understanding of the phenomena.

INTRODUCTION

There is no need to emphasize the important role that thin films play in modern technology. They appear as vital elements in integrated circuits as well as superconducting elements and solar energy devices. It has become apparent in the past several years that ion beam techniques offer unique capabilities for thin film analysis. This is because the thickness of most thin films lies in the range 0.1 to 1 μm and this is a difficult region as far as conventional analytical techniques are concerned.

A bewildering variety of phenomena can be listed under the general heading of thin films and we will concentrate in this review upon the areas where ion beams are contributing significantly to the understanding of thin films. There are two general classifications of studies: 1) the study of as-deposited films for the determination of such parameters as stoichiometry or impurity content and 2) interdiffusion and reaction between films after deposition. Ion beam techniques have been discussed in several recent reviews and compendia[1,2,3] and will not be discussed in any detail here. It is worthwhile to emphasize however that no one

technique gives the comprehensive information necessary for complete understanding of the thin film phenomena. In particular it
is often essential to combine diffraction measurements (X-ray or
electron) with the ion beam studies.

We will first list those thin film areas where ion beams have
been successfully employed and then select three distinct fields
for detailed discussion where particular progress has been made
over the past two years or so. In the general listing, references
to representative articles will be given.

Ion Beam Studies of Thin Films

I) Insulating Films. Composition and impurity content of
such films as Al_2O_3,[4] SiO_2,[5] Si_3N_4 [6] and anodic oxide films.[7]
Transport and growth mechanisms of anodic oxide films have been
studied extensively by the Chalk River[8] and École Normale
Superieure groups.[9]

II) Metal-Metal Interdiffusion. Discussed later

III) Metal-Semiconductor Reactions. Discussed later

IV) Metal-Dielectric Reactions. Mainly metal-SiO_2 reactions[10]

V) Solid State Epitaxy. Diffusion and epitaxial growth at
interfaces of Si and Ge through metal films, studied mainly by
Caltech group[11]

VI) Single Crystal Films and Heteroepitaxy. Quality and
composition of epitaxial films[12]

VII) Superconducting Films. Discussed later

VIII) Impurities at Interfaces or Surface Layers. Observation
of controlled or non-controlled impurity concentrations.[13]

SUPERCONDUCTING THIN FILMS

A parameter of prime importance in superconductivity is the
superconducting transition temperature, T_c. It would appear
desirable in thin film studies to be able to correlate T_c with other
variables such as the stoichiometry of the films and impurity content. Thin films can   easily be prepared with ranges of composition, impurity content and even grain size. But before the
application of ion beam techniques to these films there was not
much quantitative progress in correlating T_c or other superconducting properties with the actual material of the films. Perhaps the

biggest effort to date has gone into the Rutherford backscattering analysis of ion implanted films.[14],[15],[16] However Meyer et al.[17] have measured the T_c dependence on the N/Nb ratio in NbN thin films.

We have investigated in detail the superconducting properties of Nb_3Ge thin films with a variety of techniques. Gavaler[18] in 1973 obtained T_c onsets of 22.3K in films of Nb-Ge sputtered onto hot substrates in a high Ar pressure. This important result marked the first increase in the maximum known T_c in about four years. Subsequently, Testardi et al.[19] reproduced these results and obtained films with T_c onsets ~23K. The question of how sensitive the T_c of Nb-Ge is to composition near the exact stoichiometric ratio 3/1 has attracted considerable comment. Carpenter and Searcy[20] were the first to synthesize and identify A-15 structure Nb-Ge and Geller[21] was the first to recognize that the compound normally grown to be Nb_3Ge was, in fact, non-stoichiometric. Later, Matthias et al.[22] splat-cooled the approximate Nb_3Ge and obtained a very broad superconducting transition beginning at ~17K. The high T_c was assumed to result from the nearly exact stoichiometric Nb_3Ge obtained by the rapid quench preparation compared to the bulk compound where the Nb/Ge ratio is ~3.3 to 4 and T_c ~ 6K. We[23] (Testardi et al.) have examined sputter deposited Nb-Ge thin films using Rutherford scattering, nuclear reactions and glancing angle X-ray diffraction techniques.

Figure 1 shows two superimposed backscattering spectra for 1.9 MeV ^{4}He from films with radically different T_c's of 21.4 and

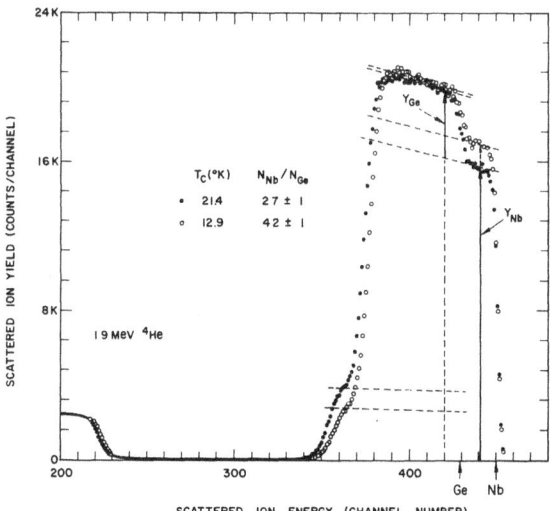

Fig. 1. Backscattering Spectra from Nb-Ge superconducting thin films (after Testardi, et al.[23]).

12.9K; the films were prepared using different targets in the
getter sputtering apparatus. It is evident that, to first approximation, the films have uniform composition with depth but that the
films have markedly different compositional ratios, Nb/Ge of
2.7 ± .1 and 4.2 ± .1 respectively. The composition ratios are
given from the ratio of Y_{Nb}/Y_{Ge} and a knowledge of Rutherford
scattering cross sections and backscattering kinematics (Although
stopping powers enter the calculation, they do so in ratio and
cancel to first approximation).

Before conclusions can be reached regarding stoichiometric
dependencies the structures of the films have to be obtained.
X-ray diffraction data were obtained by use of a wide-film Debye
camera and CuK_α radiation. This camera uses 5 in. × 7 in. film
which allows the detection of diffracting planes having preferred
orientation at a considerable angle to the plane of the substrate;
the samples were held inclined to the X-ray beam at an angle of 18°.

Figure 2 shows the X-ray diffraction pattern from a high T_c
film with a Nb/Ge ratio of approximately 3/1. The film is single
phase A15. The lattice parameters were found to decrease with
decreasing Nb/Ge ratio. For Nb/Ge ratio ~3, a_o ~ 5.13 ± 0.01 Å
which is considerably less than the value ~5.17 Å obtained for
bulk A15 Nb-Ge. However the films prepared with the bulk composition Nb/Ge ~ 4 have, in fact, considerably smaller lattice parameters than that of the bulk.

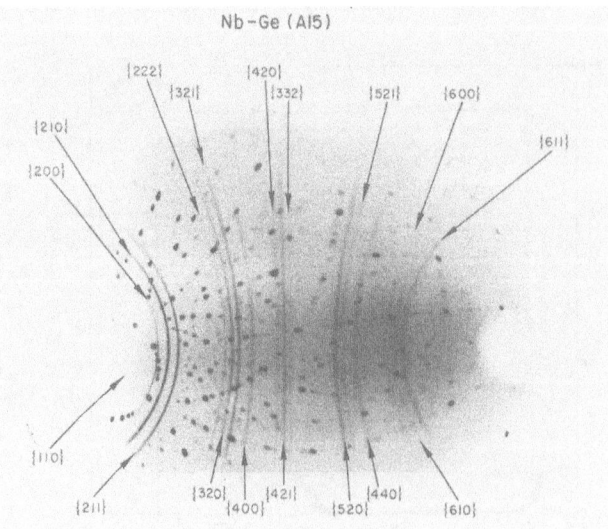

Fig. 2.   Glancing angle X-ray diffraction pattern from A15
Nb$_3$Ge (after Testardi et al.[23]).

The X-ray results show that in the compositional range
Nb/Ge ∼ 2.5 to 5.5 the films yield predominantly single phase A-15
material. The very weak second phase lines observed in these
films have been estimated to correspond to less than five percent
second phase. Thus the Rutherford determination of average Nb/Ge
composition is essentially correct in describing the A-15 phase
composition and Figure 3 shows a plot of T_c against a variety of
film compositions. The films prepared under nonoptimum conditions
were usually obtained at a substrate temperature ∼50°C lower than
optimum. These results show that although stoichiometry is important for high T_c, it is not crucial.

To determine whether impurity content was important in controlling T_c, we deposited films on C substrates thus allowing
observation of light impurities such as C, N, O. Low levels of C
and O impurities were observed, < 5 atomic percent, but no correla-
tions with impurity content and T_c could be observed. Nitrogen is
an obvious impurity candidate and to better establish the N levels
in the films we used the $^{14}N(d,\alpha)$ nuclear reaction. By this technique the N levels were similarly found to be very low, < 1 atomic
percent, with no obvious correlation with T_c.

These measurements of the Nb/Ge compositional ratios vs T_c are
surprising and intriguing as it has been widely held that stoichiometry of 3/1 is crucial for the high T_c material. Models[24] have been
proposed for the A15 materials based on the integrity of the A
(i.e. Nb) chains. Any deviations from stoichiometry should therefore drastically reduce T_c. Our results do not bear these models
out. A further point which deserves note concerns the shape of

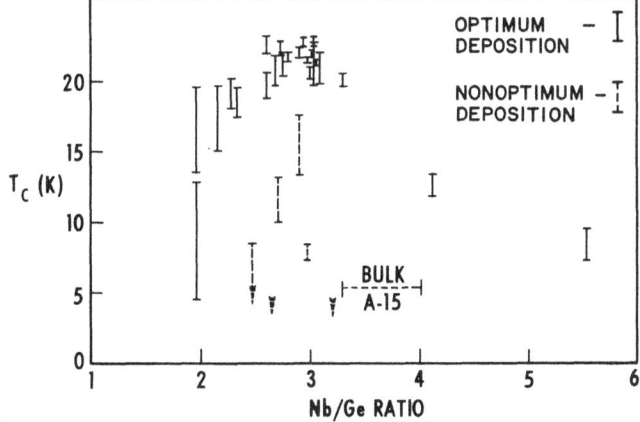

Fig. 3. T_c vs Nb/Ge compositional ratio (after Testardi, et al.[23]).

the trend shown in Fig. 3. Since, the films have mainly the A-15
phase we might assume that for Nb/Ge > 3 the excess Nb atoms are
on the Ge sublattice with the converse for Nb/Ge < 3. Assuming
then that the integrity of the A(Nb) chains is important for T_c
in this A_3B structure, one would expect that with compositional
variation T_c is reduced more drastically with disorder on A chains
(e.g., with Nb/Ge < 3) than on B sites. Figure 3 shows that this
simple concept fails in the sputtered films. Moreover it is
possible to grow low T_c films that have the 3/1 stoichiometry.
Another interesting point also is that it is possible to grow
films with the "bulk" stoichiometry (4/1) but with much higher
T_c's than bulk material. The T_c difference between thin films
and bulk material does not therefore appear to depend on stoichiometric differences.

There is one other piece of relevant information. A simple
and striking correlation between resistance ratio $\rho(300K)/\rho(25K)$
and T_c was observed which was largely independent of all sputtering conditions and composition. From this result and the previous
results we concluded that a key factor which determined whether T_c
was low or high was the presence or absence of microscopic defects.
We have pursued this argument further by introducing defects into
the film by 2 MeV ^{4}He bombardment[25] and indeed the same resistance
ratio correlation is observed.

This Nb-Ge study has been discussed at length to show how
quantitative measurements of the composition of as-deposited
superconducting thin films can lead to other physical developments.
It is hopefully evident that there are many other examples in
this area amenable to such studies. Indeed at this conference
Miller et al.[26] will also discuss detailed ion beam investigations
of thicker Nb_3Ge films and Müller et al.[27] will discuss such
studies of V, NbTi and Nb_3Sn thin films.

METAL-METAL INTERDIFFUSION

Rutherford backscattering measurements have provided the first
quantitative measurements of diffusion profiles between thin metal
films. Recent reviews[28,29] summarize the results obtained by ion
backscattering on thin metal film systems. Here we will concentrate on those experiments that have led to an understanding of
diffusion or reaction mechanisms in thin films. It is natural to
categorize the subject into those systems that form complete
series of solid solutions and those that form compound or intermetallic phases.

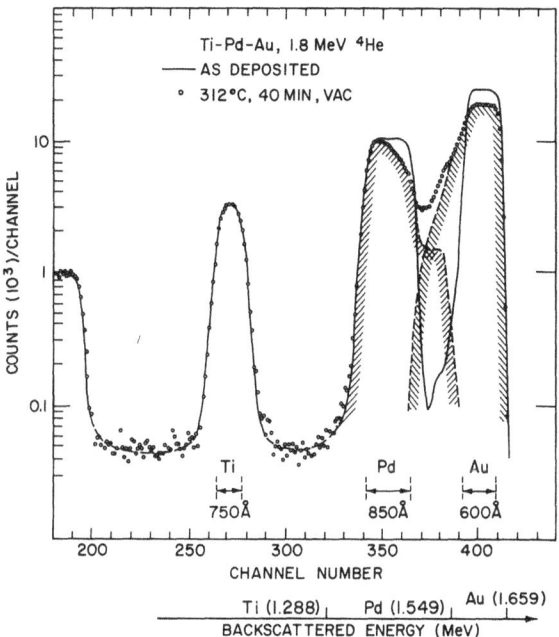

Fig. 4. Backscattering spectra of Ti-Pd-Au thin films. Au is
        outer layer and Ti is next to sapphire substrate (after
        Poate, et al.[30]).

Solid Solutions

    Examples of alloys that form complete series of solid solu-
tions are Ag-Au or Pd-Au. We (Poate et al.[30]) have studied
interdiffusion between Pd-Au thin films using backscattering in
conjunction with transmission electron microscopy for the determination of grain sizes. For many technological applications a
metallic "glue" layer such as Ti is applied between the conductor
system (e.g. Pd-Au) and the insulating substrate. Figure 4 shows
a backscattering spectrum of a Ti-Pd-Au thin film composite before
and after annealing. There is much interdiffusion between the
Pd-Au couple, as indicated by the deconvoluted scattering yields,
but no reaction between Ti-Pd.

    The Ti interactions show several interesting effects.
Figures 5 and 6 show couples of Ti-Au and Ti-Pd on sapphire substrates for air and vacuum annealing. Although the Ti-Au interdiffusion is faster than Ti-Pd the general effects are the same
for vacuum annealing. However if identical couples are annealed

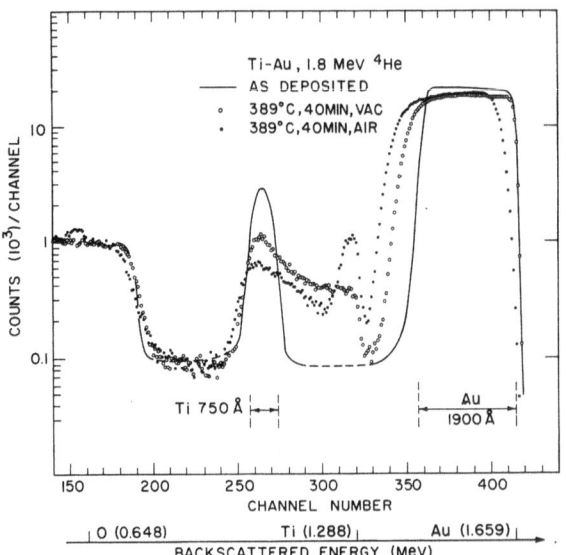

Fig. 5.   Backscattering spectra of Ti-Au couple (after Poate et al.[30]).

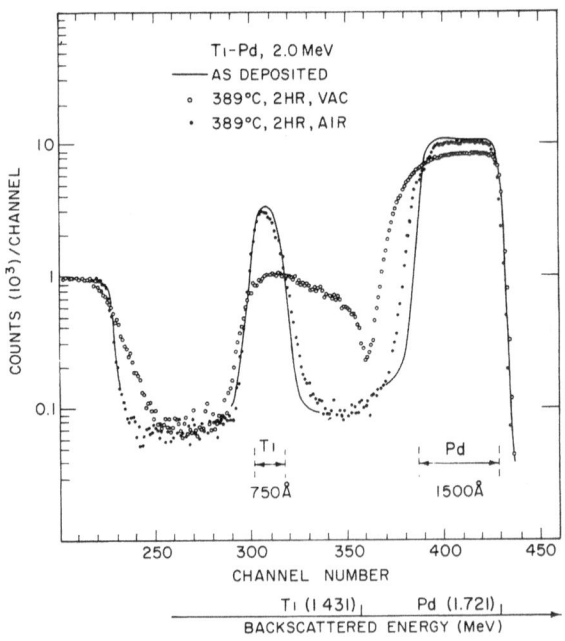

Fig. 6.   Backscattering spectra of Ti-Pd couple (after Poate, et al.[30]).

in air the behavior is markedly different. For Ti-Au, the Ti
out-diffuses to the free surface of the Au where it is oxidized.
The surface oxidation of Ti causes continued enhanced diffusion
by creating a chemical potential sink. For Ti-Pd, however, it is
suggested that several processes occur simultaneously during air
annealing: 1) interdiffusion of the Ti and Pd, 2) diffusion of
oxygen through Pd, and 3) oxidation of Ti near the Ti-Pd interface. As soon as the oxidation reaches a critical stage, most
likely when a continuous oxide layer is formed along grain boundaries near the interface, further diffusion of Pd and Ti is
blocked. Therefore for the Ti-Au couple, oxygen enhances diffusion but blocks diffusion for Ti-Pd.

We also observe that, for the case of vacuum annealing, Ti
diffusion in Ti-Au or Ti-Pd couples is much greater than in the
corresponding Ti-Au-Pd or Ti-Pd-Au ternary systems. In other
words the diffusion of Ti into Pd or Au is markedly decreased
when a third diffusing species (Au or Pd) is present in the form
of an outer film; we term this a solute effect. However the
rapid interdiffusion of the outer Pd-Au or Au-Pd couple is little
affected by the presence of the underlying Ti. The situation
can be described phenomenologically by assuming the Au-Pd interdiffusion to be so rapid that the diffusion paths for Ti are
blocked. Measurements of grain-boundary diffusion in bulk polycrystalline samples have shown that grain-boundary diffusivities
may be greatly enhanced or reduced by solute effects; this field
has been reviewed by Gleiter and Chalmers.[31] It is to be expected
therefore that solute effects will play an important role at low
temperatures in thin film interdiffusion owing to the dominance
of defect diffusion in these regimes.

Consider diffusion in the Pd-Au thin film couples. It is
clear from our results that we are witnessing a diffusion process
which is very much faster than would be observed for bulk values,
for example for the Pd/Au system bulk diffusivities are expected
to be $\sim 10^{-20}$ cm^2/sec at $\sim 250°$C. However the Pd/Au results, for
example, indicate that the operative diffusivities at temperatures
in the vicinity of 250°C should be $\sim 10^{-14}$ cm^2/sec (assuming the
simple diffusion equation $D = x^2/t$ where $x \sim 500$ Å and $t \sim 2500$
secs). We will assume that this rapid diffusion is caused by the
short-circuiting action of grain boundaries, an approach first
analyzed quantitatively by Fisher.[32] An exact mathematical solution of this problem which takes into account the simultaneous
leakage of atoms into the grains by lattice (i.e., bulk diffusion)
has been given for the case of an isolated planar grain boundary
by Whipple.[33] In the Whipple model the logarithm of the average
concentration profile, ln $\bar{c}$, should have an $x^{6/5}$ dependence where
$\bar{c}$ is the concentration at depth x. Knowing the lattice or bulk
diffusivity (D_L) the grain boundary diffusivity (D_B) can be calculated from the slopes of the concentration profiles. However

from a knowledge of the grain sizes of the films and the amount of diffusant that can be contained within the grain boundaries, it is possible to show from our grain-boundary-assisted bulk diffusion[30,34] model that diffusion within the grains is defect enhanced. Diffusion within the grains cannot therefore be treated in terms of the simple lattice diffusivity, D_L.

In a later study, Hall <u>et al.</u>[35], of diffusion in the Pd-Au system we correlated resistivity changes with Auger electron spectroscopy and Rutherford backscattering profiles. Backscattering was used to calibrate the Auger profiling with its greater depth sensitivity. Figure 7 shows diffusion profiles for Pd in Au obtained from Auger profiling. There are two distinct regions: within 400 Å of the interface erfc diffusion predominates, beyond this the Whipple model is applicable. Hall <u>et al.</u>[35] have shown how it is possible to use the erfc concentration gradient at the interface to determine defect enhanced diffusion coefficients within the grains. Values thus obtained agree quite well with those obtained independently from the grain-boundary-assisted bulk model mentioned previously and a model based on the increase of the electrical resistance. We have therefore been able to determine diffusivities not only in the grain boundaries but also within the grains themselves. We cannot, of course, identify the defects, whether dislocations or point defects such as vacancies, causing enhanced diffusion within the grains.

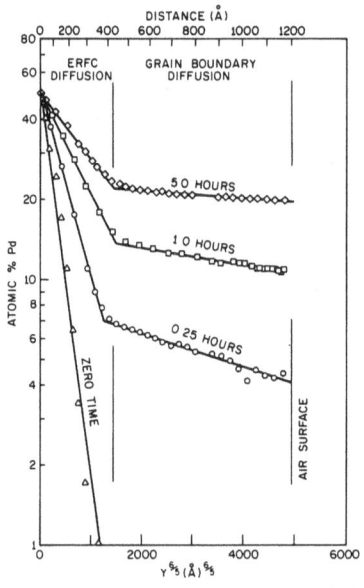

WHIPPLE PLOT OF Pd PROFILE IN GOLD AT 250°C

Fig. 7.   Pd diffusion profile in Au obtained by Auger profiling
            (after Hall, <u>et al</u>.[35]).

Intermetallic Compound Formation

Unlike the previous examples, where there appears to be a clear delineation and competition between grain boundary and bulk diffusion, there are a whole class of thin film couples characterized by the formation of discrete intermetallic layers. Moreover these layers are formed at temperatures well below their melting points. Thin film couples in which Al is an element often show such layered compounds. The Al/Ag[28,36] and Al/Ti[37] couples have been studied by backscattering and show a $(time)^{\frac{1}{2}}$ dependence for the growth of the compound phases.

Campisano et al.[38] have recently investigated the Al/Au thin film system in detail. Figure 8 shows one of their backscattering spectra following an anneal at 230°C for 10 min. of 1600 Å Au and 4800 Å Al films. The Au has been consumed and the Au_2Al and $AuAl_2$ phases are formed in discrete layers. These phases were also identified using glancing angle X-ray diffraction techniques.

The results of their investigations are schematically represented in Figure 9. They show that the resulting end phases are determined by the starting thicknesses of the films. When one of the metal layers has been consumed, the intermetallic formation is

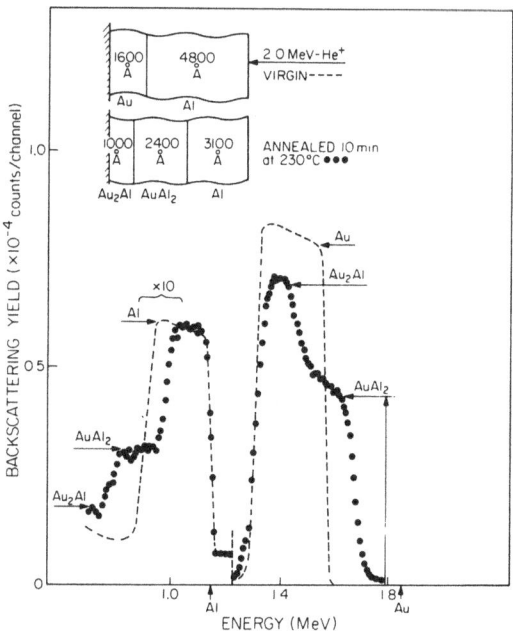

Fig. 8. Backscattering spectra of Au-Al thin film reactions (after Campisano, et al.[38]).

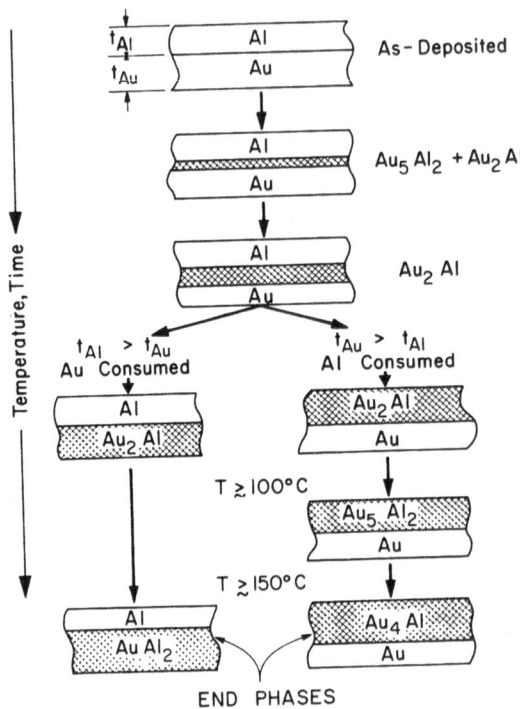

Fig. 9.   Schematic diagram showing phase formation in the Au-Al
          thin film couples (after Campisano et al.[38]).

driven towards the phases that are rich in the remaining element.
Such end phases are $AuAl_2$ and Au_4Al and indeed these are the end
phases shown in the equilibrium phase diagrams. Both the $AuAl_2$
and Au_2Al phases were observed to grow with a $(time)^{\frac{1}{2}}$ dependence.
Although this extremely rapid process is diffusion limited there
is no evidence as yet indicating whether it is taking place by
grain boundary or bulk diffusion.

     We have been able to illustrate several features of the inter-
diffusion and reaction between metal films. The most important,
perhaps, is that the interdiffusion and reactions occur at very
low temperatures due to either diffusion along defects or compound
formation. We have also been able to demonstrate chemical gradient
and solute effects. Thin films are complicated structures and
there must be other features affecting diffusion. Balluffi and
Blakely[39] have recently reviewed this area in detail and point out,
for example, the importance stresses should play in interdiffusion.
To our knowledge, there have been no experiments in metal films
which deal unambiguously with this issue.

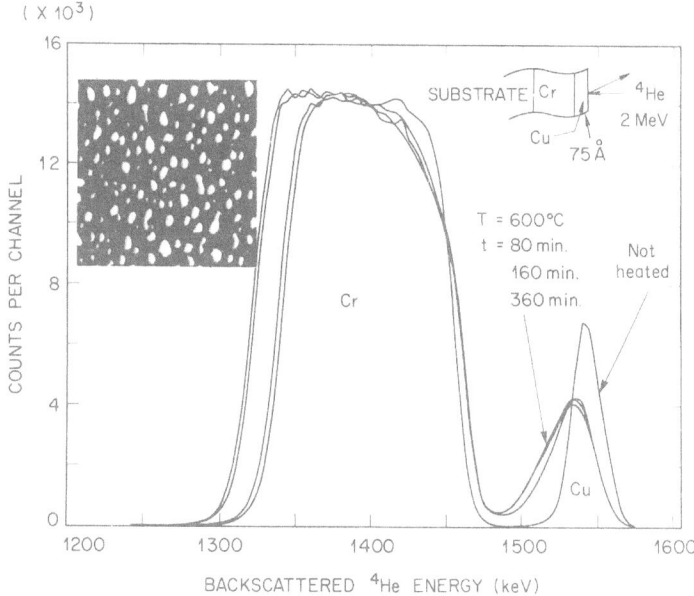

Fig. 10.   Backscattering spectra of 75 Å Cu film on Cr with SEM
            (x26,000) insert of couple after treatment at 700°C for
            2 hr. (after Baglin, et al.[40]).

It is worthwhile emphasizing once more the importance of
complementary experimental techniques. We have given examples
where X-ray diffraction and transmission electron microscopy have
been employed. Figure 10 shows backscattering spectra of Baglin
et al.[40] of a 75 Å film of Cu on Cr undergoing heat treatment.
The spectra would indicate that the Cu is diffusing into the Cr,
however, a high resolution (x26,000) scanning electron microscope
picture (see insert in figure) gives an entirely different story.
The Cu is just balling up under heat treatment with an average
globule size ∼1000 Å. By the combined using of backscattering,
X-ray diffraction and SEM Baglin and d'Heurle[41] have also investigated the Al-Ni couple. They have been able to observe nucleation and growth of the Al_3Ni phase at the interface. The Al_3Ni
grains then grow to form a continuous layer of Al_3Ni.

Many investigations, using ion beams, of thin metallic film
are in progress. At this meeting Baglin and d'Heurle[42] will discuss many of the morphological and intrinsic effects present in
thin film diffusion; Campisano, et al.[43] will discuss grain boundary diffusion in Cu thin films and Barcz, et al.[44] diffusion in
the NiCr-Au thin film system.

METAL-SEMICONDUCTOR REACTIONS

Many features of metal-semiconductor reactions parallel those of the metal-metal reactions discussed previously. Not only are there many striking examples of low temperature phase formation but also low temperature reactions without compound formation. McCaldin[45] has recently reviewed this latter area of the eutectic metal-semiconductor systems (Si-Al, Ge-Al and Si-Au) in detail and we will only mention some salient points. It was backscattering,[46] for example, that provided the first direct evidence for atom movements at the Au-Si interface. Hiraki and colleagues[47] have pursued this question of the Au-Si interfacial reactions using a variety of techniques, as well as backscattering, and postulate that the Si at the interace is metallic and forms a metallic bonding with Au.

As McCaldin[45] points out the diffusion behavior of Al or Au thin metal films on Si or Ge is quite remarkable. The diffusion occurs almost entirely in the metal phase with great rapidity. Si diffusion through Al, for example, is so rapid at low temperatures that it is more akin to diffusion in a liquid medium than to conventional diffusion in a solid Si medium.

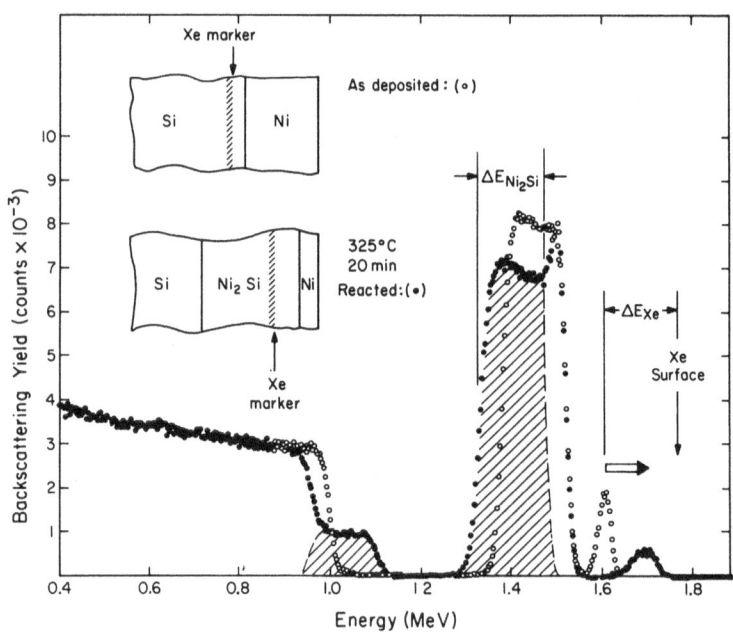

Fig. 11.   Backscattering spectra of Ni_2Si formation showing motion of the Xe marker (after Tu, et al.[52]).

Backscattering techniques have probably had the greatest
inpetus on the study of those metal-Si systems that form compounds, i.e. silicides. Tu[48] has reviewed this subject and his
table shows the silicides that have been studied using thin film
reactions, principally by backscattering in conjunction with
glancing angle X-ray diffraction techniques.

Silicides Formed by Thin Film Contact Reactions

M	M_2Si	MSi	MSi_2
Mg	x		
Pd, Pt	x	x	
Ni, Co	x	x	x
Ti, (Zr), Hf, Fe		x	x
Cr, Mo, W, V, Nb, Ta			x

The silicides are grouped into three categories: M_2Si, MSi and
MSi_2 where M represents the metallic elements listed in the left
column of the table. It should be noted that while the equilibrium phase diagrams show many of the binary M-Si systems containing more than three silicide phases only the three shown in
the Table have been observed in thin film reactions between metal
films and Si substrates. The sequence of formation is the metal
rich silicide, followed by the monosilicide and then the disilicide
i.e. $M_2Si \rightarrow MSi \rightarrow MSi_2$.

In the previous section on metal-metal interdiffusion we did
not address ourselves to the important question of the diffusing
species. Some experiments on grain boundary diffusion,[49] for
example, show that there is only one species mobile along the
grain boundaries. To determine unambiguously the diffusing species,
diffusion markers are needed. In studies of diffusion and compound formation in bulk samples, diffusion markers have played a
key role. Kirkendall markers,[50] usually fine Mo or W wires placed
at the interface between two metals, are used to determine which
of the two metals diffuses through the compound during phase
formation.

Of course such macroscopic markers cannot be used in thin
films but by the techniques of ion implantation of inert gases and
backscattering to observe motion of the markers, it is possible to
determine the diffusing species. This technique was pioneered by
Brown and Mackintosh[8] to determine the metal-to-oxygen transport
ratio in anodic oxidation studies and has been successfully
employed by Chu et al.[51] in the silicides. Figure 11 shows the
backscattering spectra of Tu et al.[52] for a Xe implant in Si with
a superimposed Ni film. On annealing the marker moves towards the
surface indicating that Ni is the diffusing species in the Ni_2Si.

In Figure 12 the two broken lines give the limiting positions of
the marker assuming that the diffusing flux is either Ni or Si.
The positions of the displaced Xe marker agree extremely well with
the line for Ni.

Considerations of the very low temperatures of formation and
low activation energies of the metal rich silicides (Ni, Pd, Pt
silicides form at 200°C with activation energies ~1.5 eV, for
example) and nature of the diffusing species has led Tu[53] to propose a model for thin film silicide growth. It is proposed that
the noble and near noble metal atoms can be incorporated on interstitial sites in the Si lattice to sufficiently high concentrations
so that the Si bonds at the interface change from covalent to
metallic-like. Since an interstitial increases the number of
nearest neighbors of its host atoms, a Si atom at the interface
will be bonded by unsaturated rather than saturated covalent bonds.
These metallic-like unsaturated covalent bonds can be broken more
easily at low temperatures than the saturated covalent bonds.
Hence the interface possesses a high mobility and enables the
silicide growth to proceed at low temperatures. The interface will
lose its high mobility whenever it is depleted of near noble metal
atoms. The growth of the metal rich silicides therefore depends
on the diffusion of the metal atom to the interface as indicated
by the above diffusion marker experiment.

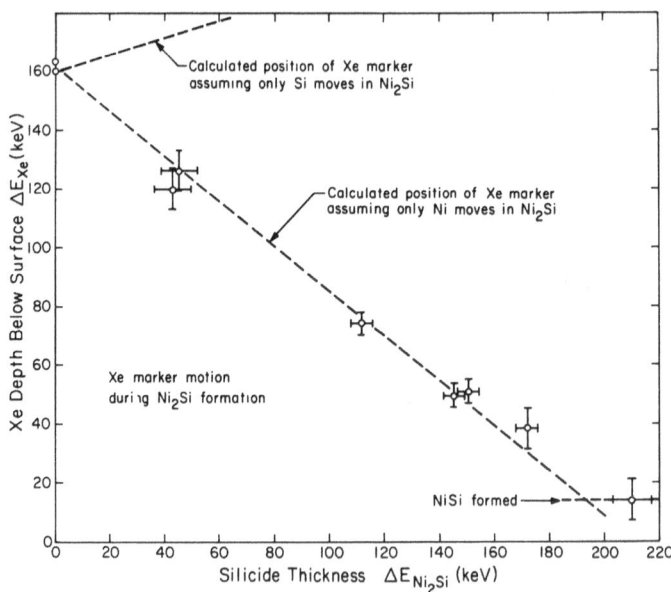

Fig. 12.   Broken lines give limiting positions of Xe marker.
            Points are the measured marker positions (after Tu,
            et al.[52]).

ACKNOWLEDGMENTS

I am indebted to my colleagues at Bell, Caltech, IBM and Sandia for discussions and collaboration on the subject of this review.

REFERENCES

1. "Ion Beam Surface Layer Analysis", Proceedings of 1973 Yorktown Heights Conference edited by J. W. Mayer and J. F. Ziegler, Thin Solid Films 19 (1973).

2. Catania Working Data "Ion Beam Analysis", Catania 1974 edited by J. W. Mayer and E. Rimini.

3. T. M. Buck and J. M. Poate, J. Vac. Sci. Technol. 11, 289 (1974).

4. I. V. Mitchell, M. Kamoshida and J. W. Mayer, J. Appl. Phys. 42, 4378 (1971).

5. W. K. Chu, E. Lugujjo, J. W. Mayer and T. W. Sigmon, Thin Solid Films 19, 329 (1973).

6. O. Meyer and W. Scherber, J. Phys. Chem. Solids 32, 1909 (1971).

   M. Croset, S. Rigo and G. Amsel, Appl. Phys. Lett 19, 33 (1971).

7. J. M. Poate, P. J. Silverman and J. Yahalom, J. Phys. Chem. Solids 34, 1847 (1973).

   P. J. Silverman and N. Schwartz, J. Electrochem. Soc. 121, 550 (1974).

8. F. Brown and W. D. Mackintosh, J. Electrochemical Soc. 120, 1096 (1973) and Abstract No. 99 in Extended Abstracts, Spring Meeting, Toronto, Electrochemical Society, 1975.

9. S. Rigo and J. Siejka, Solid State Comm. 15, 259 (1974) and Abstracts 98, 101 and 129, Spring Meeting, Electrochemical Society, 1975.

10. H. Kräutle, W. K. Chu, M. A. Nicolet, J. W. Mayer and K. N. Tu in Proc. Conf. on Applications of Ion Beams to Metals (ed. S. T. Picraux, E. P. EerNisse and F. L. Vook) p 193, Plenum Press, N. Y. (1974).

11. V. Marrello, J. M. Caywood, J. W. Mayer and M. A. Nicolet, Phys. Stat. Sol (a) 13, 531 (1972).

     C. Canali, J. W. Mayer, G. Ottaviani, D. Sigurd and W. F. van der Weg, Appl. Phys. Lett. 25, 3 (1974) and Abstracts 113, 114, 120 and 130, Spring Meeting, Electrochemical Society 1975.

12. S. T. Picraux, J. Appl. Phys. 44, 587 (1973).

13. T. M. Buck, J. M. Poate, K. A. Pickar and C. M. Hsieh, Surface Sci. 35, 362 (1973).

14. O. Meyer, H. Mann and E. Phrilingos in Proceedings "Applications of Ion Beams to Metals", p. 15 (1974).

15. O. Meyer, Proceedings "Ion Implantation in Semiconductor" (ed. S. Namba), p. 301, Plenum Press, N. Y. (1975).

16. G. Linker and O. Meyer, ibid. p. 309.

17. O. Meyer, G. Linker and B. Kraeft, Thin Solid Films 19, 217 (1973).

18. J. R. Gavaler, Appl. Phys. Lett. 23, 480 (1973).

19. L. R. Testardi, J. H. Wernick and W. A. Royer, Solid State Commun. 15, 1 (1974).

20. J. H. Carpenter and A. W. Searcey, J. Am. Chem. Soc. 78, 2079 (1956).

21. S. Geller, Acta Crystallogr 9, 885 (1956).

22. B. T. Matthias, T. H. Geballe, R. H. Willens, E. Corenzwit and G. W. Hall, Jr., Phys. Rev. 139, A1505 (1965).

23. L. R. Testardi, R. L. Meek, J. M. Poate, W. A. Royer, A. R. Storm and J. H. Wernick, Phys. Rev. B 11, 4304 (1975).

24. J. Labbé, S. Barišić and J. Friedel, Phys. Rev. Lett. 19, 1039 (1967).

25. J. M. Poate, L. R. Testardi, A. R. Storm and W. M. Augustyniak (1975) to be published.

26. J. W. Miller, B. R. Appleton and J. R. Gavaler, this conference.

27. P. Müller, G. Ischenko and F. Gabler, ibid.

28. S. T. Picraux, Proc. 6th International Vacuum Congress 1974, Japan J. Appl. Phys. Suppl. 2, Pt 1, p. 657 (1974).

29. J. A. Borders and S. T. Picraux, Proc. IEEE 62, 1224 (1974).

30. J. M. Poate, P. A. Turner, W. J. De Bonte, J. Yahalom, J. Appl. Phys. in press.

31. H. Gleiter and B. Chalmers, Progress in Materials Science, 16 (1972) "High Angle Grain Boundaries" Chapter 4.

32. J. C. Fisher, J. Appl. Phys. 22, 74 (1951).

33. R. T. P. Whipple, Phil. Mag. 45, 1225 (1954).

34. W. J. De Bonte, J. M. Poate, C. M. Melliar-Smith and R. A. Levesque "Applications of Ion Beams to Metals", p. 147 (1974).

35. P. M. Hall, J. M. Morabito and J. M. Poate, Thin Solid Films, to be published.

36. J. E. Westmoreland and W. H. Weisenberger, Thin Solid Films 19, 349 (1973).

37. R. W. Bower, Appl. Phys. Letters 23, 99 (1973).

38. S. U. Campisano, G. Foti, E. Rimini, S. S. Lau and J. W. Mayer, Phil. Mag. 31, 903 (1975).

39. R. W. Balluffi and J. M. Blakely, Thin Solid Films, 25, 363 (1975).

40. J. E. E. Baglin, V. Brusic, E. Alessandrini and J. F. Ziegler "Applications of Ion Beams to Metals", p. 169 (1974) and unpublished data.

41. J. E. E. Baglin, F. M. d'Heurle and P. S. Ho, to be published and Abstract 109, Spring Meeting, Electrochemical Society 1975.

42. J. E. E. Baglin and F. M. d'Heurle, this conference.

43. S. U. Campisano, E. Costanzo, G. Foti and E. Rimini, ibid.

44. A. Barcz, A. Turos, L. Wieluński, ibid.

45. J. O. McCaldin, J. Vac. Sci. Technol. 11, 990 (1974).

46. A. Hiraki, M. A. Nicolet and J. W. Mayer, Appl. Phys. Lett. 18, 178 (1971).

47. K. Nakashima, M. Iwami and A. Hiraki, Thin Solid Films, <u>25</u>, 423 (1975).

48. K. N. Tu, Abstract 116, Spring Meeting Electrochemical Society 1975.

49. W. J. De Bonte, J. M. Poate, C. M. Melliar-Smith, R. A. Levesque, J. Appl. Phys. in press.

50. E. O. Kirkendall, Trans AIME <u>147</u>, 104 (1942).

51. W. K. Chu, K. Kräutle, J. W. Mayer, H. Müller, M. A. Nicolet and K. N. Tu, Appl. Phys. Letters <u>25</u>, 454 (1974).

52. K. N. Tu, W. K. Chu and J. W. Mayer, Thin Solid Films, <u>25</u>, 403 (1975).

53. K. N. Tu, Appl. Phys. Lett. <u>27</u>, 221 (1975).

STUDIES OF TANTALUM NITRIDE THIN FILM RESISTORS*

R. A. Langley

Sandia Laboratories

Albuquerque, New Mexico 87115

I. INTRODUCTION

Extensive interest in tantalum as a thin film material has been stimulated by its use in the microelectronics industry. Tantalum has a refractory nature which implies that any imperfection frozen in during deposition will not anneal out during the life of the thin film. In addition, tantalum belongs to a class of metals which form tough self-protective oxides either through heat treatment or through anodic oxidation.[1] The preferred method of deposition has been sputtering rather than evaporation because of the refractive nature of the tantalum. Sputtered tantalum films have a tendency to become contaminated since it is a reactive material although some degree of contamination is desirable in order to achieve useful properties.[2]

The technology of tantalum resistors is complicated by the fact that tantalum films can exist in two separate structures or a combination of both.[3] The conventional bcc tantalum structure corresponding to the bulk is known as the α structure. The β structure is tetragonal in nature. The exact conditions under which α or β tantalum will form are not fully understood; however, it has been found that the β form does not appear in systems in which there is an appreciable degree of gaseous contamination.[4] Frequently films contain a mixture of the α and β structures. Reproducibility of the films is difficult to achieve under these conditions. Originally oxygen was used as a form of contamination which could be added in a controlled fashion. Oxygen has a profound effect on both the resistivity and the thermal coefficient of resistance (TCR) of the tantalum films.

---

*This work was supported by the United States Energy Research and Development Administration, ERDA.

The change in resistivity with % dissolved oxygen is extremely rapid and difficult to control. A more attractive contaminant for deliberate introduction during the sputtering process is nitrogen. In practice film composition close to that of Ta_2N was chosen since it was thought that resistors having this composition would display the greatest stability during life tests.[5] Tantalum nitride films are susceptible to oxidation and will change characteristics on exposure to air. It is common to lightly anodize the film or to anneal the film in air to create a thin oxide layer on the surface which stabilizes them.

The purpose of this study was to correlate the electrical properties of tantalum nitride thin film resistors with the nitrogen concentration in the films. Previous studies have used either separately or in combination Auger sputter profiling[6] and chemical analysis[7,8] to give constituent concentrations and depth distributions in a destructive analysis. In the present experiment ion backscattering was used to provide both the constituent concentrations and depth distributions in a single nondestructive test. In addition growth of the oxide during the stabilization period was studied. The present study has accomplished this correlation by using high energy (2.0 MeV) ion backscattering to characterize the thin film tantalum nitride resistors, as to the amount of surface oxide, the nitrogen content, its depth distribution, the amount of argon sputtering gas incorporated into the film and its depth distribution.

The films for these experiments were deposited on two types of substrates: alumina and beryllium. The use of a low Z substrate ion backscattering, namely Be, allows observation of film components of higher Z, e.g., C, N, O, and Ar. The films on the alumina were characterized both by ion backscattering and electrical measurements and the films on the beryllium were characterized only by ion backscattering. The electrical properties of the films measured were the sheet resistance, differential Seebeck potential, thermal coefficient of resistance, and stability. Preliminary results were given in a previous paper.[9] This paper discusses the preparation of the substrates and resistive films, the resistor definition, the stabilization procedure, and the electrical and ion backscattering measurements. The results of the electrical and ion backscattering measurements are presented, and finally the conclusions based on the experimental results are discussed in detail.

II. EXPERIMENTAL PROCEDURE

A. Sample Preparation

Substrates of both alumina and beryllium were used in the experiment. The alumina substrates had dimensions 2.54 x 2.54 x 0.068 cm and were cleaned by ultrasonically washing successively in trichlorethylene, acetone, detergent, and deionized water followed

by hot deionized water rinse and an air firing at 900°C for one hour.
The beryllium substrates were machined from high purity bulk material
which had been formed by cold isostatic pressing followed by hot isostatic pressing and whose grain sizes were of the order of 10 µm.[11]
The substrates were subsequently mechanically polished, electrically
polished and finally rinsed in deionized water. Air exposure of
these substrates at room temperature provided a stable surface oxide
layer, approximately 100 Å thick, so that deposition was accomplished
on similar oxide surfaces: namely alumina and beryllia. In the
vacuum station the substrates were heated to 400°C for two hours
before sputtering was accomplished. The films were prepared by dc
diode reactive sputtering of tantalum in nitrogen-argon mixtures in
a 15 cm diameter oil diffusion pumped system with a liquid nitrogen
trap and a stainless steel bell jar. Sputtering from a 730 cm^2
cathode was conducted at a potential of 5 kV, a current density of
0.25 mA/cm^2, an interelectrode spacing of 7.0 cm and a gas pressure
of 4.5 x 10^{-3} Torr. The nominal deposition rate was 150 Å per minute,
the argon flow rate was 20 Std cc^3/min (sccm) and the nitrogen flow
rate was varied from 0.5 to 0.9 sccm. This technique resulted in
film compositions varying from 0.25 to 0.55 at.% nitrogen. A presputtering period of 20 minutes was accomplished with a shutter
covering the substrates. Following this presputtering the shutter
was removed and the films were deposited in approximately 180 seconds.
This procedure produced tantalum nitride films of nearly the same
thickness but with varying amounts of nitrogen incorporated into the
film. Each deposition was made on a batch of 35 alumina substrates
arranged in a square configuration. To obtain films of the same
thickness only the 4 central substrates were used in this study.
This was necessary because the substrates not in the center were
found by ion backscattering to have minor variations in the thickness
of the resistor films. This variation was attributed to perturbation of the electric field caused by fixturing in the sputtering
system. Two of the four central substrates diagonal to each other
had 0.95 cm holes through them to accommodate the beryllium substrates
which were used in the ion backscattering study.

B. Resistor Definition and Stabilization

In order to make certain of the electrical tests, it was necessary to deposit conductor films on the resistor material so that
leads could be attached. This was carried out by vacuum evaporation
of approximately 300 Å of chromium followed by a deposition of
30,000 Å of gold. During these depositions the substrate temperature
remained at 375°C. Following these depositions resistor patterns
were photolithographically defined on the alumina substrate for subsequent determination of the thermal coefficient of resistance and
the long-term stability. Photolithography is discussed extensively
in both references 2 and 5 and will not be discussed further here.
Following the resistor definition, stabilization of the resistors

was accomplished by annealing in air for 2 hours at 300°C. This
formed a passivating oxide layer on the films. Both electrical and
backscattering measurements were made before and after this anneal.
Fifteen parallel resistors were defined on each substrate. They were
1.27 cm long and had a width of 0.13 cm with a nominal resistance of
10 kOhms each. Following the photolithography process and the resistor stabilization, gold-plated copper leads were attached by thermal
compression bonding.[11]

C. Electrical Measurements.

Sheet resistance and the differential Seebeck potential measurements were performed on the thin film resistors before metallization.
Measurements of the sheet resistance, R_s, were performed using an

A & M Fell Ltd. four-point probe system. The resistivity, ρ, of the
thin film resistors was determined from the sheet resistance and the
film thickness, T, as determined by ion backscattering, by the
equation

$$\rho = R_s \cdot T \quad .$$

Four resistance measurements were made on each substrate and the
results from the four substrates of each deposition run were averaged
to give a sheet resistance for the run. Seebeck thermoelectric
measurements were made using two gold-plated copper contact blocks,
13 x 13 mm, separated by 22 mm. One contact was thermostated at 10°C
using a Peltier cooler and the other contact was thermostated at 60°C.
This configuration allowed measurements of the differential Seebeck
potential. Two measurements were made on each substrate along the
diagonal axes of the resistor. Again the results from the four substrates of each deposition run were averaged.

The thermal coefficient of resistance (TCR) measurements were
performed after resistor definition and stabilization by measuring
the resistance of each resistor at 3 temperatures: -195°C, 23°C
and +150°C. From these measurements, TCR can easily be calculated.[12]
Stability tests were run at 150°C for 1000 hours. The resistance
was sampled at 30, 100, 300 and 1000 hours. In each case the 15
resistors on one substrate were averaged to give an average value
of TCR and stability. On those substrates which had the hole in
them to accommodate the beryllium substrates, only 8 resistors could
be defined. For each deposition there were values obtained for
46 resistors. These were averaged to give a value for TCR and
stability for that particular deposition run.

D. Backscattering Measurements

Ion backscattering has been used extensively to determine composition vs depth in solids.[13] A well-collimated monoenergetic beam

of He$^+$ ions was directed onto the sample. The incident ions lose
energy in the solid by two processes: (i) numerous collisions with
electrons of the sample material, and (ii) individual elastic scattering events due to the repulsive Coulomb force between the nuclei
of the incident ion and the target atom. The stopping power of the
electrons of the sample material provide information about the depth
distribution of the elements in the sample since ions scattered from
atoms deep in the sample will have lost more energy than those nuclei
scattered from atoms of the same element nearer the surface. The
nuclear scattering allows identification of the constituent elements
of a thin film by the amount of energy the incident ion loses to
recoil of the target atom. Light ions such as nitrogen and oxygen
can take up a large portion of the energy of the ion, while heavy
atoms like tantalum absorb relatively little energy in recoiling.
Beryllium is an ideal substrate to use because the low backscattering yield from this low-Z element is relatively small and occurs at
low energy. The higher energy backscattering yield from a higher-Z
element is therefore easily resolved. This backscattering technique
has the important feature of being nondestructive.[14]

The ion backscattering measurements were made with 2 MeV helium
ions obtained from a Van de Graaff accelerator. The experimental
apparatus and the basic equations pertinent to ion backscattering
have been described elsewhere.[15] The beam had dimensions of 1 x 1 mm
and a current in the range of 10-50 nA. Typically, the charge deposited per spectrum was 10 µC for the alumina substrates and 40 µC for
the beryllium substrates. Particles backscattered at an angle of
170° were energy-analyzed with a surface-barrier Si detector whose
energy resolution corresponded to ~ 120 Å depth resolution. For each
deposition, eight backscattered energy spectra were taken: one on
each of the alumina substrates, and one on each of the two beryllium
substrates before stabilization and one on each after stabilization.

III. RESULTS

A typical backscattered energy spectrum for an unannealed sputtered tantalum nitride film on a beryllium substrate is shown in
Fig. 1. For energies less than 1.7 MeV the vertical scale is expanded
by a factor of 50. The solid vertical lines indicate the energies
with which 2.0 MeV He ions are backscattered from Ta, Ar, O, N, C and
Be atoms when these elements are on the surface. These energies correspond to the half-heights of the leading edges of the respective
scattering peaks.

The ion backscattering yield from Ta is shown at high energy,
~ 1.8 MeV. The width of the peak is the measure of the film thickness. The yield from the argon incorporated into the film in the
deposition process is easily distinguished and is evenly distributed
throughout the depth of the film at a concentration of 2 at.%. This

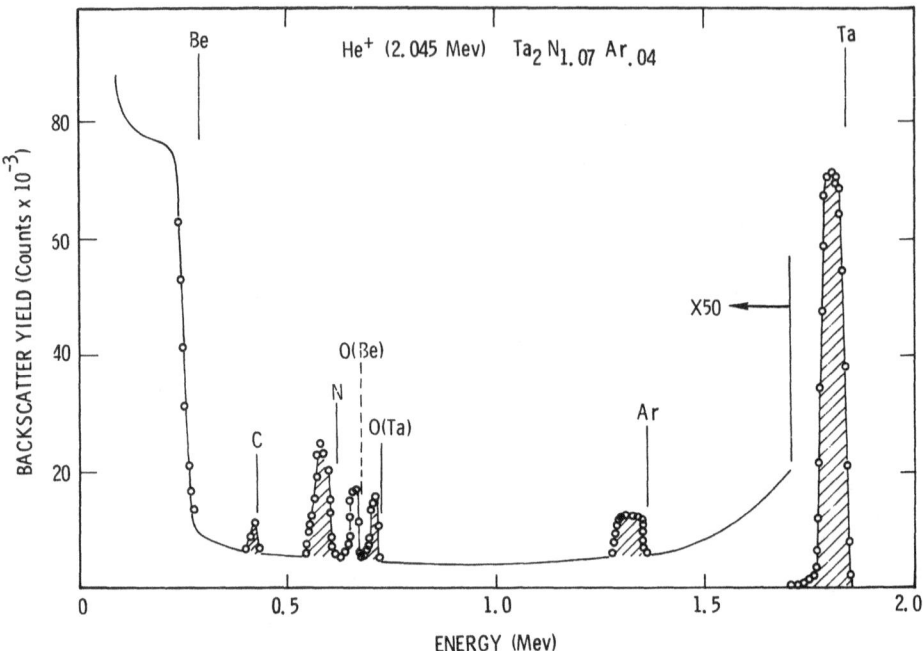

Fig. 1. Energy spectrum of 2.0 MeV ^{4}He ions backscattered from a
reactively sputtered Ta-N film about 500 Å thick on a beryllium substrate. The low-energy side of the spectrum is magnified by a factor
of 50. The shaded areas are the scattering yields of the elements in
the film. The energies with which the incident ions are backscattered
from the film elements Ta, Ar, O, N, and C are shown as well as the
substrate elements Be and O. The edge labeled O(Ta) is associated
with the Ta-N film as a surface oxide layer and the edge labeled
O(Be) is associated with the Be substrate as BeO.

is in agreement with the results of Morabito.[6] The oxygen peak
located at 0.7 MeV results from a tantalum oxide surface layer, presumably Ta_2O_5, which formed on room temperature air exposure of the
film after deposition. The high energy edge of this peak is energy-
resolution limited while the low energy edge is not, indicating that
the average spatial oxygen concentration is decreasing with greater
depths in the film. The low-energy oxygen peak is associated with
the substrate as BeO. The shape of the high-energy edge of the
nitrogen peak indicates that nitrogen is depleted from the surface
region but is probably distributed uniformly throughout the remainder
of the film. This uniformity is difficult to ascertain since the
resistive film thickness is only three times the depth resolution.
A small surface carbon peak is shown, its presence was probably due
to the breakup of residual hydrocarbons in the vacuum system caused
by the ion beam bombardment. The integrated composition of the film
is obtained from the elemental scattering cross section.[13] The
atomic percentage nitrogen, oxygen and argon in the film with respect

to the total amount of tantalum in the film shown in Fig. 1 is 38%, 11%, and 2% respectively. For very low Z targets, e.g., beryllium, the elastic cross section is enhanced over the Rutherford cross section, but this is not the case for nitrogen.[16]

Because of possible differences in films deposited on different types of substrates it was necessary to show that as near as possible the films deposited on the beryllium substrates and those deposited on the alumina substrates were the same. The backscattering spectra taken on the alumina substrates appeared similar to that in Fig. 1 except the aluminum yield from the alumina substrate occurred at 1.08 MeV and together with the oxygen in the substrate obscured all information about the surface oxygen layer and the nitrogen incorporated into the film. However, the tantalum and the argon in the spectrum are well resolved. The measured film thicknesses as determined by ion backscattering for all substrates, alumina and beryllium, were the same for each deposition and the distribution of argon and the amount of argon incorporated into the films were the same within an experimental error of 2%. It was therefore assumed that the films deposited on the beryllium substrates were the same as those deposited on the alumina substrate and had essentially the same compositional content.

Ion backscattering spectra were taken on each of the beryllium substrates before and after stabilization to measure the growth of the surface oxide layer. It was found that the amount of oxygen in the surface layer increased by approximately a factor of two for all deposition runs. The oxide layer appeared to become more dense with oxygen and only increased slightly in depth. One cannot determine the thickness of the surface oxide layer from the width of the oxygen peak since it is energy-resolution limited, but one can determine an average thickness from the area under the oxygen peak. In order to obtain a thickness in Å it is necessary to make two assumptions: (i) the stoichiometry for this oxide layer, namely Ta_2O_5, and (ii) its density, 8.73 gm/cm^3. Using these assumptions the oxide thickness layers on the films before stabilization were determined to have an average thickness of 40 Å and varied from 35 Å to 45 Å. After stabilization this oxide layer had grown to a thickness of about 80 Å and varied between 75-85 Å. For all of the films investigated, it appeared as though the nitrogen was rather uniform with depth, but depleted somewhat from the surface region, i.e., ~ 50 Å. Morabito has found that nitrogen is homogeneous beneath the surface oxide layer.[6]

It is important to point out the distinction between the film thickness and the resistor thickness, the film thickness being the resistor thickness plus the surface oxide layer thickness. The nitrogen content of the resistor film is determined not by the total amount of tantalum present in the film, but by the amount of tantalum available for nitriding, i.e., the total amount of tantalum minus

that associated with the surface oxide (assumed to be Ta_2O_5). All
backscattering data were reduced in this manner. The uncertainty
associated with the nitrogen content varies with the absolute nitrogen content and is mainly due to counting statistics: ± 5% for the
highest nitrogen content of $Ta_2N_{1.07}$ and ± 16% for the lowest nitrogen content of $Ta_2N_{0.5}$. Figure 2 shows the relationship between the
nitrogen flow rate in the deposition system and the nitrogen content

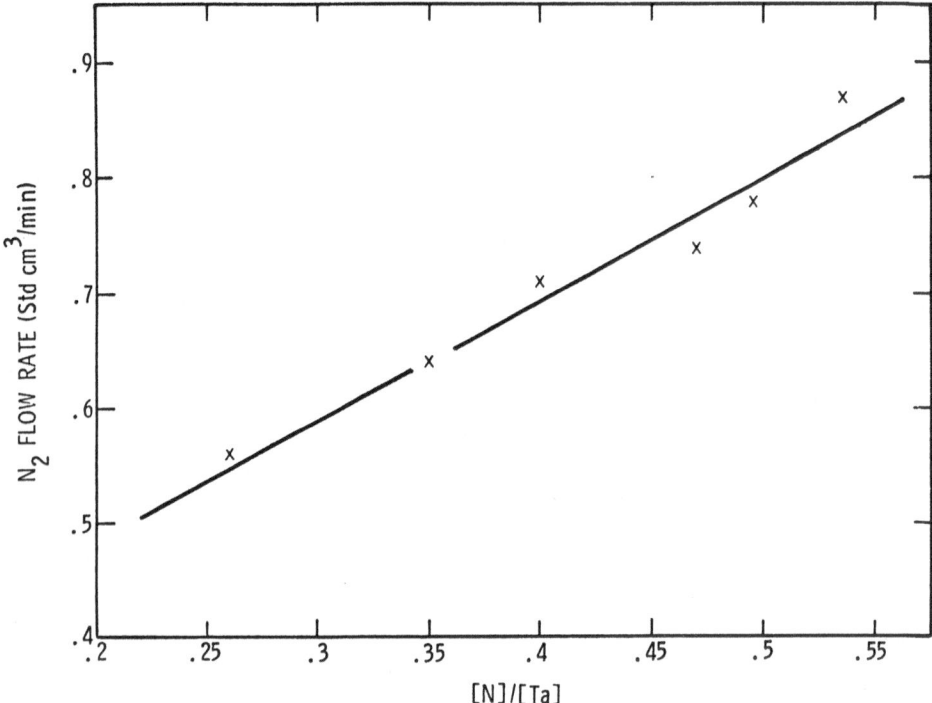

Fig. 2. Nitrogen flow rate vs nitrogen content in the resistive
film. The tantalum content is that available for nitriding (i.e.,
excluding Ta surface oxide). See text for further explanation.

in the resistor film. Within experimental error the dependence is
linear and a best-fit straight line has been drawn to guide the eye.
Figure 3 shows the sheet resistance vs nitrogen content and also the
resistance values of the defined resistors vs nitrogen content. The
correlation between the two is quite good.

In Fig. 4 is shown the dependence of the differential Seebeck potential on the nitrogen content in the resistor film. Within experimental error the dependence is linear and a best-fit straight line
has been drawn to guide the eye. The probable error for the differential Seebeck potential was set equal to the rms deviation and was
< ± 1% for all deposition runs. A recent work by Trudel correlated
nitrogen partial pressure in the deposition chamber with Seebeck
ratios.[17] Qualitative agreement with the present results and those
of Trudel was found.

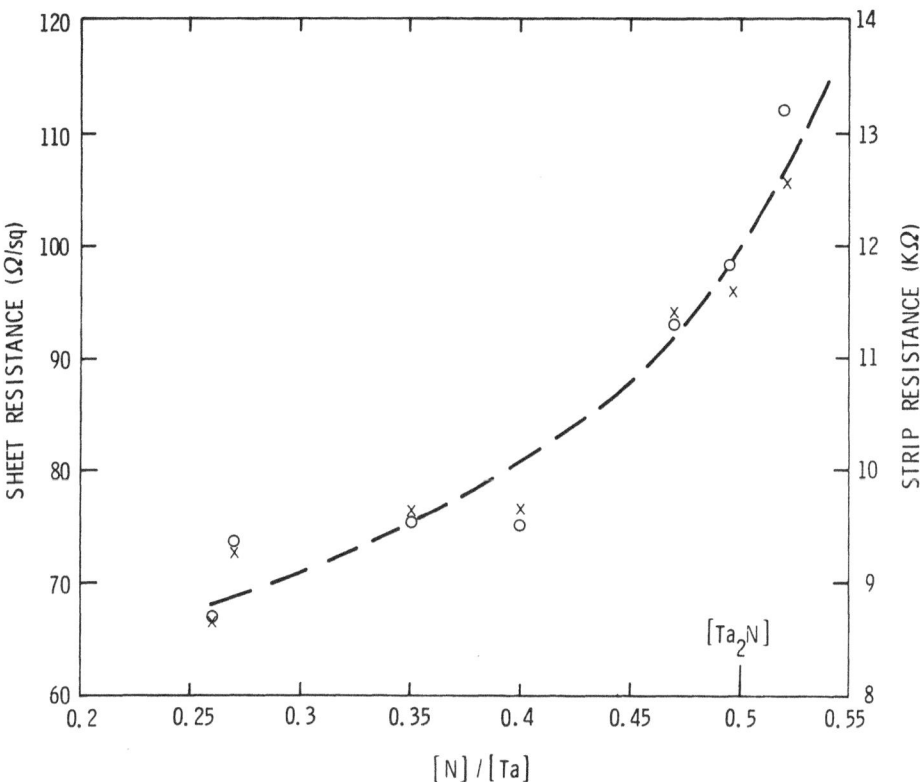

Fig. 3. Sheet resistance and strip resistance vs nitrogen content in the resistive film.

The resistivity vs nitrogen content is shown in Fig. 5. Resistivity is a bulk property and therefore to determine it from the sheet resistance, one must account for both the resistive film thickness and its roughness caused by the substrate roughness. The thickness was determined by measuring the energy width of the tantalum peak and applying Bragg's Rule[18] to the stopping cross sections[19,20] and accounting for the various constituents of the film. This yielded an areal density of tantalum in the film. Taking only that portion of the tantalum associated with the nitrogen and assuming a density for tantalum nitride of 15.8 grams/cm^2 a thickness in Å was obtained. The effect of surface roughness was taken into account by noting the increase in sheet resistance of films deposited on alumina substrates over that for single crystal Al_2O_3 (sapphire) in the same deposition. This effect was measured in a separate experiment. This correction was determined by averaging the difference obtained from 12 deposition runs and was 25 ± 1%. The resistance increased with increasing surface roughness. A measure of the surface roughness could be made by observing the

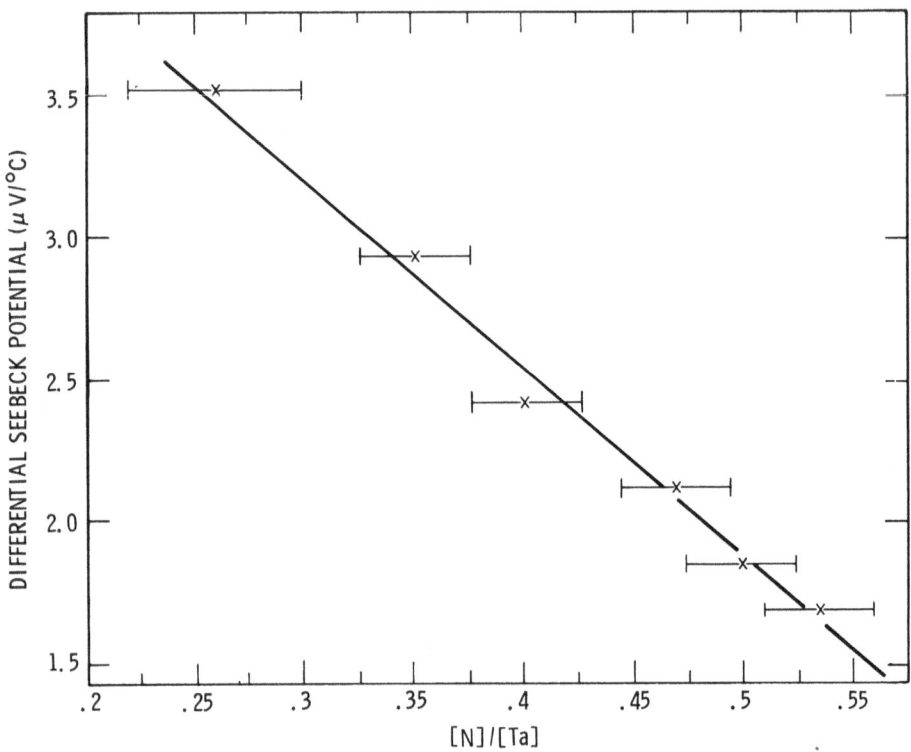

Fig. 4. Differential Seebeck potential vs nitrogen content.

slope of the trailing edge of the tantalum peak in the ion backscattering spectra. The slope was very high for both the polished beryllium and single-crystal Al_2O_3 but somewhat less for the alumina substrate. The resulting dependence of resistivity on nitrogen content is linear within the experimental uncertainty which is shown for all data points. The rms deviation for the sheet resistance was less than ±1%, but due to the assumption of a constant density, the ± 5% error for the stopping cross sections and the correction for film roughness, the experimental uncertainty for the resistivity is estimated to be ± 5%. Previous experiments by Guldner[7] and Waterhouse[8] have correlated film resistivity and nitrogen content. Both experiments used some form of chemical analysis to determine nitrogen content. The results of Guldner are shown in Fig. 5 and agree quite well with the present results while those of Waterhouse are about 60% higher and are not shown.

In most microelectronic circuit applications it is desirable to have resistors which have a small temperature coefficient of resistance, TCR, defined as $1/R \, (dR/dT)$. The TCR can be useful for film characterization and it can be important in circuit functions

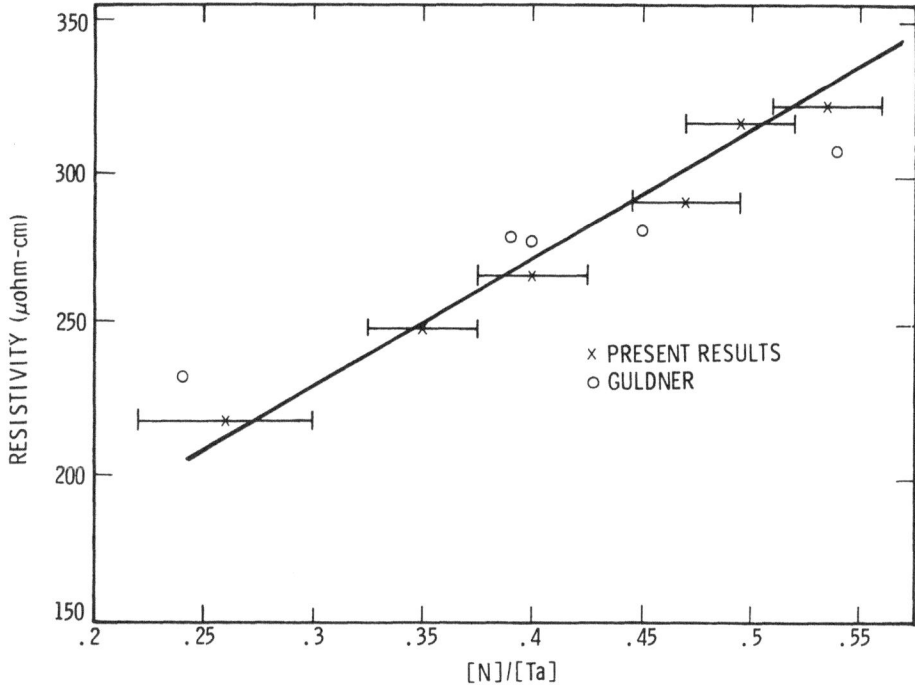

Fig. 5. Film resistivity vs nitrogen content. See Ref. 7 for
results of Guldner.

where ambient temperature changes would cause unwanted changes in
circuit performance. The studies relating TCR to nitrogen content
reveal that it is a strong function of nitrogen content below
[N]/[Ta] = 0.35, but remains almost constant throughout the range
of 0.35-0.55 as shown in Fig. 6. The units of TCR are parts per
million (ppm) per degree centigrade, and values less than 100 ppm/°C
are normally considered acceptable. Thin film resistors may drift
from their initial resistance values especially if they are operated
at higher temperatures. This effect is expressed as percent resistance change per unit time. In reality the percent drift is not
linear with time but logarithmic. Thin film resistors usually
become more stable as they age partly due to self-limiting processes
such as surface oxidation which forms a passivating layer. The
results of the stability tests also shown in Fig. 6 indicate that
the stability of the thin film resistors is independent of the nitrogen content to within the experimental error. The stability is
given as percentage change resistance for 1000 hours at 150°C.
Values for the stability less than 0.5%/1000 hr-°C are acceptable.

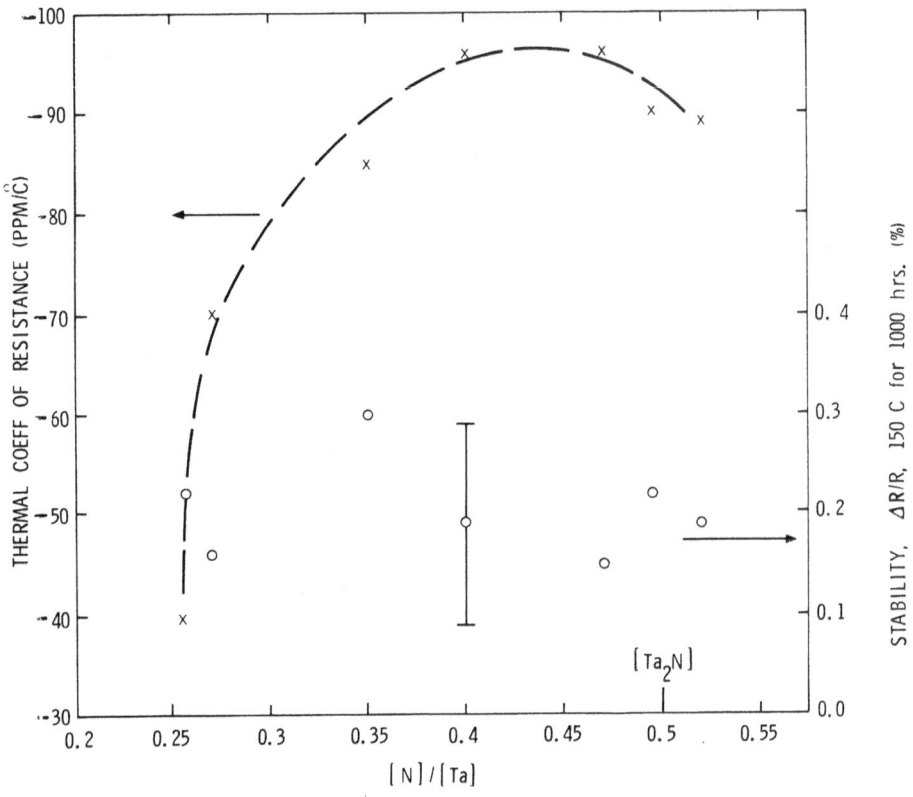

Fig. 6. Thermal coefficient of resistance and stability vs nitrogen content.

IV. DISCUSSION AND CONCLUSIONS

Brauer and Zapp[21] studied the tantalum nitride system using both x ray and chemical analytical methods and found only two stable stochiometric tantalum nitrides: Ta_2N and TaN. The solubility of nitrogen in metallic tantalum was observed not to exceed 4 at.% and from this concentration through $TaN_{0.35}$ mixed phases ot Ta_2N and metallic tantalum were observed. They found that Ta_2N is homogeneous from $TaN_{0.41}$ to $TaN_{0.50}$, while at greater nitrogen concentrations mixed phases of Ta_2N and TaN occur until TaN is reached. Based on the linear dependence of both the differential Seebeck potential and the resistivity on nitrogen content no abrupt change from a single phase to another single phase can be detected in the present study. However, the present results are not necessarily in conflict with the findings of Bower and Zapp since sputtered thin films are notorious for not being in thermodynamic equilibrium, i.e., they are multiphase and have numerous defects. Presumably a combination of at least two phases exist. The differential Seebeck voltage is an indication of the disparities of carrier concentration and their mobilities at the two temperature junctions. Thus the linear decrease of the differential Seebeck potential with increasing nitrogen content is indicative that barrier height is increasing with nitrogen content, i.e., the metallic character of the film is changing to a semiconductor character.

Our results lead to the following important conclusions: 1. A surface oxide layer grows on air exposure of the tantalum film at room temperature. Assuming the oxide is composed entirely of Ta_2O_5, then the equivalent oxide thickness is about 40 Å. After stabilization of the thin film resistors the surface oxide layer thickness increases to approximately 80 Å. Nitrogen was found to be somewhat excluded from this surface layer region. 2. The sputtering gas argon appears to be included uniformly throughout the film at a concentration of 2 at.%. 3. For sputtered thin tantalum nitride films both the resistivity and the differential Seebeck voltage ratio appear to exhibit a linear dependence on nitrogen content in the range $0.25 < [N]/[Ta] < 0.55$. Since the differential Seebeck potential is a strong function of nitrogen content, it is quite possible to use this parameter as a measure of nitrogen content when the remaining parameters in the deposition system remain constant. 4. The TCR of tantalum nitride film was found to decrease sharply below $[N]/[Ta] = 0.35$ and to be relatively constant from 0.35 to 0.55. 5. The stability of the resistors was independent of nitrogen content in the range covered by this experiment. 6. Use of ion backscattering using low Z substrates can give absolute depth concentrations of light elements in thin films nondestructively.

ACKNOWLEDGMENT

The author is indebted to D. J. Sharp for directing a substantial portion of the depositions and electrical measurements, to C. R. Peeples and to B. S. Gardner for preparation of the samples, to J. M. McDonald for running the accelerator and taking a portion of the backscattering data, and to R. E. Hampy and J. Sweet for fruitful discussions.

REFERENCES

1. L. I. Maissel, Trans. 9th National Vacuum Symposium, 1962, p. 169.

2. L. I. Maissel and R. Glang, Handbook of Thin Film Technology (McGraw-Hill, N. Y., 1970), p. 18-12.

3. D. A. McClean, J. Electrochem. Soc. Japan 34, 1 (1966).

4. J. Sosniak, W. S. Polito and G. A. Rozgonyi, J. Appl. Phys. 38, 3041 (1967).

5. R. W. Berry, P. M. Hall and M. T. Harris, Thin Film Technology
    (Van Nostrand Reinhold, N. Y., 1968, p. 343.

6. J. M. Morabito, Thin Solid Films 19, 21 (1973).

7. W. G. Guldner, Proceedings of the IEEE 14th Electronic Computer
    Conference (IEEE, New York, 1964), p. 9.

8. N. Waterhouse, Proceedings of the IEEE 22nd Electronic Computer
    Conference (IEEE, New York, 1972), p. 58.

9. R. A. Langley and D. J. Sharp, J. Vac. Sci. Technol. 12, 155
    (1975).

10. The bulk polycrystalline Be was obtained from Kaweki Berylco
    Industries, Inc.

11. D. R. Johnson, Sandia Laboratory Report No. SC-DR-71-9539 (1971).

12. T. B. Linnerooth, private communication.

13. A general review of the methods of ion beam analysis has
    recently been given by W. K. Chu, J. W. Mayer, M.-A. Nicolet,
    T. M. Buck, G. Amsel and F. Eisen, Thin Solid Films 17, 1 (1973).

14. There is always some radiation damage to the sample by the ion
    beam. However, this usually can be neglected in metal systems.
    In the present experiment, the analyzing ion beam was found to
    have no observable effect.

15. R. A. Langley and R. S. Blewer, Thin Solid Films 19, 187 (1973).

16. J. F. Ziegler and J. E. E. Baglin, J. Appl. Phys. 45, 1888 (1974).

17. M. L. Trudel, Proceedings of the IEEE 22nd Electronic Computer
    Conference (IEEE, New York, 1972), p. 431.

18. R. D. Evans, The Atomic Nucleus (McGraw-Hill, New York, 1955),
    p. 587 ff.

19. W. K. Lin, H. G. Olson and D. Powers, Phys. Rev. B8, 1881 (1973).

20. R. A. Langley, Phys. Rev., to be published (1975).

21. G. Brauer and K. H. Zapp, Zeitschr. Anorg. Chemie 277, 129
    (1954).

DISCUSSION

Q: (E. Wolicki) Is Argon still present in the film after the 300°C anneal ?

A: (R. Langley) Yes it is.

Q: (J.F. Ziegler) When we studied Rutherford scattering from nitrogen several years ago we did not find a good agreement between ^{14}N (α,α) and calculated cross-sections. What did you find about the backscattering of He ions from nitrogen ?

A: (R. Langley) If I remember your work correctly, you found agreement between the elastic and Rutherford cross section to within $\pm$ 5 % in the energy ranges around 2 MeV. I have referenced your work and included the $\pm$ 5 % from your work in the uncertainties shown in the present figures.

INVESTIGATION OF CVD TUNGSTEN METALLIZATIONS ON
SILICON BY BACKSCATTERING

P. Eichinger

Institut für Festkörpertechnologie der Fraun-

hofer-Gesellschaft, München

H. Sauermann und M. Wahl

Valvo Röhren- und Halbleiterwerke der Philips

GmbH, Hamburg

ABSTRACT

Pyrolytically deposited tungsten layers on Si have been
investigated by RUTHERFORD backscattering and some correlation to electrical properties has been established.
The parameters which have been varied were the temperature of the pyrolytic reaction and subsequent anneals
up to 900 C. Film thickness and metal-silicon interactions such as silicide formation and interdiffusion were
determined using N^+ and He^+ ions of 1.3 MeV energy. Formation of W_2Si at a reaction temperature of 600 C and
WSi at 900 C together with non-stochiometric mixing has
been observed explaining data from sheet resistivity
measurements. The best films on Si grow at a reaction
temperature of 700 C and remain stable with heat treatments up to 900 C. Thus the properties of CVD tungsten
layers on silicon are quite different from those of
sputtered layers investigated by BORDERS and SWEET /1/
the controlling factor being the reaction temperature
of the pyrolytic process.

1. INTRODUCTION

The pyrolytic deposition of tungsten on silicon has been suggested by several authors as basic process for a high-temperature resistent contact metallization system /2,3/. Tungsten seems to be advantageous because of its thermal expansion coefficient and the high temperature (1400 C) for formation of a WSi eutectic mixture according to the phase diagram for this particular system. However, as has been pointed out by BOWER /4/, compound phases from the phase diagrams represent strictly equilibrium conditions while compound formation for thin film metallizations can be entirely governed by the transport kinetics which occur when the films are heated (or grown). This has been verified for the W-Si system by BORDERS and SWEET /1/ who studied the formation of WSi$_2$ by annealing sputtered W-films on Si in the temperature range from 625 C to 750 C. Quite different kinetic processes can be expected for pyrolytically deposited films; this paper gives the experimental results for layer thickness, electrical resistivity and compound formation of W-layers grown on Si by WF$_6$ reduction as a function of the process parameters, mainly the reaction temperature.

2. THE PYROLYTIC PROCESS /2/

The process which will be investigated is based on the reaction formula

$$2 \ WF_6 + 3 \ Si \xrightarrow{\ T > 300C\ } 2 \ W + 3 \ SiF_4$$

This process occurs selectively with Si(i.e. not with SiO$_2$)and consumes Si. Theoretically for the growth of a 0,55 micron thick W-layer a 1 micron thick Si-layer is removed. The reaction has been performed in a diffusion furnace using N$_2$ as carrier gas. Special care has been taken to avoid oxidation of the W-layer by spurious amounts of oxygen during the reaction and subsequent cooling. All silicon samples were oriented in the 111 plane.

3. PROPERTIES OF THE TUNGSTEN LAYERS

a) Film Thickness and Electrical Resistivity

The thickness of the layer has been measured by com-

paring the step heights between unreacted areas of the
samples (masked with SiO_2 during the reaction) and areas
on which the reaction took place, before and after the
removal of the tungsten layer by etching with a H_2O_2/
/NH_4OH mixture (one part of 25 percent NH_4OH to one part
of 30 percent H_2O_2 to give an etch rate of 500 Å per minute at room temperature). In this way the tungsten layer
thickness and the thickness of the removed silicon can be
determined; typical results are shown in fig. 1 for reaction temperatures between 300 C and 1000 C. (Taking into account the possibility of compound formation, d has
to be interpreted as the thickness of the layer affected
by the etchant rather than simply as the tungsten layer
thickness, see section 3 b.) The occurence of a maximum
at ca. 500 C and a minimum at ca. 700 C clearly indicates
three different mechanisms for the layer growth in the
temperature range considered.

Sheet resistivity measurements and specific resistivity values calculated with the thickness taken from fig
fig. 1a are shown in fig. 1b. The variation of the specific resistivity over more than a factor of 20 stimulated the compositional analysis by backscattering of He^+
and N^+ particles.

b) Backscattering of 1.3 MeV He^+ and N^+ Particles

Spectra were taken for W-layers deposited at 600 C,
700 C and 900 C. Furthermore the minimum W-concentration
in Si has been determined which is necessary for the
etchant used for thickness measurements in order to remove material. W-layers grown at 700 C have been annealed up to 900 C. For each process a reference spectrum
taken with the same ion dose from a pure tungsten layer
on Si is included for comparison of the backscattering
yields.

600 C: Fig. 2 and fig. 3 show the energy spectra
of He^+ particles backscattered to an angle of 165
degrees from a tungsten layer deposited on n-type (n =
= 5 x 10^{18} cm^{-3}) resp. p-type (p> 10^{20} cm^{-3}) Si. For weaker doped concentrations (n, p = 10^{14} cm^{-3} and p = 10^{18}
cm^{-3}) essentially the same spectra resulted clearly
showing compound formation by the occurence of a well
defined step in the W-signal with a corresponding step
in the Si-signal. In order to establish the composition
of the compound the total Si-yield can be compared with
the total W-yield if the compound on the step height

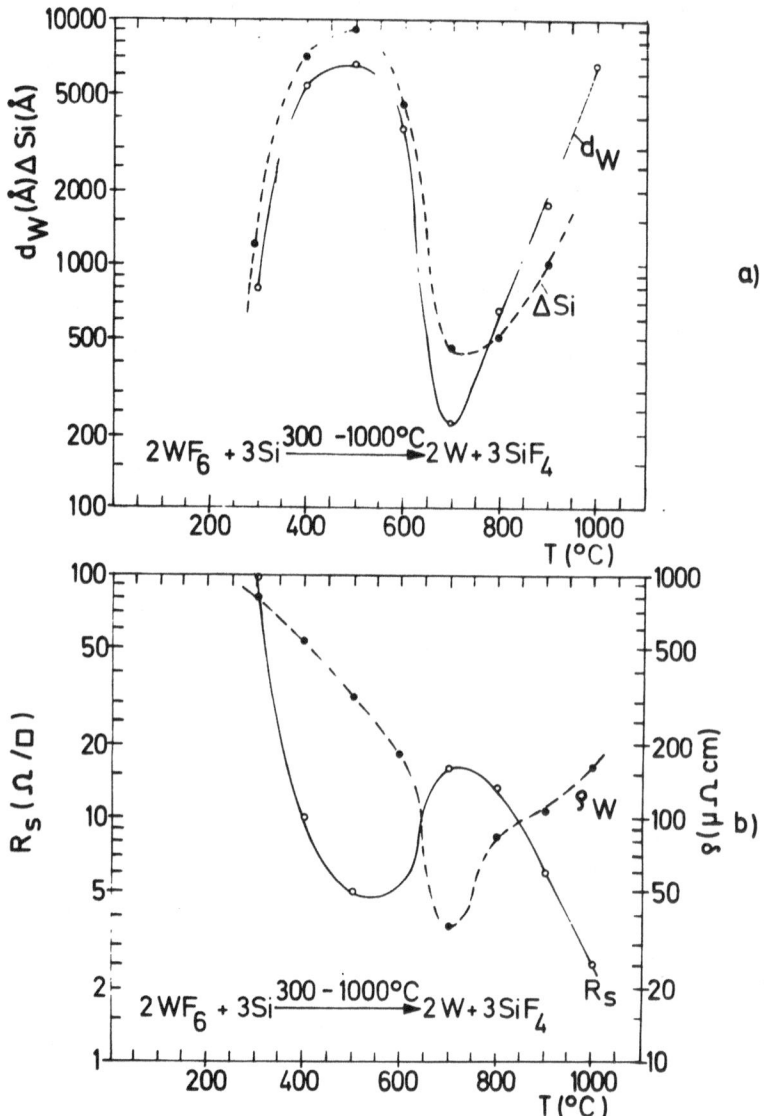

Fig. 1: a) Tungsten layer thickness d_W and thickness
         of the silicon removal Δ_S, as a function of
         the reaction temperature.
         b) Sheet resistivity R_S and specific resisti-
         vity ρ_W of the tungsten layer.

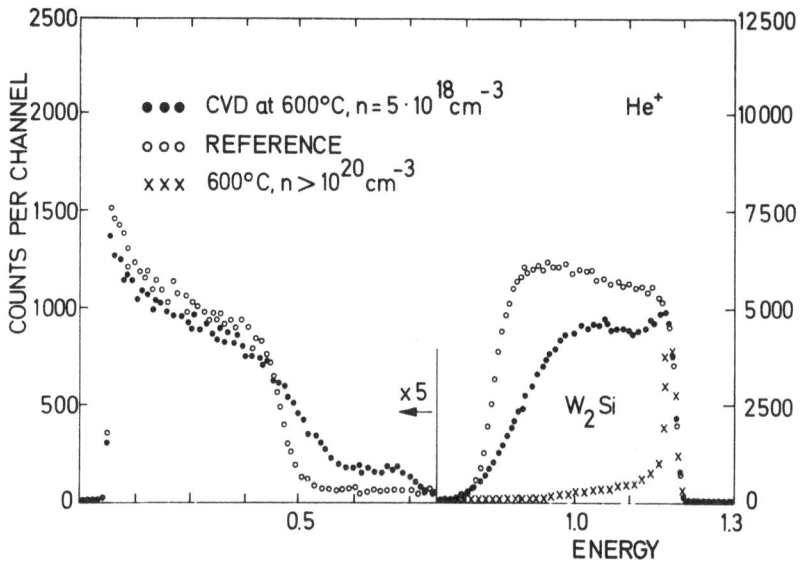

Fig. 2: Spectra of 1,3 MeV He$^+$ backscattered from 600°C
n-type CVD samples (n = 5 x 10^{18} cm^{-3}, n>10^{20} cm^{-3})
and from a reference sample (pure W on Si).

between compound and pure material in the W-signal or in
the Si-yield can be evaluated /5/. The latter method has
the disadvantage that the ratio of the atomic stopping
powers   and   is needed, but is easier to apply, esp.
when the layers are relatively thick so that the two signals overlap. In our spectra results consistent with both
methods could be achieved with an $\varepsilon_W/\varepsilon_{Si}$ ratio of 2 in
agreement with the values of ZIEGLER /6/ and BORDERS /7/:
The W:Si ratio in the compound is 2:1, indicating that

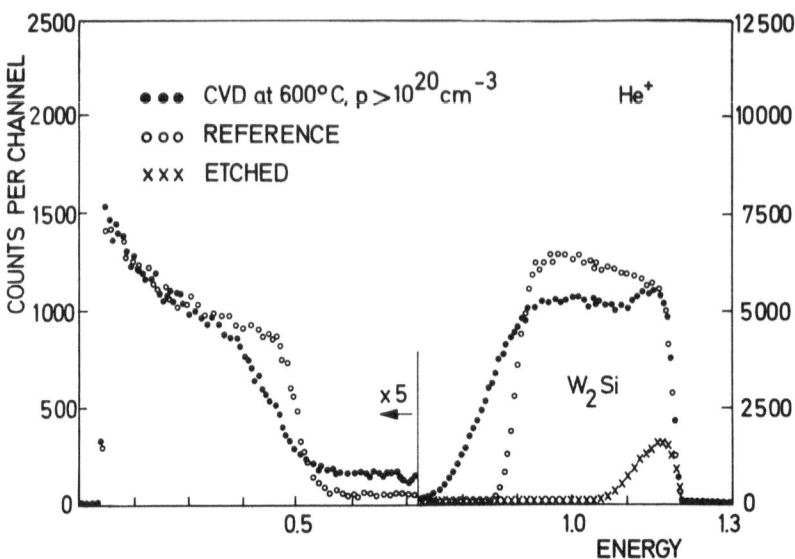

Fig. 3: Spectra of 1,3 MeV He$^+$ backscattered from a 600$\overset{o}{C}$
        p-type CVD sample (p>10^{20} cm^{-3}) before and after
        etching.

a W Si compound has been formed with a thickness of several thousand Å. At the surface several hundred Å of
pure W are present. Assuming a relatively low conductivity of W$_2$Si compared to pure W the high specific resistivity of fig. 1 can be explained if the compound
layer is removed with the H$_2$O$_2$ /NH$_4$OH solution because
in this case the silicide thickness is included in the
layer thickness measurement. That this is really so can
also be seen from fig. 3 where the sample has been etched
for more than 30 minutes giving a lower limit for the
etchant to be effective of about 20 percent W in Si. A
different layer develops at the same temperature for
heavily (n>10^{20} cm^{-3}) doped n-type samples (fig. 2):
No silicide formation can be observed, instad a small
concentration W-tail - hardly reproducible for different
samples - extends from the pure W surface into the Si.

    700 C: Fig. 4 shows the layer resulting from a re-
action temperature of 700 C. The yield in the W signal
indicates that the layer is pure W with a thickness of

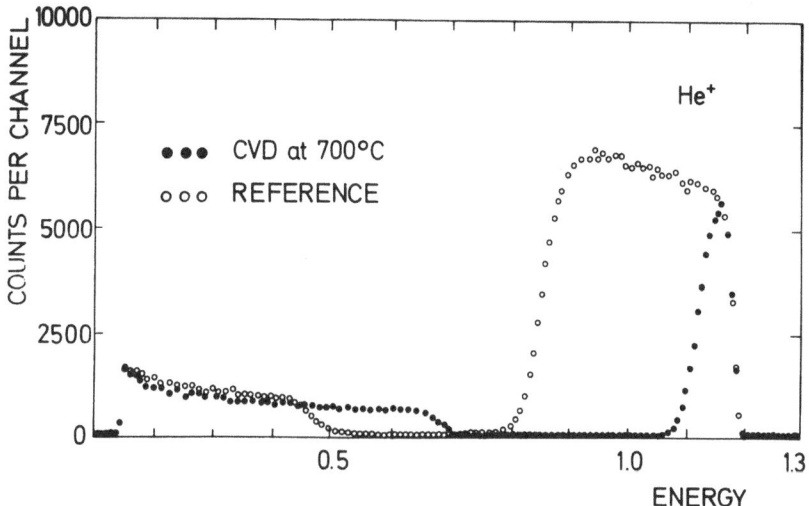

Fig. 4: Spectrum of 1,3 MeV He$^+$ backscattered from a
700°C CVD sample.

some hundred Å. This time the thickness taken from the
etched layer corresponds to the metal thickness and the
calculated resistivity is accordingly low. A comparison
of the pyrolytically grown layer at 700 C with sputtered
layers annealing at 700 C /1/ is of interest: According
to BORDERS and SWEET a WSi$_2$ layer of at least 800 Å is
formed on sputtered layers within 5 min which was the
reaction time for the pyrolytic process. This difference
becomes still more pronounced if the 700 C pyrolytic
layers are annealed at temperatures of 800,850 and 900C
like it is shown in fig. 5 for an anneal time of 20 minutes. Because the layer was in this case only about
300 Å thick N$^+$ions were taken to improve depth resolution. The result is that no compound is formed even at
850 C while with sputtered tungsten the layer should be
transformed into WSi$_2$ almost immediately (within much
less than a minute according to an extrapolation of the
data of BORDERS and SWEET).
Only at 900 C some decomposition starts to develop.

        900 C : At this temperature the compound WSi is
formed as can be seen from fig. 6 with a layer of pure
W at the surface. Similar to /4/ continuous variation
from pure Si to pure W was also found due to nonunifor-

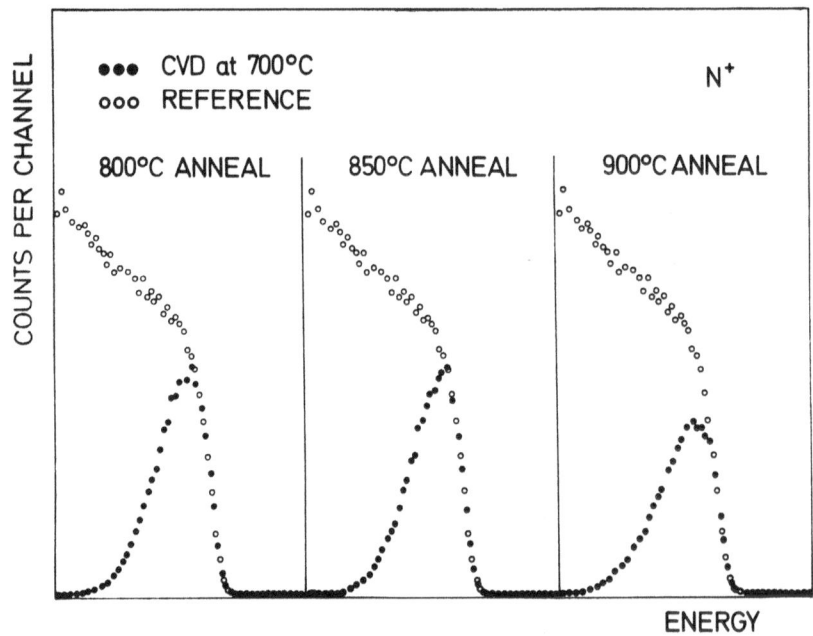

Fig. 5: W portions of the spectra of 1,3 MeV N$^+$ from
700 C CVD samples annealed at 800 C, 850 C
and 900 C.

mities of the layers over the area of the analyzing
beam. Again the high specific electrical resistivity
of the total reacted layer can be explained by the
formation of a low conductivity compound.

4. DISCUSSION

The results for pyrolytically grown W-layers on Si
differ in several points from those observed on sputtered layers - most markedly in the formation of W_2Si at
600 C which is not found with sputtered layers, and the
stability of 700 C layers against silicide formation at
anneals up to 850 C. Any theory explaining these results
must include the reaction and transport kinetics of the
pyrolytic process which seems to be a very difficult
task. From a practical viewpoint our results show that
pyrolytic tungsten deposited with a reaction temperature

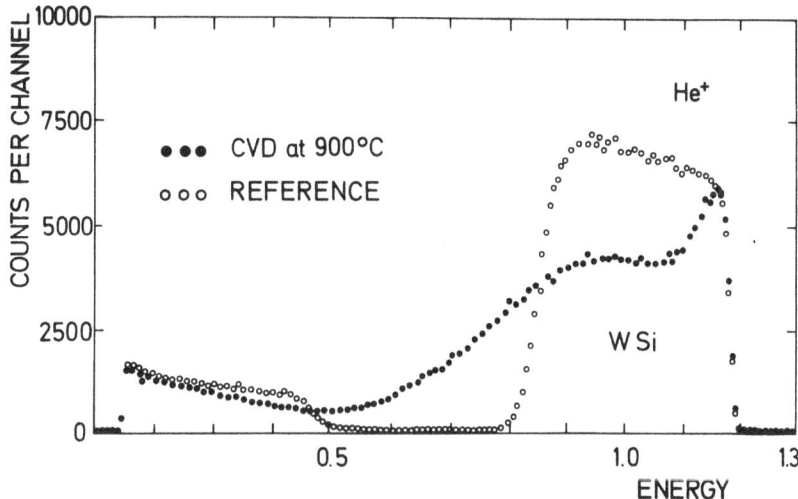

Fig. 6: Spectrum of 1,3 MeV He$^+$ backscattered from a
        900 C   CVD sample.

of 700 C is a promising metallization for obtaining
high temperature resistent contacts on Si.

5. REFERENCES

/1/ J. A. Borders and J. N. Sweet : Application of
     Ion Beams to Metals, S. T. Picraux, E. P. Eer
     Nisse, and F. H. Vook edts., Plenum Press, New
     York 1974, p.179

/2/ J. M Shaw and J. A. Arnick: Solid State Technology,
     Dec. 1971, 53

/3/ D. M. Melliar-Smith, A. C. Aclams, R. H. Kaiser
     and R. A. Kushner: J. Electrochem. Soc. : Solid
     State Science and Technology Febr. 1974, 298

/4/ R. W. Bower: Thesis, Cal. Inst. of Technology,
     April 1973.

/5/   J. A. Borders and J. N. Sweet : J. Appl. Phys.
      Vol. 43, No. 9, 3803 (1972)

/6/   J. F. Ziegler and W. K. Chu : Thin Solid Films,
      19, 281 (1973)

/7/   J. A. Borders : Radiation Effects 21, 165 (1974)

DISCUSSION

Q: (A. Hiraki) Is there any explanation about the effect of impurity concentration-dependence on the silicide (W_2Si) formation? And do you know other examples which show similar effects ?

A: (P. Eichinger) It is not yet clear, maybe the lattice stress introduced by heavy P doping can affect the reaction kinetics.

Comment: (J.A. Borders) Phosphorus precipitation may have something to do with the phosphorus impurity concentration dependence.

Q: (W.K. Chu) I did not understand the reaction kinetics. You mention that W_2Si will form at $600^{\circ}C$ but at $700^{\circ}C$ it does not form during CVD. Why ?

A: (P. Eichinger) In view of the complexity of the CVD process including parasitic reactions and transport phenomena in the solid and gaseous phases an explanation is beyond the scope of this investigation. But I would like to point out that we found complete reproducibility of this temperature dependence over a great number of W layer deposition runs where reaction time, WF_6 concentration and gas flow have been varied.

ION BEAM ANALYSIS OF ALUMINIUM PROFILES IN HETERO-EPITAXIAL $Ga_{1-x}Al_xAs$-LAYERS

P. Bayerl, W. Pabst and P. Eichinger

Institut für Festkörpertechnologie der Fraun-

hofer-Gesellschaft, München

ABSTRACT

RUTHERFORD-backscattering, ion-induced X-rays and nuclear reaction techniques were used to measure aluminium
profiles in GaAlAs. While backscattering is mainly useful to give information about profiles near the surface
with a sensitivity of about 5 percent of aluminium and
a depth resolution of about 360 Å at 0.3 μm depth, and
X-rays give the mean concentration in the surface layer,
the $^{27}Al(p,\gamma)^{28}Si$ reaction is generally more applicable
for deeper multilayer structures. The sensitivity is
a few percent of aluminium and the depth resolution is
limited by energy straggling to about 0.2 μm at 2 μm
depth. For profiles extending over more than 3 μm corrections must be made due to the presence of other nuclear resonances at energies different than 0.992 MeV;
the resonance curve is deduced from the thick target
yield. Aluminium profiles are shown for single layers
and a multilayer heterolaser structure.

I. INTRODUCTION

Heteroepitaxial layer structures of GaAs and GaAlAs
are of major interest in the development of solid state
optoelectronic emitters such as light emitting diodes
and lasers. Generally, the optically active GaAs layer
with a submicron thickness is sandwiched between GaAlAs
layers to achieve optical and minority charge carrier
confinement due to the layer bandgap of GaAlAs. Multi-

layer heteroepitaxial structures are grown by liquid
phase or molecular beam epitaxy. For the determination
of aluminium profiles in epitaxial layers a variety of
analytical techniques has been applied including wavelength analysis of the recombination light with electron beam or optical excitation /1/, RUTHERFORD-backscattering of He^+ particles /2/ and AUGER-spectroscopy
together with ion milling /3/. This paper compares the
analytic possibilities of MeV ion beams for the determination of aluminium in GaAs by discussing results
obtained with backscattering, ion induced X-rays and
the $^{27}Al(p,\gamma)^{28}Si$ nuclear reaction.

2. He^+ INDUCED X-RAYS

The principle of the method consists in comparing
the intensity of the Ga K_α emission from the GaAlAs
sample with the intensity from a pure GaAs sample. The
lower intensity from the ternary system is attributed
to Al atoms occupying Ga lattice sites. The Ga K_α peak
area at 9.25 keV is normalized to the As K_β peak at
11.8 keV as was done by FELDMAN et al. /4/ who investigated outdiffusion of the GaAs components into anodic
oxides. For pure single crystal GaAs we found a ratio
of the Ga to the As intensity of R = 9.2 at 1.3 MeV
He^+ particle energy, independent of crystal orientation. Fig. 1 shows the X-ray spectra taken with a Si(Li)-
detector for pure GaAs (R = 9.2) and a 4 micron thick
GaAlAs layer grown on a GaAs substrate (R = 6.1). From
these numbers an atomic aluminium concentration of
x = 0.34 can be calculated. For accurate measurements
the fluorescence yield of Ga quanta excited by As quanta
in the substrate must be subtracted; the difference in
the fluorescence contribution for the pure GaAs sample
and the epilayer is negigible. The fluorescence yield
can in principle be determined by comparing the yield
from a GaAs and a GaP sample.

The measured aluminium concentration is an average
value over the stopping path (4.2 μm) of the He^+ particles. The strong energy dependence of the excitation
cross section, however, results in an effective probing
depth of ca. 1 micron with 1.3 MeV He^+ particles at
normal incidence.

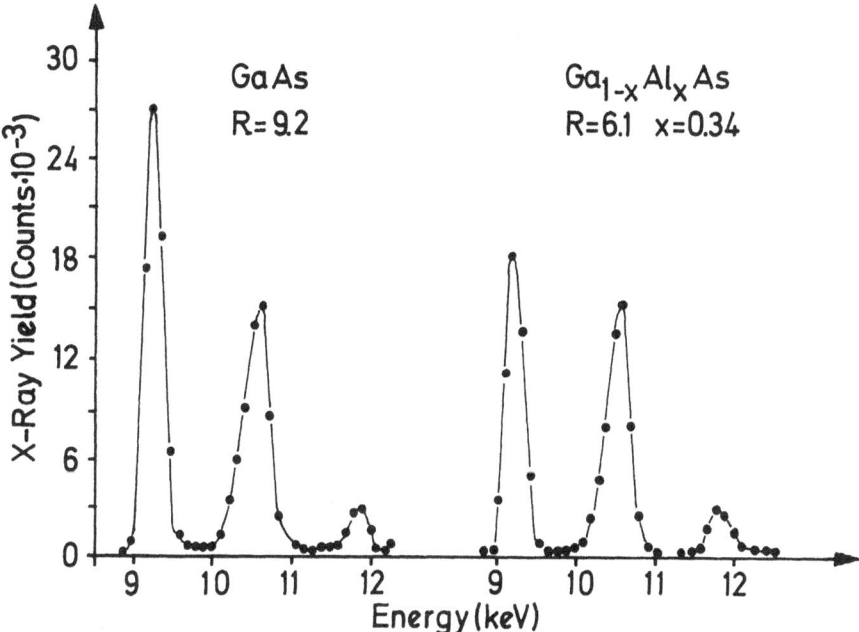

Fig. 1: X-ray spectra from a GaAs and a GaAlAs sample

3. RUTHERFORD BACKSCATTERING WITH He$^+$-IONS

Rutherford backscattering offers the possibility to determine concentration profiles of multilayer structures. The method consists in comparing the counts from GaAlAs with those from a pure GaAs target and was first demonstrated by J. W. Mayer et al. /2/. Fig. 2 shows the backscattering yields obtained from a layered structure, consisting of a GaAs substrate, a 4 μm thick GaAlAs layer and a top layer of 3200 Å GaAs. The energy of the incident particles was 1.3 MeV and the resolution of the detecting system 14 keV, therefore the backscattering contribution from Ga and As could not

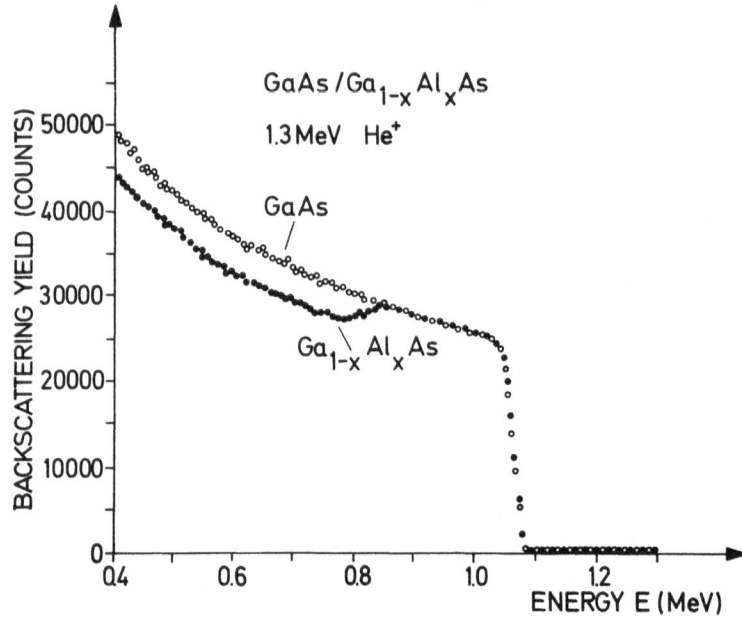

Fig. 2: Backscattering spectra from a pure GaAs (upper
curve) and a GaAs-GaAlAs layer (lower curve)
which were continuously tilted with respect to
the ion beam.

be resolved; this results in a broadening of the front
edge of the spectra at 1.05 MeV. At the beginning of
the GaAlAs layer the yield decreases, showing the replacement of Ga by Al.

In a single crystal the stopping powers are dependent on target orientation. Therefore the target was
continuously tilted with respect to the incident beam
in order to average over many crystal orientations. To
show the reproducibility of the method, the yield of
the GaAs substrate with that of the 3200 Å epilayer was
compared. The agreement was within 3 %.

Fig. 3 shows the Al-concentration dependence on
layer depth, calculated from the ratio: counts (GaAs)/

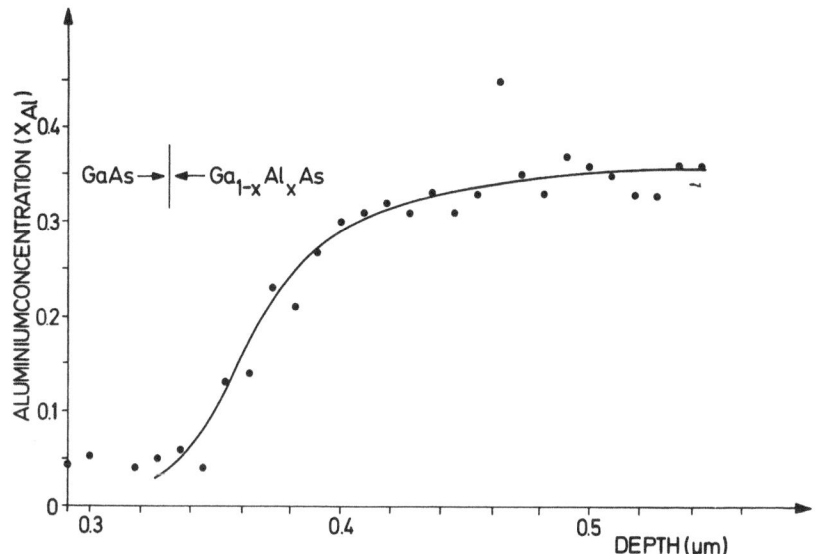

Fig. 3: Al-concentration at a GaAs-GaAlAs interface, showing the depletion of Al in the GaAlAs layer.

counts (GaAlAs), using the stopping power data from J. F. Ziegler /5/. At a depth range from 3300 Å to 4500 Å the atomic aluminium concentration increases from $x_{Al} = 0$ to $x_{Al} = 0.32 \pm 0.03$.

Energy straggling sets a limit in determing a concentration profile. From the backscattering yield of a 3200 Å polycrystalline Ge-layer, evaporated on sapphire, a depth resolution of 360 Å was measured. This value represents therefore a lower range limit for a profile that can be measured. At 1 μm depth this minimum value is 630 Å, if a detailed calculation is done, regarding the following points: different scattering cross sections for Ga and As, a thick target calculation, differences in the stopping power data introduced by the replacement of Ga by Al, an increase of energy straggling with the square root of the thickness.

4. NUCLEAR REACTION TECHNIQUES

The nuclear reaction $^{27}\text{Al}(p,\gamma)^{28}\text{Si}$ has already
been applied by Amsel /6/ and extensively discussed
by Dunning /7,8/ for profiling Al in Al_2O_3, SiO_2 and
SiC layers, for depths up to 0.8 μm. This technique
uses the narrow (100 eV) resonance in the reaction cross
section at a proton energy of 992 keV. A proton beam
passing through the sample loses energy and excites
the reaction when the resonance energy is reached. ·
Counting the γ-quanta yields the Al-concentration in
the corresponding depth region. Fig. 4 shows the energy
spectrum of the gamma rays emitted from a pure GaAs
sample and from a GaAlAs sample. Variation of the initial beam energy yields the profile. The thicker the
Al layer is, the more additional resonances in the
reaction cross section will contribute. Energy straggling of the beam limits depth resolution.

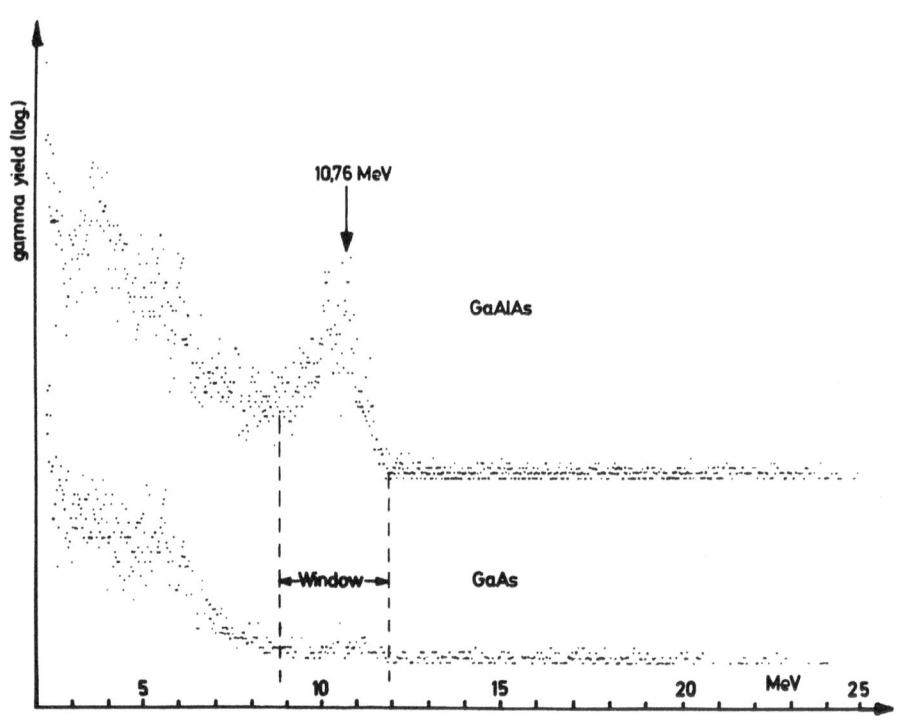

Fig. 4: γ-ray emission from a pure GaAs sample and
a GaAlAs sample.

The reaction cross section was measured with a
thick Al target so that the technique could be applied
to thick layer profiling up to 4 μm. The γ-rays were
measured by a 3'' x 3'' NaI(Tl) detector. Depth resolution and sensitivity are discussed.

a) Reaction cross section

Fig. 5 shows the reaction cross section measured
with a thick Al target (thickness > beam range). The
step at 992 keV results from the well-known strong resonance. The depth scale shown in the figure was calculated from tabulated stopping power values /9/. For
the measurement of Al layers thicker than 3 μm, the
weaker and broader resonance at about 780 keV and the
increasing cross section above 1130 keV must be taken
into account. The quantitative yield from each resonance

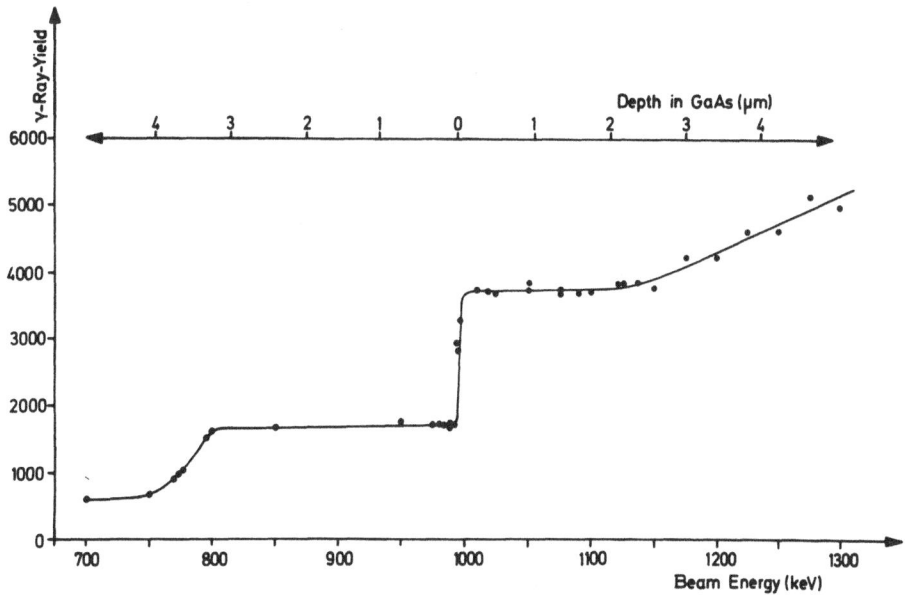

Fig. 5: Integrated cross section for the ^{27}Al(p,γ)
^{28}Si reaction, measured with a thick Al target

can equally be taken from fig. 5. At the 992 keV resonance, the yield is increased only by a factor of 2.

b) Depth resolution

Depth resolution is limited at the surface by the
energy distribution in the beam (in our experimental
set-up: 2 keV). As the beam is slowed down in the sample
by statistical processes, this energy distribution
broadens. At 2 μm the depth resolution is about 0.2 μm.
The straggling data were taken from Bohr theory /11/
(which is an overestimation compared to experimental
results, see /10/).

c) Sensitivity

The sensitivity of the method is best when the
10.7 MeV γ-rays (and the corresponding escape peaks)
are measured. Using a 3'' x 3'' NaI (Tl) detector, a
window of about 3 MeV width was found to give an optimum signal to noise ratio. In this energy region beam
independent background (mainly cosmic rays) was found
to be 10 counts in the usual measuring time of 4 min.
Background excited by the beam (in the GaAs substrate,
the target chamber and the beam line) was additional
30 counts for an integrated charge of 120 μCb. This led
to an actual sensitivity limit of 1 % Al in GaAs. Sensitivity can be improved by increasing the solid angle
of the detector and thus the counting rate.

d) Profiles

Fig. 6 shows an Al profile measured by the nuclear
reaction technique. The 4 μm thick profile is corrected
for the contributions of resonances other than the
strong 992 keV resonance, using the result of fig. 5.
Absolute concentration is given by comparison with the
yield from a pure Al standard. Fig. 7 shows the γ-yield
from a multilayer heterolaser structure. The investigated thickness was 4 μm. At a depth of about 2.8 μm
there is a thin layer of GaAlAs of 0.25 μm thickness
which could be resolved.

It should be remarked that the high proton densities required, and long measuring times to get a profile, can cause crystal degradation and heating of the
sample.

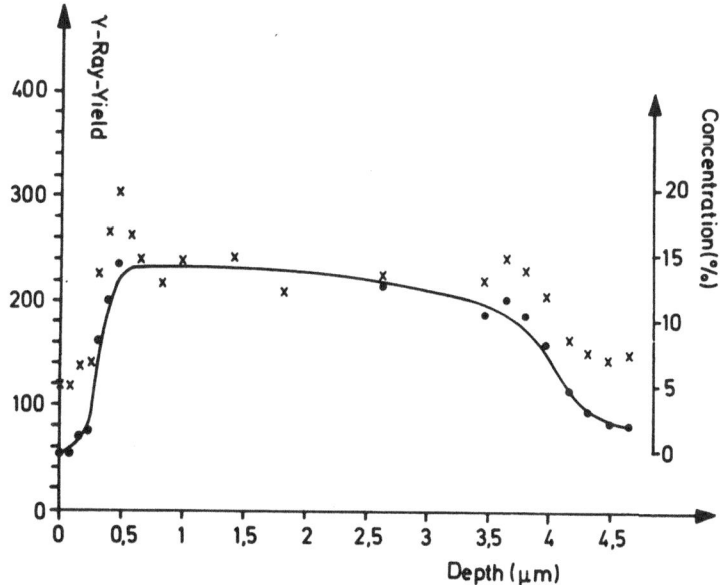

Fig. 6: Al profile in GaAlAs measured by the nuclear
        reaction technique:
        x   measured points
        •   correction for contributions from resonan-
            ces other than the 992 keV resonance.

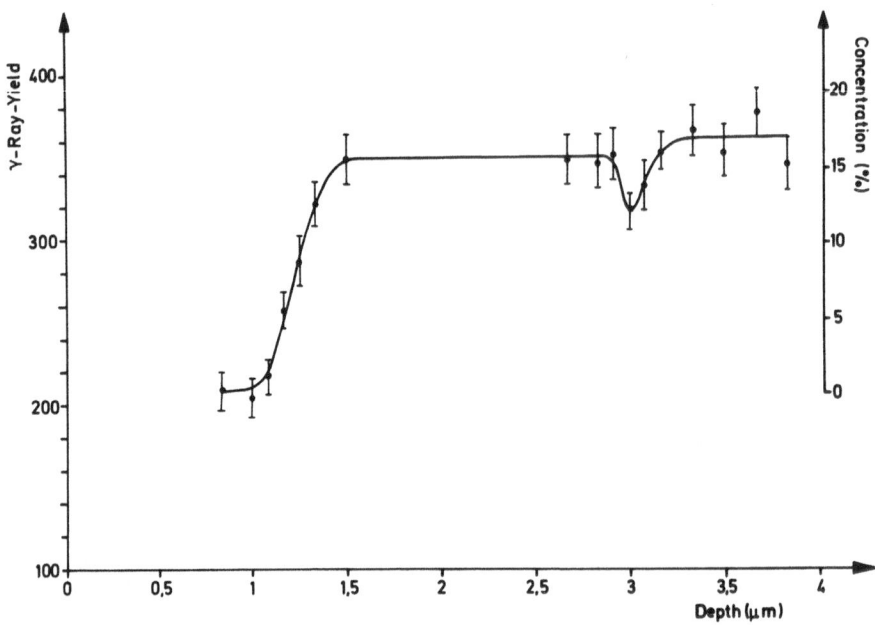

Fig. 7: γ-ray yield from a multilayer heterolaser structure. The structure consists of a GaAs top layer
and three GaAlAs layers having different atomic
aluminium concentrations.

REFERENCES

/1/   H. B. Bebb and E. W. Williams in Semiconductors
       and Semimetals Vol.8, R. K. Willardson and
       A. C. Beer edts. Academic Press New York and
       London 1972, p. 182.

/2/   J. W. Mayer, J. F. Ziegler, L. L. Chang, R. Tsu
       and L. Esaki : J. Appl. Phys. 44 (5), 2322 (1973)

/3/   R. Ludeke, L. Esaki and L. L. Chang: Appl. Phys.
       Letters 24 (9), 417 (1974)

/4/   L. C. Feldman, J. M Poate, F. Fermanis and B.
        Schwartz: Thin Solid Films 19, 81 (1973)

/5/   J. F. Ziegler, W. K. Chu: Thin Solid Films 19,
        181 (1973)

/6/   G. Amsel et al., Nucl. Instr. & Meth. 92 (1971)
        481

/7/   K. L. Dunning et al., Thin. Solid Films 19 (1973)
        145

/8/   K. L. Dunning and H. L. Hughes, IEEE Transactions
        on Nucl. Science NS-19 (1972) 243

/9/   L. C. Northcliffe and R. F. Schilling, Nuclear
        Data A7 (1970) 233

/10/   J. M. Harris et al., Thin Solid Films 19 (1973)
        259

/11/   J. Bohr, Phil Mag. 30 (1915) 581

DISCUSSION

Q: (D. Simons) Comment to question from Jim Hirvonen about where the origin of background below 992 keV resonance in the ^{27}Al (p,γ) ^{28}Si reaction.

A: There is a lower energy resonance. Thus from a thick target this would be a source of background.

ANALYSIS OF $Ga_{1-x}Al_xAs$-GaAs HETEROEPITAXIAL LAYERS BY PROTON BACKSCATTERING*

K. Gamo;* T. Inada;** I. Samid, C.P. Lee and J.W. Mayer

California Institute of Technology

Pasadena, California 91125

ABSTRACT

Proton backscattering at < 1 MeV has been used to measure $Ga_{1-x}Al_xAs$ heteroepitaxial layers on GaAs that are used in optoelectronic applications. By evaporating Ge on Al, we have obtained relative values of the stopping cross-section of Al to Ge. Deposition of Ge layers on $Ga_{1-x}Al_xAs$ layers removed problems associated with secondary electron suppression. The experimental data on composition compare well with analysis by other techniques.

I. INTRODUCTION

There has been continued interest in GaAlAs layers on GaAs for applications involving optical properties of the structures [1]. In these applications thicknesses of the epitaxial layers are typically between 0.2 and 4 μm. It has been previously shown [2] that MeV He backscattering techniques can be used to determine the composition profiles of periodic structures (GaAs-GaAlAs) for thicknesses up to 1 μm. The purpose of the present paper is to investigate the applications of MeV proton backscattering techniques to analyse thicker layers of GaAlAs.

One of the problems in utilizing proton backscattering is that there has not been as extensive an evaluation of stopping cross-sections as for He ions. In the present case this problem can be circumvented by depositing Ge on Al and GaAs to obtain relative stopping cross-sections. The stopping cross-section of Ge is very close [2] to that of GaAs for He ions and suggests that Ge can be used as a reference for proton analysis.

II.   ANALYSIS PROCEDURES

In the analysis, we have assumed that the stopping cross-section ε of GaAlAs is the simple addition of the stopping cross-section for the individual elements weighted according to their atomic composition. To calculate the height of the spectra as a function of energy loss we have followed the procedures of Behrisch and Scherzer [3]. Recently this analytical approach has been shown to be in good agreement with the exact computation for cases when the backscattering kinematic factor K has values near unity [4]. This condition is satisfied for proton in our cases, since K_{Al} = .86 and K_{Ge} = .95.

Following the procedures of reference 3 and using conventional notation [5], we obtain

$$R_1(E_1) = \frac{N_{Ga}^{GaAlAs}\sigma_{Ga}(E) + N_{As}^{GaAlAs}\sigma_{As}(E)}{N_{Ga}^{GaAs}\sigma_{Ga}(E) + N_{As}^{GaAs}\sigma_{As}(E)} \cdot \frac{\delta t_{Ge}^{GaAlAs}}{\delta t_{Ge}^{GaAs}}$$

$$= \frac{\sigma_{Ga}(1 - x(t)) + \sigma_{As}}{(\sigma_{Ga} + \sigma_{As})[1 - 0.5x(t)\{1 - R_\varepsilon(E_1)\}]} \tag{1}$$

where $R_1(E_1)$ is the ratio of the heights of Ga + As components of $Ga_{1-x}Al_xAs$ to that of GaAs at the detected energy E_1 and $R_\varepsilon(E_1) = \varepsilon^{Al}(E_1)/\varepsilon^{Ge}(E_1)$ with the approximation that $K_{Ga} = K_{As} = K_{Ge}$. The results of this calculation are shown in Fig. 1(a) for different values of R_ε at E_o = .987 MeV. (This energy was chosen because of the Al resonance at 992 keV.) It can be noted that the errors in determination of the Al concentrations are of the same magnitude as any errors in R_ε. Also the backscattering yield ratio depends upon the detected particle energy since $\varepsilon^{Al}/\varepsilon^{Ge}$ varies with energy. For example, when E_1 changes from .94 to .6 MeV, $\varepsilon^{Al}/\varepsilon^{Ge}$ changes from .54 to .58.

In the analysis of GaAlAs with thick layers with protons, the Al component of the backscattering yield is superimposed on

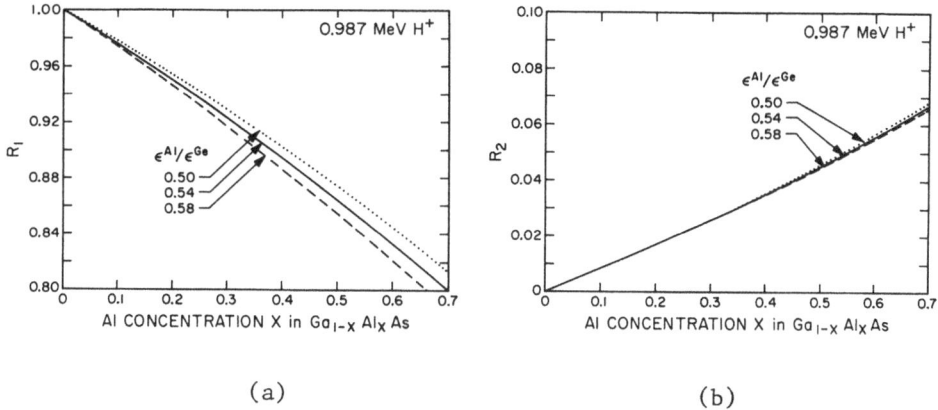

Figure 1.   Calculated ratios of (a) Ga + As counts from
            Ga$_{1-x}$Al$_x$As to those from GaAs (R$_1$) and (b) Al counts
            from Ga$_{1-x}$Al$_x$As to Ga + As counts from GaAs (R$_2$) as
            a function of Al concentration x.

that from the Ga and As components of the yield as shown in the
inset of Fig. 2.   The Al component of the yield can be subtracted utilizing iterative procedure.   In order to perform this
subtraction one must relate the energy E_1' of the signal from the
Al to the corresponding energy E_1 of the signal from the Ga + As
at a given depth t within the sample.   The relation is

$$E_1' = [K_{Al}^{1-n} E_o^{1-n} - (K_{Ge}^{1-n} E_o^{1-n} - E_1^{1-n}) \frac{K_{Al}^{1-n} + \cos\alpha/\cos\beta}{K_{Ge}^{1-n} + \cos\alpha/\cos\beta}]^{\frac{1}{1-n}} \qquad (2)$$

where α and β are the angles of incidence and reflection to the
normal of the surface and the stopping cross-section is des-
cribed by $\varepsilon = \varepsilon_o E^n$ [3].   The results of this calculation with
$n = -1/3$, $\alpha = 5°$ and $\beta = 10°$ are shown in Fig. 2.   At the energy
E_1, the total heights of the spectrum is the sum of the height of
Al, $H_{Al}^{GaAlAs}(t)$ ($H_{Al}(t)$ in Fig. 2), and that of Ga + As,
$H_{Ge}^{GaAlAs}(t)$ ($H_{Ge}^{Al}(t')$ in Fig. 2).   The component $H_{Ge}^{GaAlAs}(t')$
corresponds to backscattering from Ga + As at a different depth

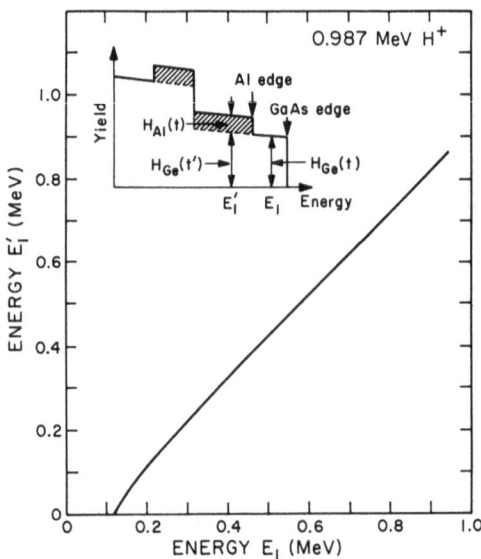

Figure 2.   Relation of the energy E_1' of the signal from the Al
to the corresponding energy E_1 of the signal from the
Ga + As at depth t.

$t' > t$. The ratio $R_1(E_1') = H_{Ge}^{GaAlAs}(t')/H_{Ge}^{GaAs}(E_1')$ can be obtained by subtracting $R_2(E_1') = H_{Al}(t)/H_{Ge}^{GaAs}(E_1)$ from the ratio of
the heights of GaAlAs spectrum to GaAs spectrum at the energy E_1.
The ratio $R_2(E_1')$ is given by

$$R_2(E_1') = \frac{x(t')\sigma_{Al}(E)}{\sigma_{Ge}(E') + \sigma_{As}(E')} \cdot \frac{K_{Ge} + \cos\alpha/\cos\beta}{K_{Al} + \cos\alpha/\cos\beta}$$

$$/[1 - 0.5x(t')\{1 - R_\varepsilon(E_1')\}] \qquad (3)$$

where E and E' are the energy before the collision for Al and Ge
and are

$$E = \{\frac{E_1'^{1-n} - (\cos/\cos)E_0^{1-n}}{K_M^{1-n} + \cos\alpha/\cos\beta}\}^{1/(1-n)}, \quad M = Al, Ge \qquad (4)$$

The results of this calculation are shown in Fig. 1(b).

The procedures of Behrisch and Scherzer are exact if the stopping cross-section can be expressed as $\varepsilon = \varepsilon_0 E^n$. As shown in the appendix the stopping cross-section of Ge and Al can be approximated by the power law with good accuracy in a wide range of energy. One notices that the stopping cross-section of GaAlAs can also be approximated by the power law with the power which depends on the composition. For $Ga_{0.7}Al_{0.3}As$, for example, the stopping cross-section can be approximated with n = - 0.346. The procedures, however, give reasonably accurate results independent of n if the kinematic factor is near unity as the present case [4]. In fact, for the above example, the errors in R_1 and R_2 produced by taking n = - 1/3 instead of n = - 0.346 are estimated to be ~ 0.17%. Therefore, we can neglect the change for n for most cases.

III. RESULTS

The application of this procedure is shown in Fig. 3 for a 4 μm thick layer of $Ga_{1-x}Al_xAs$ with x = .55 as determined by electron microprobe measurements [6]. The solid line is the yield from the GaAs reference samples and the dashed line represents the yield from GaAlAs with the Al yield subtracted. In order to determine the dashed curve the composition at depth 1 is found from the measured height ratio $R_1 = 0.859$. From Fig. 1(a) this ratio corresponds to a composition x = 0.52. For this value, the relative height of the Al components is given by $R_2 = 0.047$ in Fig. 1(b). This Al height is then subtracted from the spectrum at energy position $E_1'(t = 1)$ as determined from the solid

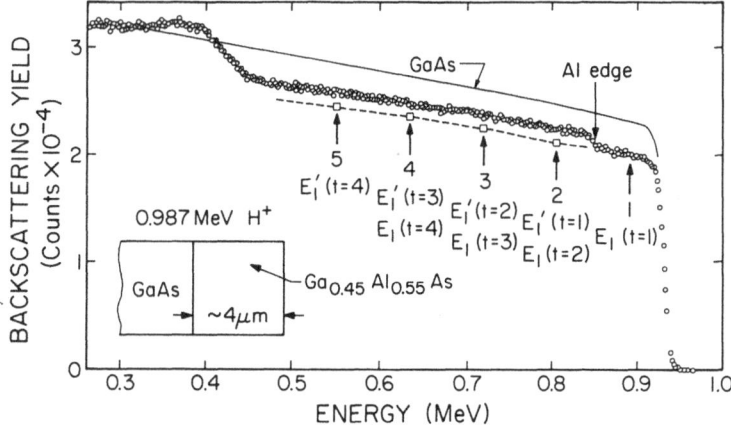

Figure 3.   Backscattering spectrum from a GaAlAs grown on a GaAs substrate.

curve in Fig. 2. Then the composition at the depth 2 is determined from the height of dashed curve to that of the solid line
giving x = .51 and a Al component that appears at position
$E_1'(t = 2)$. The reminder of the curve can be generated in the same
fashion. The composition of the film was found to vary between
0.49 and 0.53 as compared to the microprobe value of 0.55.

As shown in Fig. 1, the determination of the Al concentration in GaAlAs is an extremely sensitive function of the
ratio R_1 of the yield from the GaAlAs sample to that of GaAs
reference. In order to remove any systematic changes in measured yield due to inadequate secondary electron suppression, we
have vacuum-deposited a layer of Ge on the samples. The height
(upper curve in Fig. 4) of the deposited Ge spectrum closely
matches that for the underlying GaAs samples as would be expected
from similar results found in the analysis with He ions [2]. Contamination of low mass impurity in the film was checked by simultaneously depositing Ge on a carbon substrate and using backscattering analysis. In order to eliminate channeling effects,
the samples were rotated during the analysis. This rotation
procedure was tested by forming amorphous layers on GaAs by
implantation of 400 keV Se ions at LN_2 substrate temperature. No
steps were noted in the backscattering spectra between the amorphous layer and the single crystal substrate.

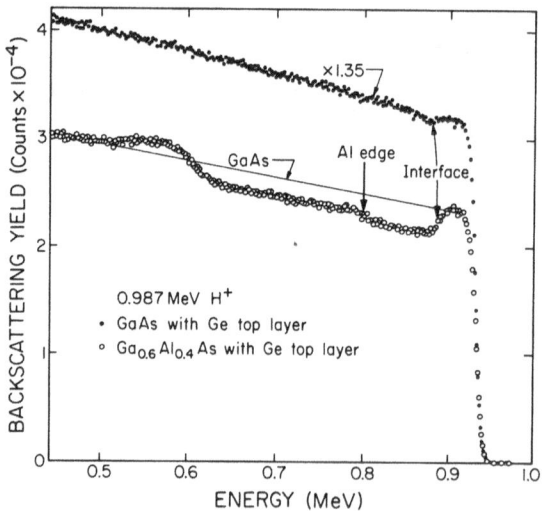

Figure 4. Backscattering spectra from a GaAlAs grown on a GaAs
           substrate with Ge top layer and from a GaAs.

As shown in the lower curve of Fig. 4, the deposited
layer of Ge about 0.4 µ thick provides a sufficiently broad
plateau to establish a reference height for the surface yield.
Analysis of the ratio of backscattering yield indicates that the
composition of the Ga$_{1-x}$Al$_x$As is between x = .40 near the surface
to .38 near the GaAs substrate. The heights of the Al component
at the GaAlAs/GaAs interface (at E$_1$ = 0.55 MeV) correspond to a
composition of x = .40. These values of the composition agree
with those determined by photoluminescence results.

The analysis of a more complicated four layer structure
is shown in Fig. 5. The nominal composition is shown in the
inset of Fig. 5. Backscattering measurements for the region near
the interface between reference Ge layer and Ga$_{1-x}$Al$_x$As are not
reliable (see dip in the upper curve of Fig. 4). Below this
interface region the spectra indicate a value of x between 0.06
and 0.11 followed by a layer of GaAs and then a layer of Ga$_{1-x}$Al$_x$As
with x = .27. This latter value is in good agreement with independent measurements of the Al concentration on photoluminescence
measurements. The upper curve is an analysis with 2 MeV He ions
on the same sample but without the Ge reference layer. The composition in regions 4 and 2 are in agreement with those with
protons.

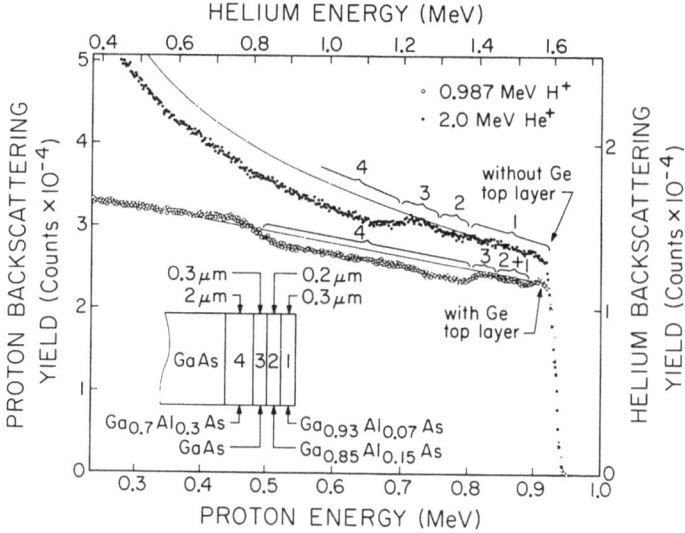

Figure 5.   Comparison of H and He backscattering spectra from
GaAlAs grown on a GaAs substrate with 4 layer structure.

IV. CONCLUSION

We have found that proton backscattering measurements can be utilized to measure the composition profile of $Ga_{1-x}Al_xAs$ layers. In order to determine ratio of the stopping cross-section we found it useful to deposit Ge on Al and measure the relative yield at the interface. As shown in the appendix, the relative yield measurements are in reasonable agreement with the published values by Northcliffe and Schilling. We also found that the deposition of a 0.4 μm thick layer of Ge on the top surface of $Ga_{1-x}Al_xAs$ samples provide a reference height. The height of the backscattering yield from the deposited Ge is found to correspond closely with that of GaAs.

REFERENCES

*Work supported in part by National Science Foundation.
**Permanent Address: Faculty of Engineering Science, Osaka
        University, Toyonaka, Osaka, Japan.
***Permanent Address: College of Engineering, Hosei University,
        Koganei, Tokyo, Japan.

1. V. Evtuhov and A. Yariv, IEEE Trans. on Microwave Theory and
    Techniques MTT-23, 44 (1975).
2. J.W. Mayer, J.F. Ziegler, L.L. Chang, R. Tsu and L. Esaki,
    J. Appl. Phys. 44, 2322 (1973).
3. R. Behrisch and B.M.U. Scherzer, Thin Solid Films 19, 247
    (1973).
4. B.M.U. Scherzer, P. Borgesen, M-A. Nicolet and J.W. Mayer,
    Proc. of this conference.
5. W.K. Chu, J.W. Mayer, M-A. Nicolet, S.U. Campisano and E.
    Rimini, Catania Working Data, Proc. US-Italy Seminar.
6. Sample and microprobe analysis furnished by J.S. Harris,
    Rockwell International Science Center.
7. L.C. Northcliffe and R.F. Schilling, Nuclear Data Tables A7,
    233 (1970).
8. J.S.Y. Feng, W.K. Chu and M-A. Nicolet and J.W. Mayer, Thin
    Solid Films 19, 195 (1973).
9. J.S.Y. Feng, W.K. Chu and M-A. Nicolet, Thin Solid Films
    19, 227 (1973).

APPENDIX

The tabulated values [7] of stopping cross-sections in Ge and Al are shown by solid lines in Fig. 6. The ratio of the stopping powers of Al to Ge at the same energy before scattering can be measured from the ratio of the height of Al to that of Ge at the interface between the deposited Ge and the Al substrate

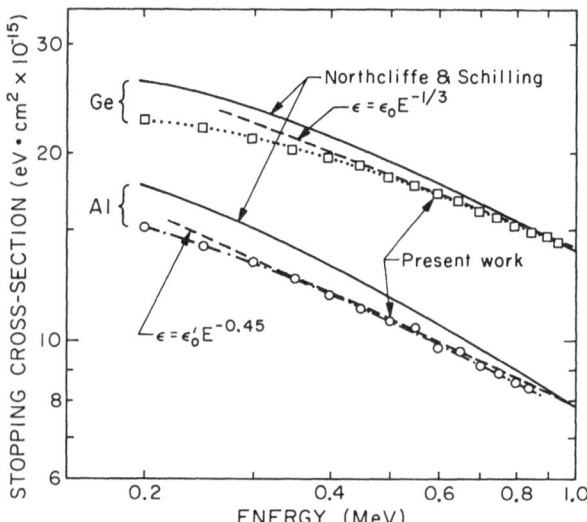

Figure 6.   Stopping cross-sections given by the present measurements (normalized at .94 MeV to reported values for
Ge [7], by power law approximation and by Ref. [7].

[8].   The ratio at 0.94 MeV is found to be 0.54 in reasonable
agreement with tablulated values.

    The energy dependence of the stopping cross-sections of
Al and Ge is estimated from backscattering spectrum by following the procedures by the reference [3].   In our case the power n
is taken to be - 1/3 for Ge and - .45 for Al.   The stopping
cross-sections for Ge and Al obtained by this calculation is
shown in Fig. 6.   The values are normalized at .9 MeV to the
stopping power of Ge by Northcliffe and Schilling.   The absolute
values are difficult to obtain due to secondary electron suppression problem and unknown detector solid angle.

    In the present case the analytical   procedure gives
reasonable values because the backscattering kinematic factor
fulfill the condition to be near unity and the stopping power
can be approximated in good accuracy with the power law in wide
range of energy [4].   In fact the synthesized spectra [9] obtained by using the calculated stopping power are in agreement
with the experimental spectra within the errors of < 2%.   The

synthesized spectra and the experimental spectra are shown in
Fig. 7.

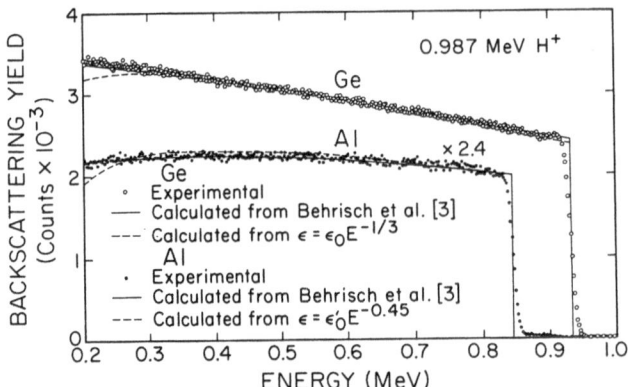

Figure 7.   Comparison of observed and calculated backscatter-
            ing spectra.

DISCUSSION

Q: (J.K. Hirvonen) What is your estimated error for the Al content
of the layers shown in your last slide ?

A: (K. Gamo) The errors are 50 to 100 % in the very low concentration region ($\chi \simeq 0.05$) and are better than 10 % at the Al concentration of $\chi \geq 0.2$.

Q: (G. Foti) Have you found some derectional effects in the analysis
of your samples ?

A: (K. Gamo) No. We measured random spectra by rotating samples a-
round an axis 6° off the channelling direction. This gave a good
reproducibility within statistical errors.

INTERDIFFUSION KINETICS IN THIN FILM COUPLES

J. E. E. Baglin and F. M. d'Heurle

IBM Thomas J. Watson Research Center

Yorktown Heights, New York 10598

ABSTRACT: We have evidence which indicates that processes of
nucleation and growth of grains can sometimes dominate the interaction between thin metal films. The evaluation of backscattering
profiles for such samples is discussed.

INTRODUCTION

The usual presumption in examining Rutherford backscattering
(RBS) profiles of interacting metal films has been that the original
films and the interaction layer remain continuous, homogeneous and
planar; or we may allow that grain boundary diffusion perturbs this
slightly. In numerous cases, however, it appears that grain growth
and nucleation effects may be the main cause of apparent "diffusion"
in observed profiles. In such cases, actual diffusion phenomena
may be obscured by complex structural changes.

It is important that such effects be recognized so that we
know what intrinsic process the observed "activation energy" and
"diffusivity" describe. Useful diagnostic methods include examining the heat-treated films (unetched and etched) with an SEM,
checking the time-dependence of the migration of each metal,
examining the shape of the backscattering profiles at the interfaces and where possible using TEM. These techniques all help in
deciding whether the migration is diffusion limited or interaction-
limited.

Interpenetration of films in which the interaction depends
strongly on grain boundary diffusion and/or grain growth effects
will be especially sensitive to the integrity, purity, grain size

385

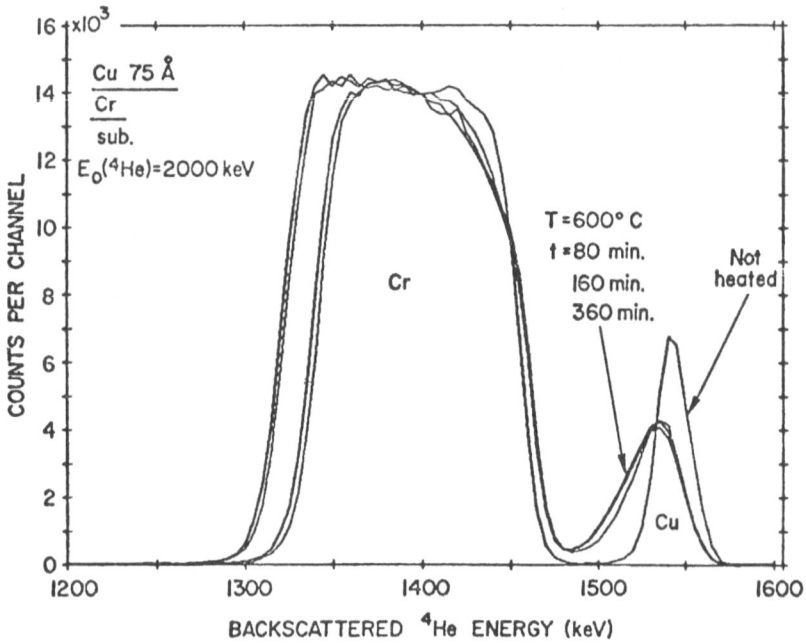

Fig.1.   RBS profiles for 75Å Cu/Cr after heating at 600°C.

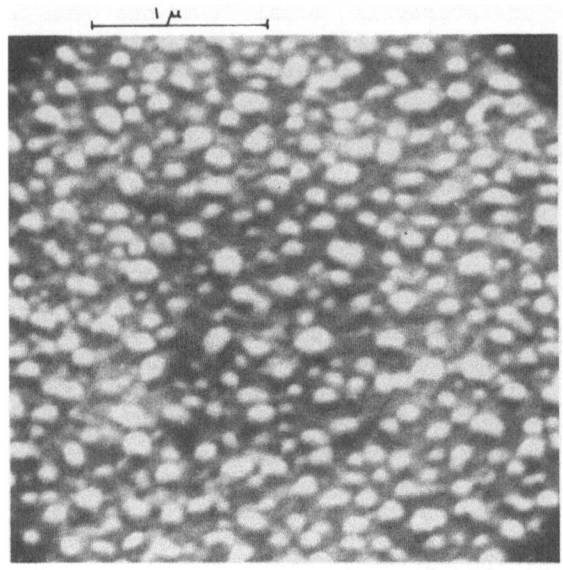

Fig. 2.   Cu(75Å)/Cr after 600°C, 20 min heat treatment.

and intrinsic stress of the initially deposited metal films. It is
important to recognize such cases whose behavior under heat treatment is strongly dependent on the preparation conditions of the films.

We proceed to cite a few examples from our recent experience.

Cr - Cu

Cr and Cu have mutual solubilities less than 0.1 atomic
percent for T < 1000°C.

a) <u>Cr(2000Å) + Cu(75Å)</u>: Fig. 1 shows RBS profiles for a
75Å layer of e-beam deposited Cu on thick Cr. After brief heating
at 600°C, the Cu profile broadens and stabilizes. Here, it turns
out, we are seeing the result of the Cu re-growing into small
beads or grains on the surface of the Cr. Supposedly the Cu grains
reach a stable size ($\sim$1200Å) consistent with minimum surface
energy at 600°C. The interpretation is supported by the SEM
photograph (Fig. 2).

b) <u>Cu(1000Å)/Cr(1000Å)</u>: (Polycrystalline, as indicated by
electron diffraction; initial grain sizes Cu:600Å Cr:200Å). As
shown in Fig. 3, this sample displays extensive intermixing of Cr and
Cu upon heating at 700°C in a purified He ambient. SEM photographs
taken of the 'reacted' sample after partial etching of the Cu
indicate a final mixture of coarse ($\sim$1200Å) grains of Cr and Cu
throughout the sample. Clearly a grain growth process has begun
at the film interface and extended into each film, possibly
augmented by nucleation and growth at high energy junctions of
grain boundaries. This is consistent with the apparent activation
energies for the "diffusion" which are similar to those for lattice
diffusion. (In impure metals, one usually observes that the activation energies for grain growth are similar to the values for lattice diffusion, not grain boundary diffusion.)

It is worth pointing out the characteristic change in the Cu
profile during the first short heat-treatment. An interface effect
like this seems to occur in many other cases. Supposedly, kinetics
computations for heat treatments of longer duration should be
referenced to the second curve representing an "initial" configuration of interface grains.

c) <u>Cu(1000Å)/Cr(1000Å)</u> ("single-crystal"): When the Cr and
Cu films were deposited on Al_2O_3 substrates aligned normal to the
c-axis, they displayed larger grain sizes (>2000Å), and showed
"single-crystal" type electron diffraction patterns. In these cases,
as shown in Ref. 1 no macroscopic interdiffusion occurred, although
surface/interface layers did appear rapidly. Their behavior is
consistent with grain boundary diffusion supplying nucleated islands
of the metal at the surface/interface. It would appear that the

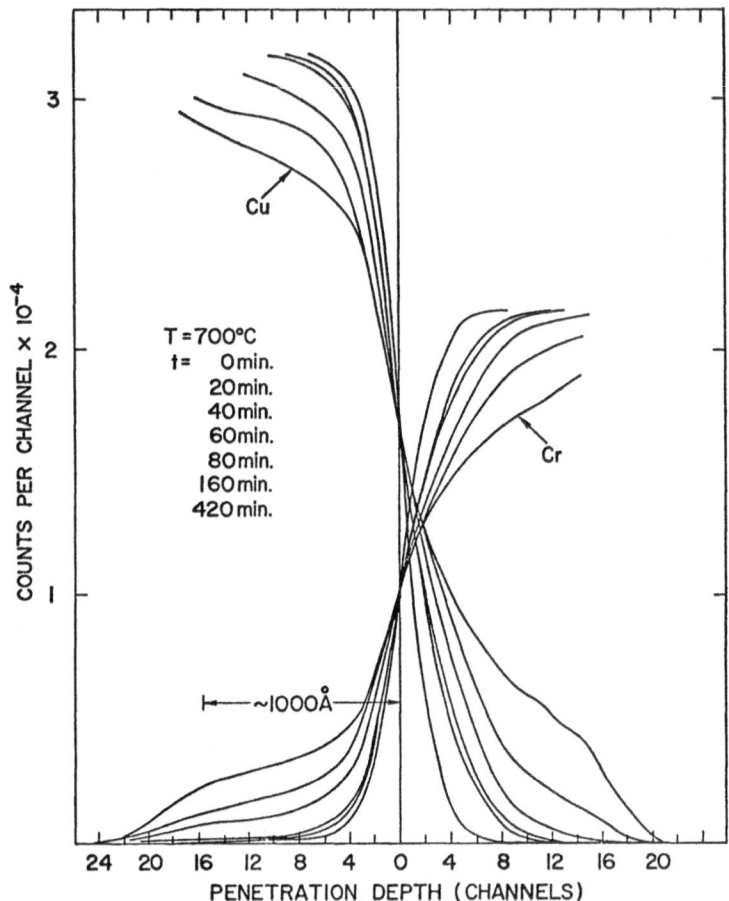

Fig. 3. Intermixing of polycrystalline films of Cu(1000Å)/Cr(1000Å). Initial grain sizes were: Cu:600Å, Cr:200Å. Final grain sizes up to 1500Å.

alignment of grains in the as-deposited Cu film inhibits the grain growth processes seen in Fig. 3.

d) Grain size effects are illustrated in Fig. 4. It is evident that initial grain sizes can strongly influence the subsequent interaction. While the effects could readily be duplicated (and are thus real), the systematics are not clear. The grain growth process evidently involves a delicate balance of surface- and interface-energies.

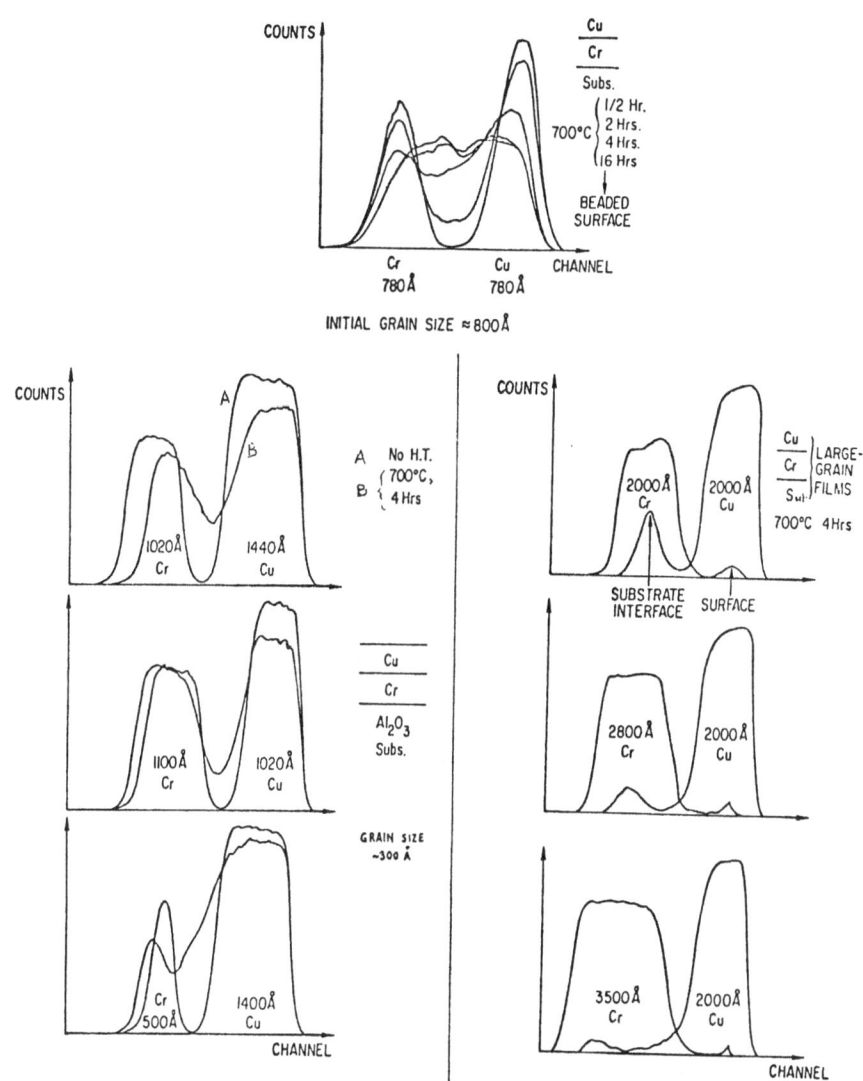

<u>Fig. 4</u>. Effects of grain size on intermixing of Cr-Cu films of various thicknesses at 700°C.

Bi - Cu

Bi and Cu likewise have very small mutual solubilities at low
temperatures. Fig. 5 shows the result of heat treatment of a
Cu(2000Å)/Bi(100Å)/Cu(1500Å) sample. The Bi rapidly forms a stable
covering at the upper surface, equivalent to about 2 monolayers.
The Bi then accumulates in the upper Cu layer. The activation

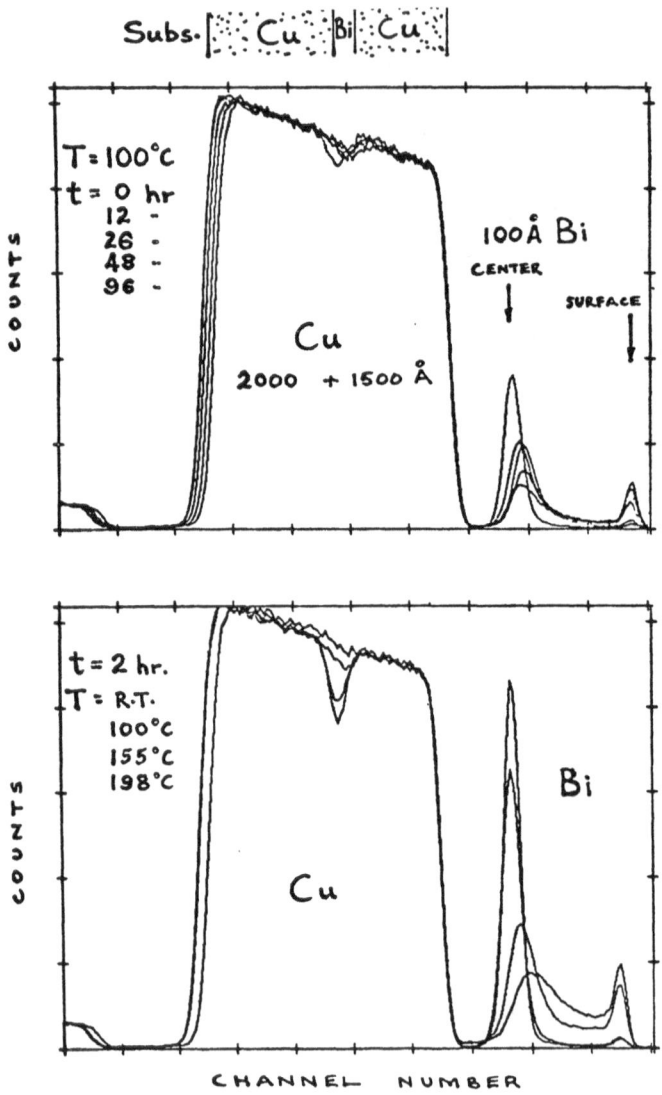

Fig. 5. Cu(2000Å/Bi()/Bi(100Å)/Cu(1500Å) after various heat treatments

energy for the process populating the surface peak was found by
obtaining sets of (T, t) for which the surface profiles were identical. The value $Q \sim 23,500$ cal/mole °C indicates a grain boundary
diffusion. One may surmise that the subsequent Bi accumulation in
the Cu involves grain growth, since there is certainly more Bi shown
than the grain boundaries alone ($\sim$5A wide) could accommodate. Our
SEM work is yet to be completed. The startling feature of these
curves is the negligible movement <u>downwards</u> of the Bi, since the
two Cu layers were deposited under nominally identical conditions.
The effect may be partly due to the excellent adhesion of the lower
Cu layer to the rf sputtered Al_2O_3 substrate.

Al - Ni

At temperatures below 1000°C, Al and Ni can form a variety of
compounds, the most Al-rich being Al_3Ni. This appears to form
exclusively in heat treatments below 300°C, as indicated by our
RBS data and supported by X-ray diffraction spectra. (At 400°C,
Al_3Ni_2 is also seen.)

Fig. 6 shows one sequence of our RBS curves for Ni(1500Å)/
Al(1500Å) after 275° heat treatment in forming gas. The sample
finally exhausts all the Al as it forms Al_3Ni and the interaction
ceases. We notice the initial "jump" in profile shapes during the
shortest heating period, which probably indicates the rapid onset
of nucleation of Al_3Ni grains along the interface. After that, the
Al_3Ni forms almost linearly with time, and the Al profile develops
a change of <u>slope</u>. This is consistent with the growth of Al_3Ni
grains in an interface-limited process. Eventually, the new Al_3Ni

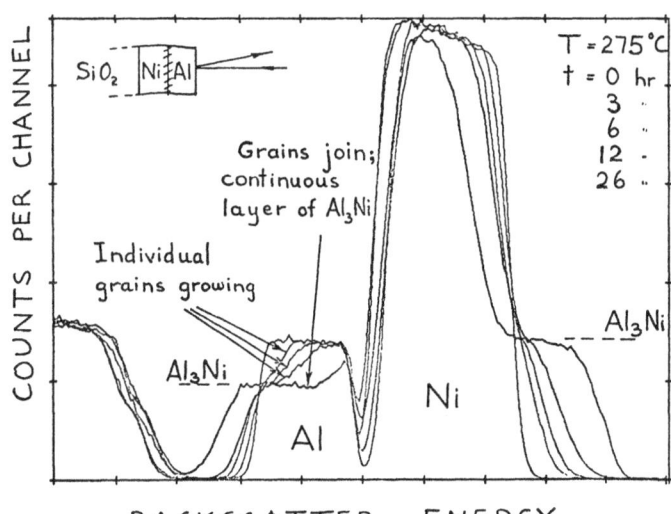

<u>Fig. 6.</u> Stages in the growth of a layer of Al_3Ni.

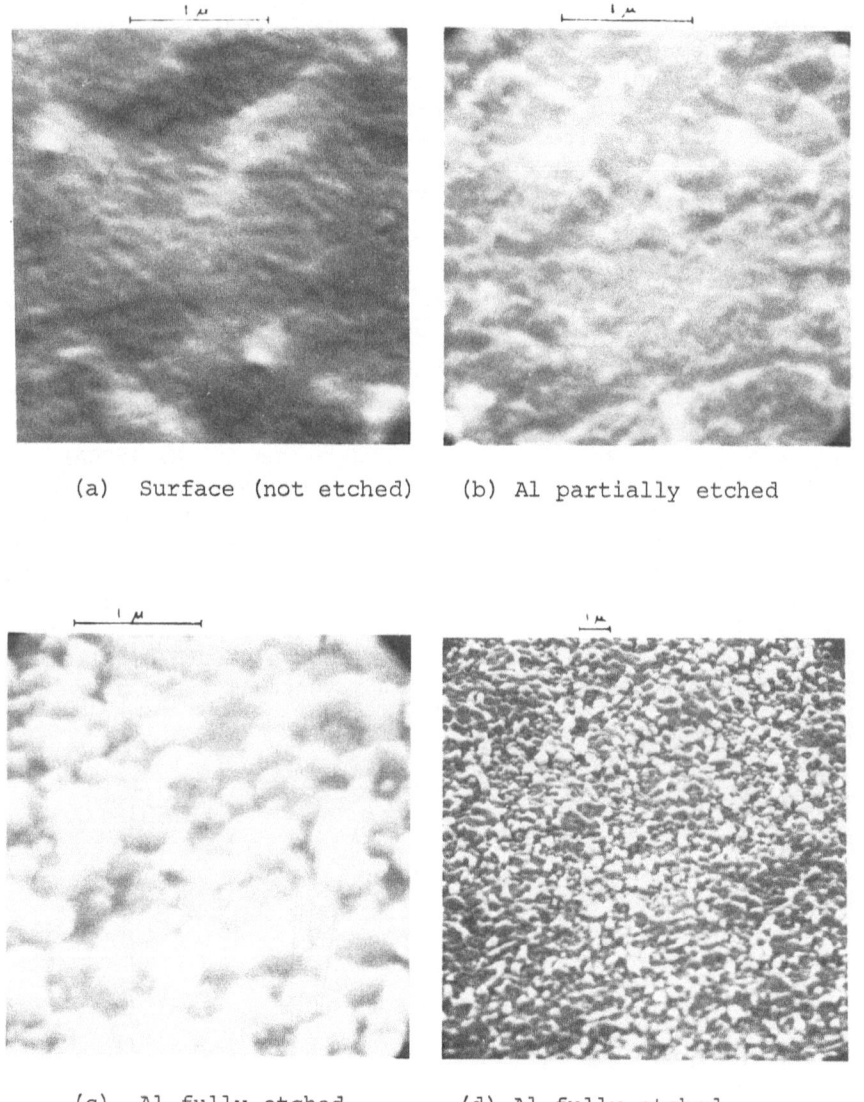

(a) Surface (not etched)   (b) Al partially etched

(c) Al fully etched   (d) Al fully etched.

Fig.7. Nucleation and growth of Al₃Ni grains after heat treatment
of Ni(1500A)/Al(1500A) at 275°C for 21 hr.

'islands' at the interface begin to link up and form a continuous layer. Then the Al profile shows a plateau, and the kinetics are governed by penetration through the continuous Al_3Ni layer (which seems to be very slow.) The apparent activation energy has been found for the first half of the process by collecting the (t, T) parameters for samples showing identical RBS profiles. We obtain 39000 cal/mole °C.

Fig. 7 shows a series of SEM photographs supporting this interpretation of the spectra. Heat treated samples were examined unetched (a) and following partial (b) and full (c), (d) etch of the Al using NaOH. The developing grains of Al_3Ni are clearly visible.

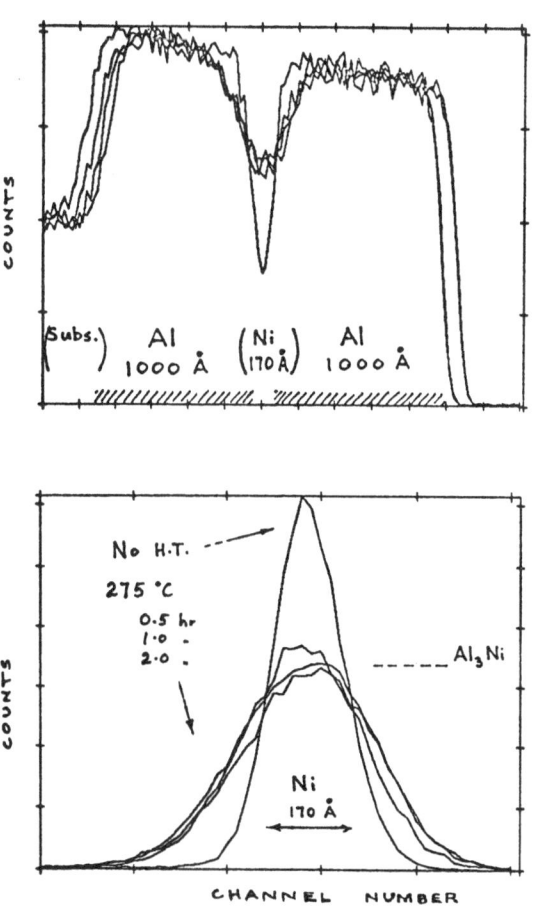

Fig.8. RBS profile for Al_3Ni grains formed in Al(1000Å)/Ni(170Å)/Al(1000Å) after heating at 275°C.

Fig. 8 shows RBS profiles for an Al-Ni-Al "sandwich" sample following a series of heat treatments at 275°C. The 170A Ni layer is seen to spread initially. It stabilizes at a point where its peak concentration corresponds to Al_3Ni. We infer that Al_3Ni grains are forming at the interface region, producing the broadened (but not flat-topped) RBS profiles. The effect was verified with better resolution using an etched upper Al layer to permit RBS at 5° grazing incidence. SEM photographs of the exposed interface show a network of beaded structures apparently being the Al_3Ni.

SUMMARY

We have seen a few examples of the possible interactions for films which have (or develop) grain structure. In most cases, nucleation and growth of grains during heat treatment can be expected to play a role. It may dominate the "diffusion" as in the Cr-Cu case, or it may be only a subtle overlaying process. In any event, we need to use all possible means of investigation to ascertain whether such effects are important in any given case, since that will determine our understanding of the kinetics, and will indicate the sensitivity of a sample to the conditions of its deposition - which affect its grain size, integrity, intrinsic stress, texture or preferred orientation, and possible contaminants. All of these factors can strongly affect thin film couple interactions of the kind discussed here.

One may exclude from these considerations cases where a layer of contaminants (oxide, or otherwise) between films prevents the occurrence of a smooth, continuous diffusion process. Beyond this it is clear that in a number of cases the free energy gain which is the driving force for diffusion is likely to be diminished, and/or overshadowed, by factors such as the grain boundary and/or surface and interface energies of the initial film couples, the interface energies of forming new phases, and the stress energy terms due to the usually anisotropic dilation (or contraction) of forming compounds with complex structures. The meaning (or lack thereof) of apparent activation energies depends on our ability to sort out the role played by such secondary factors in the overall diffusion process.

We wish to acknowledge the invaluable collaboration of V. Brusic in much of this work.

Reference

1. J. E. E. Baglin, V. Brusic, E. Allesandrini, J. Ziegler, Proc. Int. Conf. on Application of Ion Beams to Metals (Plenum Press, 1974), p.169.

DISCUSSION

Q: (G. Dearnaley) In confirmation of your suggestion that internal stress may influence interdiffusion, I refer attention to classic experiments on the thermal oxidation of crystalline and poly-crystalline Ag (see K. Lawless, Rep. Prog. in Phys. 1974). This is a gas-solid interdiffusion, and the application of biaxial stress markedly affected oxidation rates. I suggest that there is a parallel with your system.

A: (J. Baglin) Yes. Excellent !

BACKSCATTERING AND T.E.M. STUDIES OF GRAIN BOUNDARY

DIFFUSION IN THIN METAL FILMS[*]

S.U.Campisano, E.Costanzo, G.Foti and E.Rimini

Istituto di Struttura della Materia - Università di Catania

Corso Italia,57 - I95129 CATANIA,Italy

Intermixing processes in Cu-Pb thin film couples have been studied by MeV He$^+$ backscattering and transmission electron microscopy techniques. Because of the nonmiscibility of these two atomic species the only migration which can occur is via the grain boundaries. A preferential migration of Pb atoms into Cu thin film structure and their pile up at the Cu free surface has been observed. The migration has been measured as a function of annealing temperature, film thickness, oxide contamination and evaporation sequence. T.E.M. and diffraction patterns show that grain boundary diffusion and grain growth are competitive processes and this last process is inhibited as grain boundary filling occurs. The preferential migration of Pb into Cu thin film may be attributed to the larger grain boundary surface to volume ratio of the copper with respect to lead evaporated films.

INTRODUCTION

Interdiffusion processes in thin film couples differ from those occurring in bulk samples because of the high ratio of grain boundary surface to grain volume and of the large density of structural imperfections present in an evaporated film (1). Among the several experimental methods which detect composition changes over short distances (of the order of several hundreds Å), the backscattering

---

[*]Work supported in part by Centro Siciliano di Fisica Nucleare e di Struttura della Materia and by Gruppo Nazionale di Struttura della Materia del Consiglio Nazionale delle Ricerche.

technique is a well suited tool for its mass and depth percep-
tion (2).

Recently this technique has provided interesting information on silicide formation (3) between bulk silicon and deposited thin metal
films, on the compound formation and growth in Au-Al thin film couples (4) and on the interdiffusion processes in Au-Cu (5,6) and
Au-Cr (7) systems. Several conclusions can be drawn from all these
investigations which are of relevance for the metallurgical behaviour of thin films. Of the several compounds predicted by the phase diagrams only a few have been found in thin film couples, and
their formation occurs at temperatures lower than those required in
bulk specimens. This last statement is also true for miscible systems like Au-Cu or Au-Cr where a fast migration, related to low temperature short circuits for diffusion, has been evidence in the first
stage of annealing.

In non-miscible systems solid phase intermixing can occur only
via grain boundary diffusion. It seems then interesting to extend,
using backscattering and transmission electron microscopy techniques,
the investigations to non-miscible systems, to detail the role of
grain boundary diffusion and grain growth in mixing processes (8).
Cu-Pb system has been chosen because it is formed by elements well
separated in atomic masses which eases the analysis of backscattering data.

EXPERIMENTAL

The thin film couples have been obtained by vacuum evaporation
(10^{-6} torr) of pure (99.999) materials onto formvar covered aluminum
and polished carbon substrates. The substrates have not cooled during
evaporation so any heating of the substrate itself is due to irradiation from a point source 12 cm distance. The two evaporation have
been performed without breaking the vacuum and the sequence of the
evaporated materials has been interchanged. The film thicknesses
ranged from a few hundreth to a few thousand Å. Thermal processing
was carried out in a tube furnace under purified argon atmosphere in
the temperature range 100-300°C, and the temperature was held ± 1°C.

Backscattering measurements have been made using the 2.5 MeV
Van de Graaff. Helium particles backscattered through an angle of 150°
were detected by a surface barrier detector; conventional electronics were used to amplify and to display the signals with an overall
resolution of 18 keV. Secondary electron suppression techniques were
used and the dead time during analysis was less than 5%. Single element film or coupled layer have been observed by transmission electron microscopy at 100 keV* to obtain information on the structure

---

*Centro di Microscopia Elettronica dell'Università di Catania.

of the deposited films, mainly grain sizes and their dependence
on annealing temperature and mixing processes.

RESULTS AND DISCUSSION

Backscattering and T.E.M. techniques have been used to detail
the role played by grain boundaries and grain growth in the mixing
of the Cu-Pb not-miscible (9) system in the solid state. Fig.1 reports energy spectra of 2.0 MeV He$^+$ backscattered from a couple
Cu(1200 Å)-Pb(2200 Å) as deposited and after 40 min anneal at 250°C.
The two arrows indicate the surface energy location of Pb and Cu
respectively. The copper signal is shifted below the surface energy

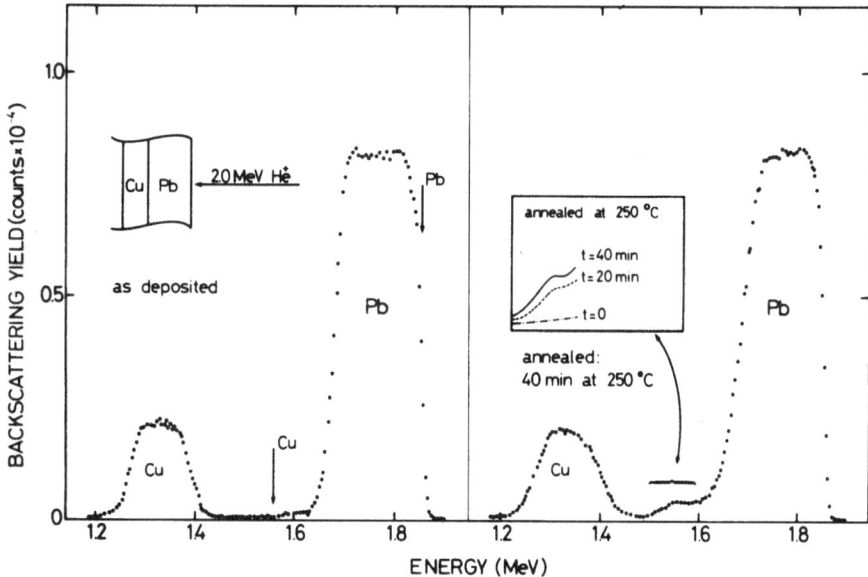

Fig.1 - Energy spectra of 2.0 MeV He$^+$ particles backscattered from
a 2200 Å Pb thin film evaporated on the top of a 1200 Å Cu thin
film, as deposited (left hand side) and after 40 min. annealing at
250°C(right hand side). The insert reports, in an expanded scale,
the yield due to diffused Pb atoms for different annealing times at
250°C.

position because of the energy loss experienced by the analyzing
beam in traversing the overlaying Pb film. Samples with the sequence Cu-Pb will be referred in the following as direct samples. The
analysis of the aged sample, shown in the right hand side of the
same figure, exhibits a step in the energy region in between the Cu
and the Pb signals. The signal which appears at energy larger than
the Cu surface backscattering, must be attributed to scattering from

Pb atoms inside the Cu film. The height of the signal is clearly higher than that corresponding to the background in the unannealed sample. Moreover it depends on the annealing time as shown in the insert, where the diffused Pb atoms yields are reported for several aging times.

To check further this Pb atom migration samples with a reverse sequence of evaporation i.e. Pb-Cu have been prepared. In this case Pb atoms migrated through the Cu film will give rise to a signal in the background free region of the backscattering energy spectra. The analysis of a couple Pb(1900 Å) - Cu(2000 Å) as deposited and after annealing at 220°C is illustrated in Fig.2. The Pb and Cu yields overlap because of thickness, kinematics, and energy loss factors; for clarity the two signals have been indicated by

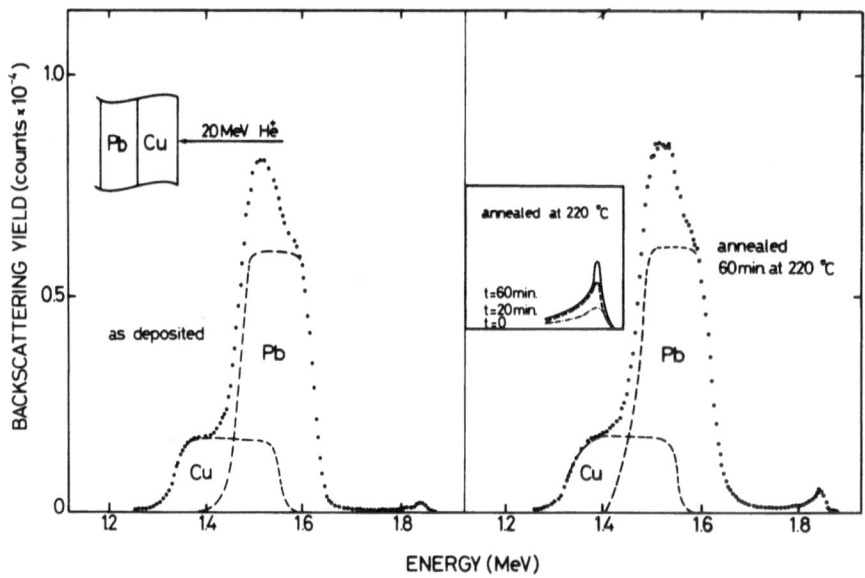

Fig.2 - Energy spectra of 2.0 MeV He⁺ beam scattered from a 2000 Å Cu thin film evaporated on the top of a 1900 Å Pb thin film, as deposited (left hand side) and after 60 min. annealing at 220°C (right hand side). The insert reports in an expanded scale the yield due to migrated Pb atoms for different annealing times at 220°C.

dashed lines. In the as deposited films a small tail is present in the high energy side and it extends up to energy corresponding to Pb surface backscattering where a small peak appears. The height of the tail increases by aging the sample as the peak area does but with a larger rate. These results, as confirmed also by the trend shown in the insert, point out a Pb migration through the copper

film and an accumulation, for reversed samples, of Pb on the copper
outer surface.

     In order to understand the mechanisms which influence the prefe-
rential Pb atom migration,the structural properties of the deposited
films have been investigated by transmission electron microscopy ob-

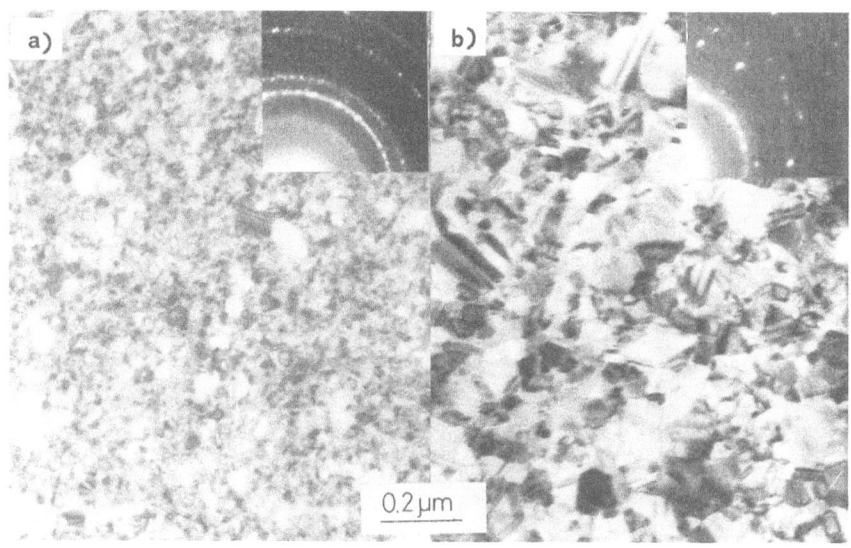

Fig.3 - Transmission electron micrograph of 700 Å thick copper film
after deposition (a) and after 20 min. annealing at 250°C (b).
The grain growth can be also inferred from the diffraction patterns
reported as inserts.

servations. A micrograph of a typical portion of an as deposited
copper film is shown in Fig.3a, the diffraction pattern of the same
area is also reported in the upper part of the figure. The micro-
grain diameter is of the order of 100 Å and the spread in the diffraction rings confirms the polycrystalline nature of the deposited
film. After 20 minutes annealing at 250°C grains grow up to few
thousand Å, as shown in Fig.3b and spots appear in the diffraction
rings. The deposited lead film is characterized instead by grain
sizes larger than the copper ones as evidenced in Fig.4, both in
the micrograph and in the diffraction pattern. The average grain
size is few thousand Å, and practically no change occurs after annealing in the temperature range 100-300°C. The island structure
is due to the small thickness (<500 Å) of the Pb layer which we used
for T.E.M. observations because of the high absorbtion coefficient
of lead.

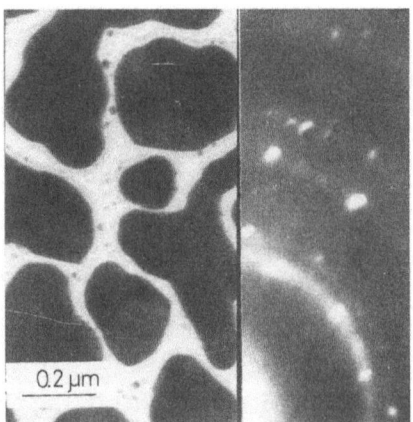

Fig. 4 – Transmission electron micrograph of 400 Å thick lead
film after deposition. The diffraction pattern shown in the
right hand side confirms the large average grain size.

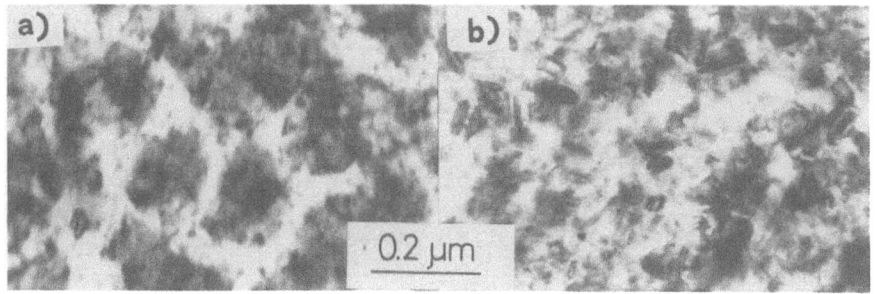

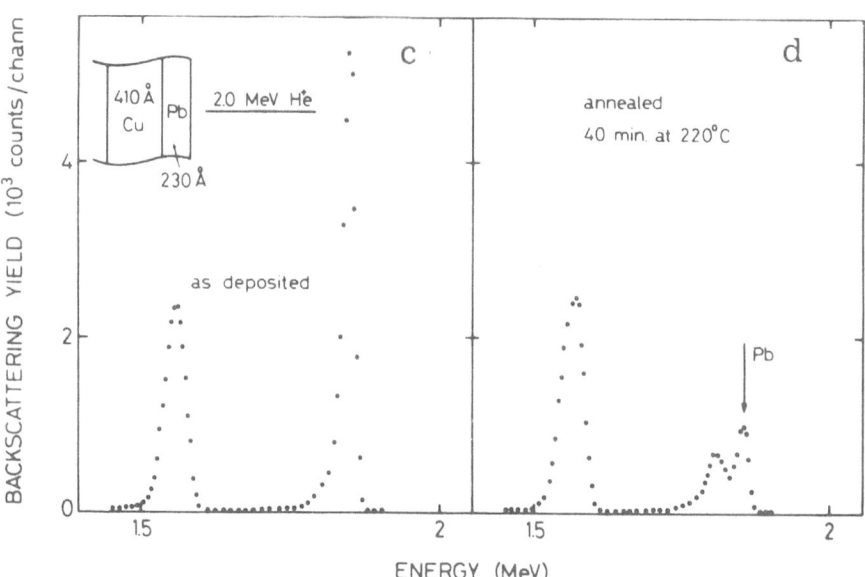

Fig.5 - Analysis of an as deposited couple of 230 Å of Pb layer on the top of 410 Å of Cu layer by transmission electron microscopy (a) and by 2.0 MeV He⁺ backscattering (c). The corresponding micrograph and energy spectrum after .40 min. annealing at 220°C are reported in (b) and (d) respectively.

The backscattering analysis indicates a preferential migration
of Pb through the Cu layer in the same temperature range at which
the T.E.M. observations show a Cu grain growth. It is then interesting to follow the thermal behaviour of the copper film structure
in the presence of a thin lead layer. A couple of 230 Å Pb on top
of 410 Å Cu layer has been analyzed by backscattering and T.E.M.
As it appears in Fig.5c and 5d the Pb signal changes strongly following 40 min. at 220°C, the double peak corresponds to Pb atoms on
the front and rear surfaces of the Cu film, and the amounts on
each surface is roughly the same. The corresponding micrographs,
shown in Fig.5a and 5b respectively, indicate that the copper average grain size increases without reaching the characteristic di-
mension of the single copper layer (see Fig.3b). Pb migration
occurs in reversed samples as illustrated by the backscattering
energy spectra for 460 Å Cu on top of 320 Å Pb layer reported in
Fig.6. In the as deposited couple a small amount of Pb is present
on the sample surface, as also previously found for thicker film
(see fig.2). This initial migration which is more evident for the
reversed samples may be attributed to the higher temperature reached

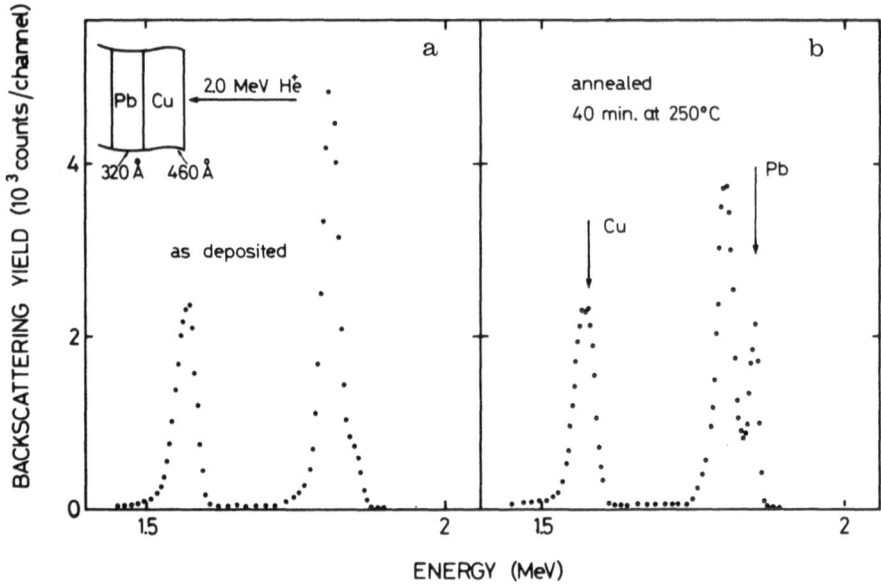

Fig.6 - 2.0 MeV He[+] backscattering energy spectra of 460 Å thick Cu
film onto 320 Å thick Pb layer as deposited (a) and after 40 min.
annealing at 250°C (b).

by the substrate during the evaporation of the higher melting point
element. The accumulation of Pb atoms at the Cu top surface, as
evidenced by the surface peak in the spectra of reversed sample,may
be due (10) to the growth of a new phase (lead oxide which could
act as a sink for Pb atoms.

MECHANISM

The preferential migration of Pb atom through the copper layer
can be attributed to the different grain sizes of the evaporated
materials. As general trend there is a correlation between grain
size and melting point of the deposited element: the lower is the
melting point the higher is the grain size (11). For the Cu-Pb
system $Tm(Cu)>Tm(Pb)$, then the Cu grain sizes are smalller than
those of Pb films (see Fig.s 3 and 4). The relevance of grain di-
mension in determining the moving species has been recently pointed
out by backscattering measurements in Au-Cu (12) and Ti-Mo-Au (13)
systems. Grain boundary and defect assisted diffusion occurs
usually in the same temperature range of grain growth in thin film
structure. In non-miscible systems these two processes become competitive: the grain growth reduces the allowed short circuit paths
for diffusion while impurity atom segregation on the grain surface can inihibit the grain growth (14). (Fig.s 3b and 5b).

The amount of Pb atoms migrated through the copper films decreases with increasing Cu layer thickness, compare for instance
Figs . 1 and 2 with Figs.5 and 6. Grain growth occurs in all the
volume of the copper film while diffusing atoms leave from interface. In thick copper films the growth is then inhibited for grains
near the interface. Growth occurs for the remaining ones thus reducing the number of allowed boundary path for diffusion.

SUMMARY AND CONCLUSION

Backscattering and T.E.M. techniques have been applied to the
investigation of grain boundary diffusion in the non-miscible Cu-Pb
system. Pb atoms migrate through the copper film and segregate on
the opposite surface. The average grain size of Pb film are nearly
temperature independent and they are larger (about one order of magnitude) than those of as deposited copper film. The Cu grain growth
occurs in the same temperature range at which Pb migration takes
place. The copper grain growth reduces the allowed short circuit
paths for lead diffusion while lead segregation on the Cu grain
boundaries inhibits the grain growth. The simultaneous occurrence
of the two processes allows an equipartion of Pb at the Cu surfaces
in a very thin film ($\sim$ 400 Å) after 40 min. annealing at 220°C,
while for thicker film ($\sim$ 2000 Å) only a few percent of Pb is allowed to migrate.

Thanks are due to Mr.A.Strano for the electron microscope observations.

REFERENCES

1) C.Weaver Physics of Thin Films, Vol.6 p.315 (1971).

2) W.K.Chu, J.W.Mayer, M.A.Nicolet, T.M.Buck, G.Amsel and F.Eisen, Thin Solid Films 19, 423,(1973).

3) J.W.Mayer and K.N.Tu, J.Vac.Sci.Technol. 11, 86,(1974).

4) S.U.Campisano, G.Foti, E.Rimini, S.S.Lau and J.W.Mayer, Phil.Mag. 31, 903,(1975)

5) S.U.Campisano, G.Foti, F.Grasso and E.Rimini, Thin Solid Films 19, 339, (1973).

6) J.A.Borders Thin Solid Films, 19, 359, (1973).

7) J.K.Hirvonen, W.H.Weisenberger, J.E.Westmoreland and R.A.Maussner, Appl.Phys.Lett. 21, 37, (1972).

8) R.W.Balluffi, Phys.Status Solidi 42 (1970) 11.

9) M.Hansen, Costitution of Binary Alloys (McGraw-Hill, New York 1958) p.609.

10) R.W.Balluffi and J.M.Blakely, Thin Solid Films 25, 363, (1975).

11) J.P.Hirth and K.L.Moazed in Physics of Thin Films, Vol.4 p.60 (1969).

12) S.U.Campisano, G.Foti, F.Grasso and E.Rimini,Japan.J.Appl.Phys. Suppl.2 Pt.1, 637, (1974).

13) J.M.Harris, E.Lugujjo, S.U.Campisano, M.A.Nicolet and R.Shima, J.Vac.Sci.Technol. 12, 524, (1975).

14) J.G.Bryne, Recovery, Recrystallization and Grain Growth (Macmillan, New York, 1965) p.77.

DISCUSSION

Comment:(J. Baglin) I am pleased to see that your Cu-Pb reaches stability with equal amounts of Pb each side of the Cu layer. In Cu-Bi we have observed the same general behaviour after prolonged heat treatment.

THE ANALYSIS OF NICKEL AND CHROMIUM MIGRATION

THROUGH GOLD LAYERS

A. Barcz, A. Turos, L. Wieluński

Institute of Nuclear Research

00-681 Warsaw, Hoża 69, Poland

ABSTRACT

Elastic backscattering of MeV He ions and the $^{16}O(d,\alpha)^{14}N$ nuclear reaction have been used to study the composition of NiCr-Au resistance films. In order to investigate the mobility of nickel and chromium atoms separately, both NiAu and CrAu couples were measured as a function of their thermal treatment. Films were deposited by the evaporation technique and subjected to heating at low temperatures in air. Significant ammounts of Ni and Cr atoms have been observed on the external surface of the Au layer. The mechanism of thermal processes cannot be treated in the case of a three-component NiCr-Au layer as a simple superposition of the processes in the two-component systems.

INTRODUCTION

Backscattering and nuclear reactions have attracted interest in solid state surface analysis owing to that these methods allow to determine in a non-destructive way the depth distribution of a given kind of atoms. Furthermore these techniques provide absolute values of elemental concentrations, and are complementary as regards their application to different solid state problems [1]. Nuclear reactions are mostly used for detection of light elements [2,3] whereas backscattering is a valuable tool for analysis of surface layers containing medium- and heavy-mass atoms [4].

The aim of this paper is to study the diffusion of nickel and chromium in gold. The intermixing of thin metallic films at low temperatures (300-450°) has important consequences in electronic applications. NiCr alloys of 100-500 Å form ohmic resistors, whereas gold layers of 500-2000 Å are used as electrical contacts and for chemical protection. The presence of chromium on the gold outer surface, as a result of annealing, [5] makes soldering difficult. On the other hand, the small chromium contents within the gold layer diminishes the gold solubility in tin and thus improves the soldering conditions.

EXPERIMENTAL

Polished Si wafers with a thermally grown layer of SiO_2 of the order of 5000 Å thick were used as sample substrates. Metal films were deposited in a sequence of evaporations under typical vacuum of $5*10^{-6}$ torr. The film thicknesses were estimated using a quartz-crystal monitor placed in the evaporation set-up.

The following metal deposition sequences were studied:

1. Cr   500 Å - Au 1000 Å
2. Au   700 Å - Cr 500 Å - Au 700 Å
3. Au   700 Å - Ni 100 Å - Au 700 Å
4. Au   700 Å - Ni 500 Å - Au 700 Å
5. NiCr   300 Å - Au 1000 Å

The annealing was performed in air at normal pressure in the temperature range of 300-450°C for 0.5 to 5 hours.

The films were analysed using 1.5 and 2.0 MeV He^+ beams backscattered.

The determination of the surface oxygen concentration from the backscattering spectra is not accurate because of the small oxygen cross section and overlapping with the silicon background. Therefore the nuclear reaction $^{16}O(d,\alpha)^{14}N$ has been used for this purpose [3].

Charged particles emerging from the target were detected by a silicon surface detector. The beam intensity was 10 nA/2mm² spot. The detector energy resolution was less than 16 keV FWHM.

RESULTS

Chromium migration through Au films has been observed in all annealed samples produced in the 1,2,5 sequences. Figure 1 shows variation of the energy spectra with annealing temperature. The step visible on the rear edge of the Au peak (Fig.1b) corresponds to the Cr_3Au phase formed at 350°C. The chromium concentration within the evaporated Au layers is nearly uniform and increases from 3 at% 320°C to 20 at% 450°C , but certain deviations from this regularity were observed. The ammounts of chromium and oxygen at the sample surface are proportional to each other, which may suggest that the oxidizing process is responsible for Cr appearance on the front of the Au layer. Other measurements with Au-Cr-Au "sandwiches" have supported this idea, since no Cr atoms appeared at the SiO_2-Au interface.

Similar experiments were performed to study nickel mobility in gold. It is seen from Fig.2 that all nickel atoms have diffused through gold to the outer surface (Fig.2c). The oxygen peak which appears in the low energy part of the spectrum(not shown in the figure)again explains the mechanism of such a migration. Nickel concentration within both Au layers is small and does not exceed 2 at%.

The energy spectra of the NiCr-Au systems subjected to thermal treatment are illustrated in Fig.3. A large part of chromium has moved to the surface from its original position in the NiCr alloy, whereas Ni atoms remain in their initial region (the energy shift of the Ni peak is due to the growth of the effective depth of the nickel layer).

Figure 4 shows the energy spectra of α -particles produced in $^{16}O(d,\alpha)^{14}N$ reaction. The surface oxygen peak is then well separated from the remaining part of the spectrum and enables measurement of the absolute ammount of oxygen concentration by the comparison with the reference standard.

DISCUSSION

The depth distribution of chromium and nickel within the Au layer appears to be constant in all samples investigated. The concentration level is strongly related to the heating temperature and hardly depends on the annealing time. This could be explained by very rapid grain boundary diffusion and practically no volume diffusion during the anneal [6].

The surface concentration of Cr varied from 5×10^{15} to 3×10^{16} at/cm² under the conditions used and generally increased with temperature. Hirvonen et al.[5] have obtained results similar to those presented, using argon for the heating atmosphere.

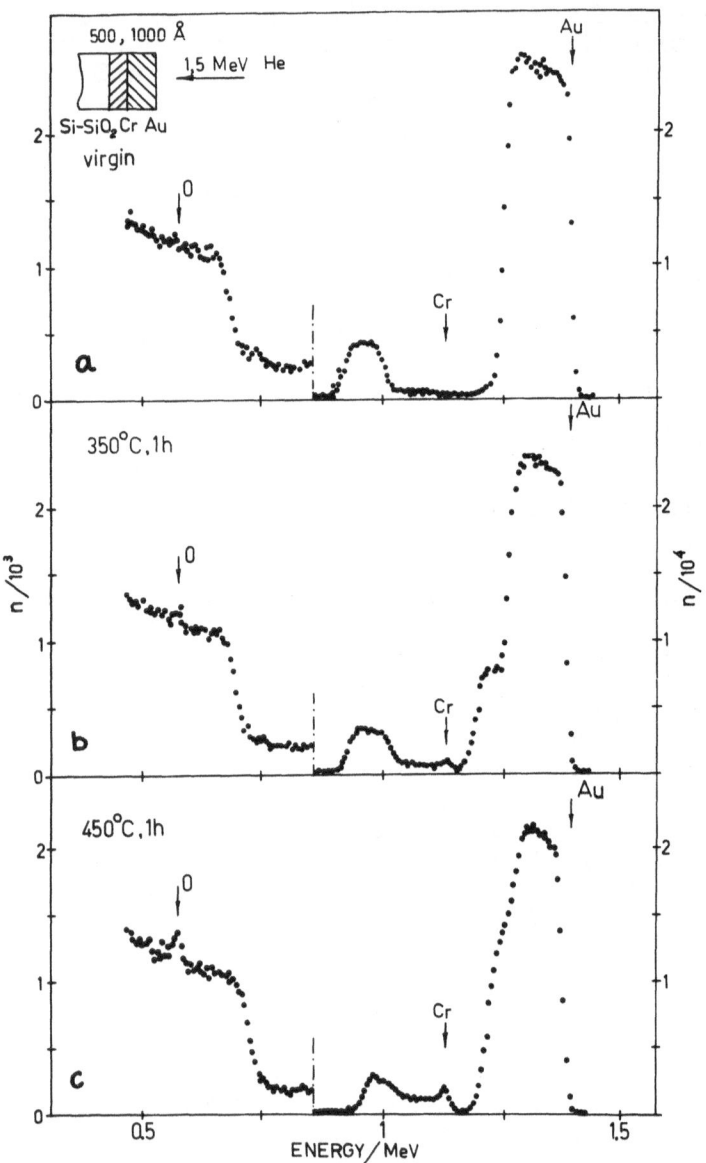

Fig.1. Energy spectra of 1.5 MeV He⁺ bombardment of
Si-SiO₂-Cr 500 Å - Au 1000 Å : (a) as deposited,
(b) annealed 350° C, 1h, (c) 450° C, 1h.

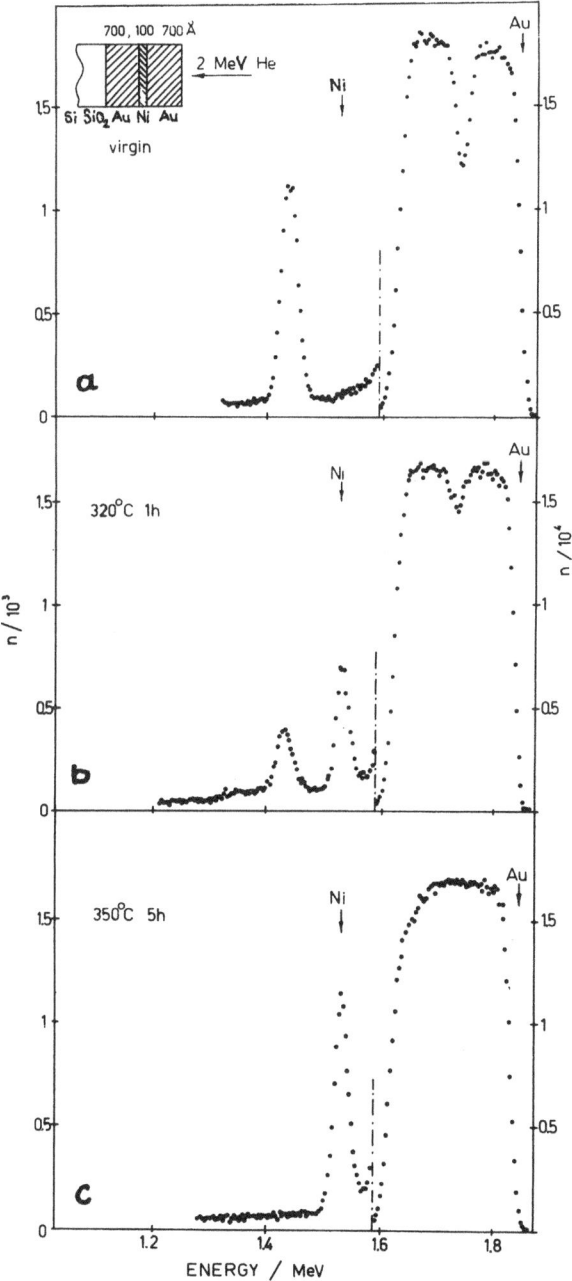

Fig.2. Energy spectra of 2 MeV He[+] bombardment of Si-
SiO$_2$-Au 700 Å - Ni 100 Å - Au 700 Å : (a) as
deposited, (b) 320° C, 1h,(c) 350° C, 5 h.

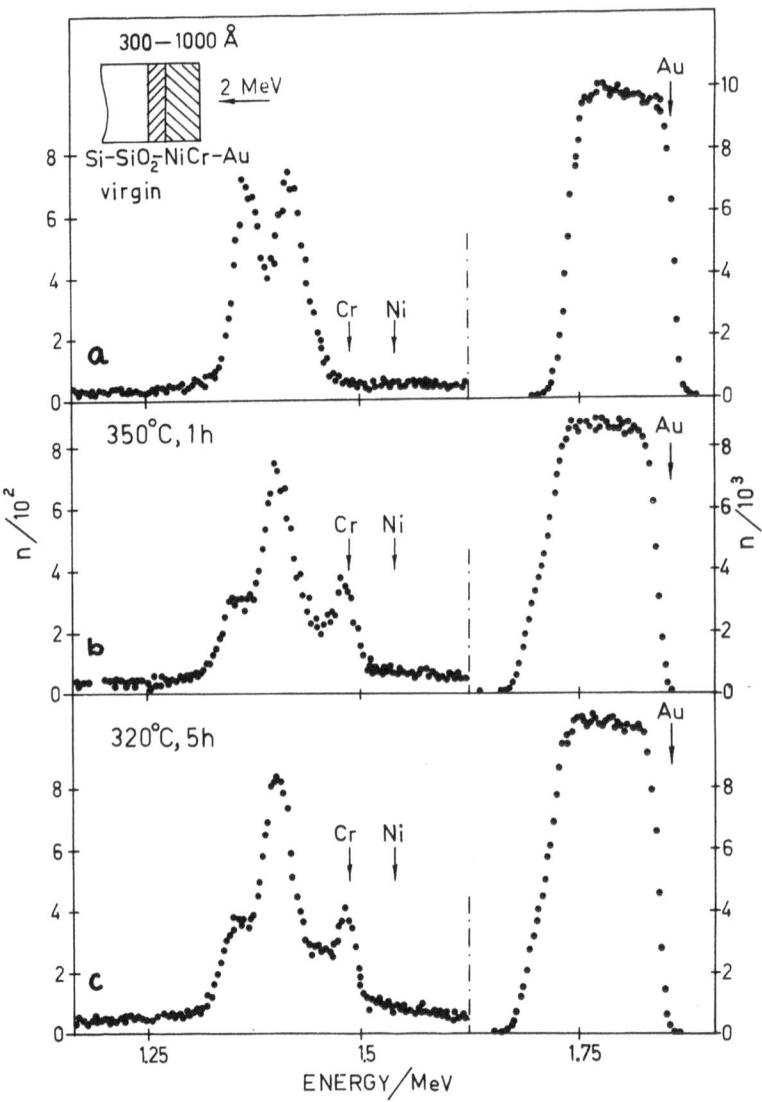

Fig.3. Energy spectra of 2 MeV He[+] backscattered from
Si-SiO$_2$-NiCr 300 Å - Au 1000 Å : a as deposited, b 350° C, 1h, c 320° C, 5h.

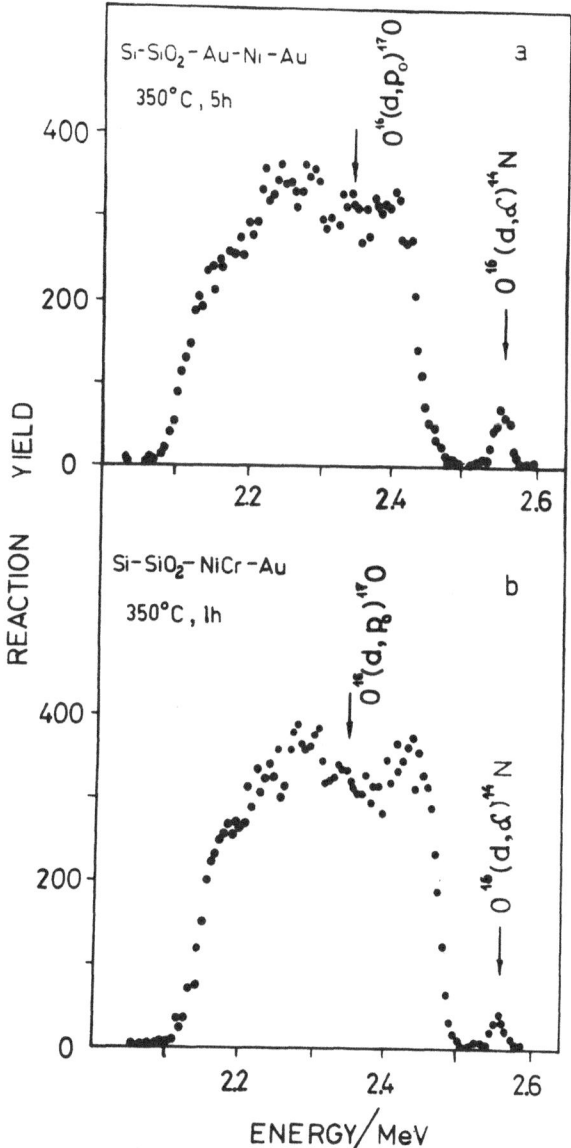

Fig. 4. Energy spectra from 900 keV deuteron bombardment of: a Si-SiO$_2$-Au 700 Å-Ni 100 Å-Au 700 Å, annealed 350° C, 5h, b Si-SiO$_2$-NiCr 300 Å-Au 1000 Å, annealed 350° C, 1h.

The appearance of Ni or Cr at the surface was always accompanied by a surface oxygen peak. The calculated stoichiometries point to NiO and Cr_2O_3.

It can be concluded that the pronounced migration of Ni and Cr to the surface is a result of their chemical activity with oxygen. The thickness of the oxide is less dependent on the oxygen concentration in the atmosphere (air, argon) than on temperature. Thin surface films of NiO and Cr_2O_3 are thus barriers for diffusion of Ni, Cr and/or O, inhibiting further oxidation.

It has been pointed out that Ni atoms are not mobile when Cr is present in the film (NiCr-Au, type 5), although nickel migration was observed in the Au-Ni-Au film. This indicates that chromium oxide, formed during annealing, makes nickel oxidation impossible.

AKNOWLEDGMENTS

The authors wish to thank Mr. B.Stępień and his team (PIE, Warsaw) for valuable discussion and samples preparation.

REFERENCES

1. J.W.Mayer and A.Turos, Thin Solid Films 19/1973/1.
2. G.Amsel, J.P.Nadai, D'Artemare E., D.David, E.Girard and J.Moulin, Nucl.Instrum.Meth. 92/1971/481.
3. A.Turos, L.Wieluński and A.Barcz, Nucl.Instrum.Meth. 111/1973/605.
4. M.A.Nicolet, J.W.Mayer and I.V.Mitchell, Science 177 /1972/841.
5. J.K.Hirvonen, W.H.Weisenberger, J.E.Westmoreland and E.A.Meussner, App.Phys.Lett. 21/1972/37.
6. J.A.Borders, Thin Solid Films 19/1973/359.

APPLICATIONS OF ION BEAM ANALYSIS TO INSULATORS*

J. A. Borders and G. W. Arnold

Sandia Laboratories

Albuquerque, New Mexico 87115

ABSTRACT

This paper reviews the use of ion analysis and ion analysis combined with ion implantation in insulators. New data are presented illustrating compositional analysis of multicomponent glass systems. Studies of diffusion in insulators is illustrated with data from Ag-implanted lithia-alumina-silica glass. Finally problems such as sample charging and effects due to energy deposited into electronic processes are discussed.

I.   INTRODUCTION

Energetic ion backscattering and ion channeling have become accepted techniques for the analysis of the near-surface layers of solids. Much of the work carried out using these techniques has been on semiconductors, particularly Si, although recently workers have begun to apply these methods to metals as well.[1] However, there have been almost no applications of energetic ion-analysis techniques to the study of insulators. Here we are excluding the analysis of insulator films of importance to the semiconductor industry (SiO_2, Si_3N_4) which have been quite extensively studied. Ion implantation can be usefully applied to insulators as well, particularly in conjunction with ion analysis, but investigations have been confined mainly to studies of refractive index changes induced by ion bombardment.[2] This paper will review various areas where energetic ion analysis and ion analysis combined with ion

*This work was supported by the United Stated Energy Research and Development Administration, ERDA.

implantation can yield information on the properties of insulators. New data on amorphous insulators will be presented illustrating these areas and some of the problems encountered when using energetic ion beams on insulators will be discussed.

II.   COMPOSITIONAL ANALYSIS

Glasses are amorphous solids and may be composed of many constituents. SiO_2 is probably the most familiar glass forming oxide with Li, Na, Al, P, K, Ti, As, Ce and Ba commonly incorporated either as network formers or as network modifiers. In addition, photoreducible ions such as Au^+, Ag^+, Cu^+ and Pt^{++} are often used to nucleate glass ceramics and may be present as well. Ion backscattering is an ideal technique for compositional analysis of such glasses and glass ceramics. The common constituents are generally far enough apart in mass for their individual surface scattering edges to be easily distinguished, and the heavier elements, to which backscattering is more sensitive, are usually present in smaller concentrations than are the lighter elements. Linker et al.[3] have used ion backscattering for compositional analysis of phosphosilicate glass layers, but there are no published data on compositional analysis of bulk glasses.

In Fig. 1 is shown a 2 MeV He^+ backscattering spectrum from a sample of silver-activated phosphate glass commonly used as a radiation dosimeter. The six atomic components of the glass are clearly shown by their surface scattering edges and the compositions indicated on the figure were calculated from the scattering yield at the

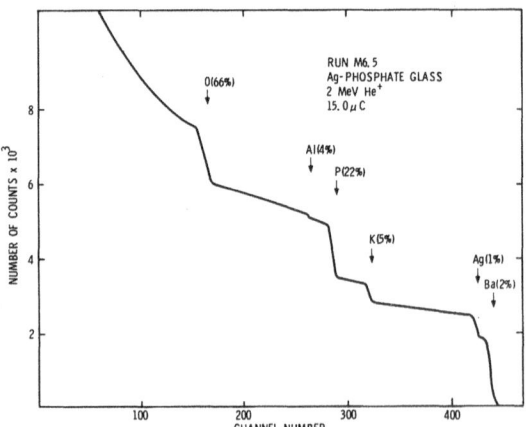

Fig. 1 - Backscattering spectrum of 2.0 MeV He^+ ions from a silver-activated phosphate glass. The atomic percentage of each component is listed beside the elemental label.

surface for each component. The methods of obtaining the composition from the scattering yield are well known and will not be discussed further here. The reader is referred to a recent compilation[4] of working data for ion beam analysis. Another example is shown in in Fig. 2 for a lithia-alumina-silicate glass which had been implanted with 10^{16} cm^{-2}, 275 keV Ag ions. Noting that the backscattering cross section is proportional to the square of the target

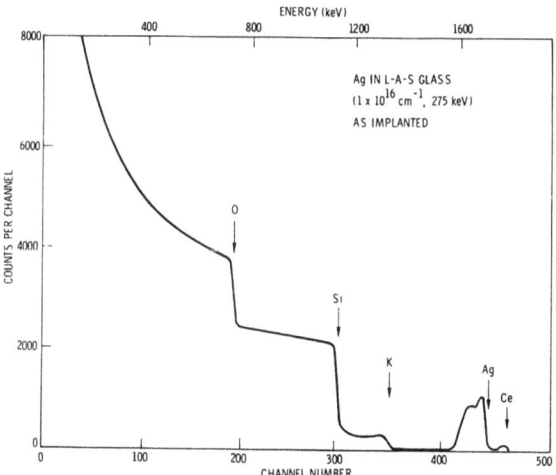

Fig. 2 - Backscattering spectrum from a lithia-alumina-silicate glass implanted with 10^{16} cm^{-2} 275 keV Ag$^+$ ions. The atomic composition of the glass before implantation was 59.3% O, 25% Si, 12.2% Li, 2% K, 1.4% Al, 2 x 10^{-2}% Sb, 2.2 x 10^{-3}% Ce.

atomic number, analysis of the O, Si and K components agree well with the melt composition and a 2.2 x 10^{-2} at.% Sb level is easily seen when the data in Fig. 2 are plotted on an expanded vertical scale. The lithium concentration is too small and the scattering is at too low an energy to be measurable with the technique used here, but analysis using proton backscattering or a Li nuclear reaction would easily resolve it. The 1.4 at.% Al cannot be seen as it is very close to the scattering edge of the 25 at.% Si. The Ce at the surface is due to the CeO$_2$ abrasive used as a final polishing compound which remained imbedded in the surface. The scattering due to the implanted silver will be discussed in Section III below.

Figure 3 shows the backscattering spectrum from a sample of As-Te-I semiconducting glass with an electrically conductive surface layer. Previous experiments[5] have shown that the average atomic composition does not change significantly with depth either with or without the surface layer and the workers were able to conclude that the electrically conductive layer formed by devitrification of the glass surface did not have a stoichiometry which was significantly altered from the bulk composition.

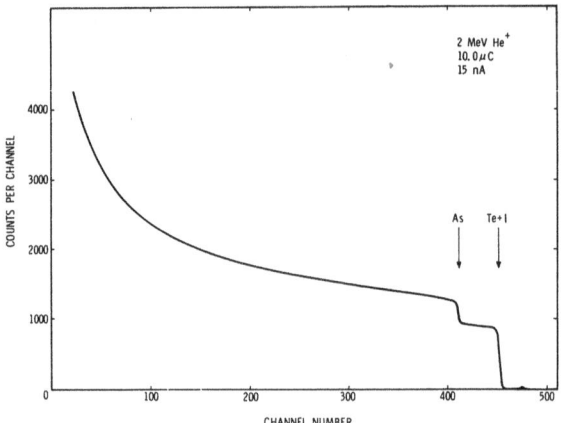

Fig. 3 - Backscattering spectrum from a $As_{50}Te_{45}I_5$ sample with an electrically conductive surface layer induced by annealing in vacuum at 160°C for 5 hr. These results indicate no change in stoichiometry to a depth ≈ 2 µm and an accuracy of ≈ 3-5% at the surface.

III.   DIFFUSION MEASUREMENTS IN GLASSES

Because ion backscattering measures the depth distribution of atomic composition, it can be used as a unique tool for measuring diffusion profiles over distances of a few thousand angstroms. This technique is especially useful for measuring the diffusion of heavy impurities in lighter hosts. Little data exist for impurity diffusion in glasses, particularly for heavy metallic impurities and it is here that backscattering can make a major contribution.

It has been demonstrated[6] that implantation of Au into the lithia-alumina-silicate (LAS) glass previously mentioned can cause nucleation and crystallization of a glass ceramic surface layer after the proper thermal treatment. Investigations of samples of this glass implanted with either gold or silver have been investigated[7] by combined ion backscattering and optical spectroscopy measurements. Figure 2 shows the backscattering spectrum of a LAS glass sample implanted with $10^{16} cm^{-2}$ 275 keV Ag^+. The profile of the silver is non-Gaussian for this ion energy, fluence and dose rate of $\sim 1 \mu A/cm^2$, and the profile contains a narrow peak on the surface side of the distribution. This distribution is very fluence-dependent as is seen in Fig. 4 where the expanded silver depth profiles are shown for implant fluences ranging from $1 \times 10^{15} Ag cm^{-2}$ to $3 \times 10^{16} Ag cm^{-2}$. The peak at about channel #463 is due to CeO_2 imbedded in the surface of the various samples. Note that for the 1×10^{15} and $5 \times 10^{15} cm^{-2}$ implantations the Ag profile is approximately Gaussian but the peak of the measured distribution is 150-200 Å deeper than the LSS projected range[8] of 1075 Å.

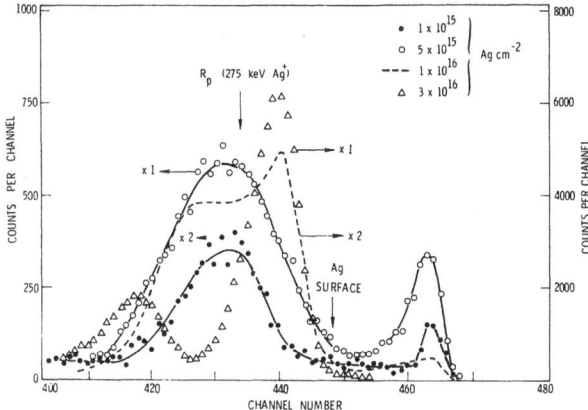

Fig. 4 - Backscattering data showing the depth profiles of implanted 275 keV Ag for fluences of 1×10^{15} cm^{-2}, 5×10^{15} cm^{-2}, 1×10^{16} cm^{-2} and 3×10^{16} cm^{-2}. The energy corresponding to Ag at the sample surface is indicated by the arrow at channel #448. Scattering from Ag at increasing depths occurs at progressively lower channel numbers. The peak near channel #463 is due to CeO_2 embedded in the glass surface.

As the fluence is increased to 10^{16} cm^{-2}, a peak ($\sim$ channel #440) begins to form about 500-600 Å from the surface. At higher fluences two distinct peaks are observed, the near-surface one corresponding to the peak observed in the 10^{16} cm^{-2} implantation. It should be noted, however, that lateral non-uniformity can influence the distributions. The 3×10^{16} cm^{-2} sample was visibly discolored due to formation of Ag colloids and the discoloration was quite non-uniform. In an attempt to understand these unusual depth distributions, samples were implanted to 10^{16} cm^{-2} under varying implant conditions; annealing experiments also were performed. The results of annealing the 10^{16} cm^{-2} sample, whose Ag profile was shown in Fig. 4, are seen in Fig. 5. Note that the vertical scale is logarithmic for these spectra. No change is observed after annealing to 100 C for 15 min, but at 200 C a small peak is observed to form corresponding to Ag on the surface of the glass. There is also a movement of Ag into the sample but at low concentrations. The concentration of surface Ag continues to increase until about 350 C; for higher annealing temperatures the surface film is observed to decrease until it disappears at $\sim$ 550 C. These changes have been related[9] to Ag agglomerate size through optical absorption measurements. In Fig. 6 are compared the 1×10^{16} cm^{-2}, 275 keV Ag depth distributions for 1) a room temperature implant at $\sim$ 1 µA/cm^2 (normal), 2) a room temperature implant at $\sim$ 0.2 µA/cm^2 (low dose rate) and 3) a 77 K implant at $\sim$ 0.2 µA/cm^2 (low temperature). Also shown is the LSS range distribution (solid line). The difference in integrated area is presumably due to 1) incomplete charge collection during

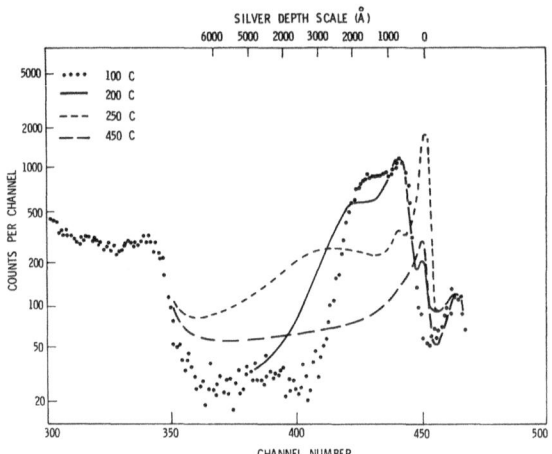

Fig. 5 - Depth profiles of implanted Ag for the 1 x 10^{16} cm^{-2} implant shown in Figs. 2 and 4 after various thermal annealing treatments of 15 minutes at 100 C, 200 C, 250 C and 400 C. The sharp edge in the scattering yield at channel #350 is due to the K component of the glass.

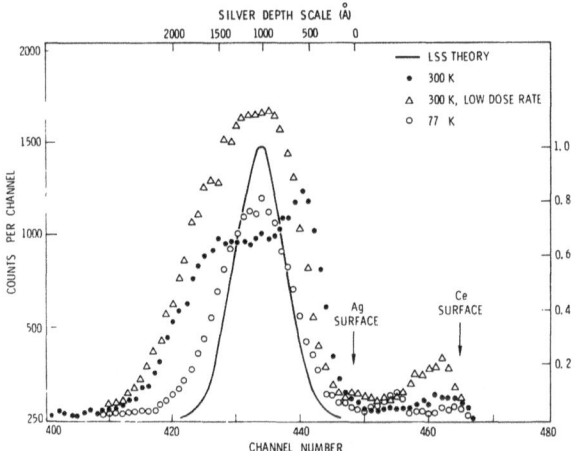

Fig. 6 - Depth profiles of 1 x 10^{16} cm^{-2}, 275 keV Ag implants performed at 1) room temperature and ~ 1 μA/cm^2, 2) room temperature and 0.2 μA/cm^2 and 3) 77 K and 0.2 μA/cm^2. The right-hand vertical scale is for the LSS distribution.

implantation and 2) lateral nonuniformity of the implant. The low temperature implant distribution agrees fairly well with the theoretical prediction whereas the low dose rate sample shows a much wider distribution although the projected range is the same as the LSS prediction. We interpret these results in terms of diffusion (possibly damage assisted) and aggregation of Ag colloidal particles. After the low temperature implant, optical measurements show no evidence for the presence of Ag colloids (of radius > 5 Å) whereas colloids are present in the low and normal dose rate $10^{16} cm^{-2}$ samples after implantation at room temperature. This suggests that, at 77 K, the Ag is present as atomically dispersed atoms but during implants at higher temperatures Ag atoms diffuse and aggregate to form colloids. These colloids tend to dissolve when annealed to higher temperatures ($\sim 200\text{-}300$ C) with some of the Ag migrating into the bulk, but most of it migrating to the surface where it forms small islands of metallic silver. In Fig. 4 the unusual peak in the $10^{16} cm^{-2}$ data and the double-peak structure in the $3 \times 10^{16} cm^{-2}$ data may be indicative that the aggregation formed during implantation is enhanced by the damaged layer created by the implantation. These studies are continuing and being extended to other photoreducible metal ions (Au, Cu and Pt).

IV. STOICHIOMETRY CHANGES DUE TO ION BOMBARDMENT

In addition to the well-known doping effects ion beams may have on materials, it is possible for damage created by the ion bombardment to induce changes in stoichiometry. In metals, dissolution of second-phase precipitates is a well-known example of such changes on a microscopic scale.[10] Picraux[11] has demonstrated enhanced outgassing of As after ion-bombardment of GaAs. Recently Naguib and Kelly[12] have reviewed evidence and formulated criteria for ion-induced stoichiometry changes in non-metallic solids. Although this subject has not been studied in detail, the effects may be particularly large in insulators due to the possible effects of damage caused by energy deposited into electronic processes as well as atomic processes. To illustrate this subject, we present data on movement of the K component of the LAS glass described earlier. A sample of LAS glass was implanted to a fluence of 10^{16} Xe cm^{-2} at 285 keV. In Fig. 7 are shown the scattering spectra from the implanted Xe distribution (top) and the K distribution (bottom) before and after the Xe implantation. The three spectra are shown on a common depth scale. After implantation, the K component of the glass is depleted (max. depletion $\sim 50\%$) near the end of the Xe range but is enhanced near the surface. If the sample is thermally annealed, the K redistributes itself between 150 and 250 C. We have also seen these effects for Au and Ag implantations, but the effect appears to depend on the implanted species as a 10^{16} Ag implantation showed little K redistribution whereas 10^{16} Xe and Au show large changes.

V. PROBLEMS ASSOCIATED WITH ION ANALYSIS OF INSULATORS

There are some unique problems associated with ion analysis of insulators, which are not encountered for metals and semiconductors. The most serious of these is probably that associated with the very low electrical conductivity of insulators. As an ion loses its charge

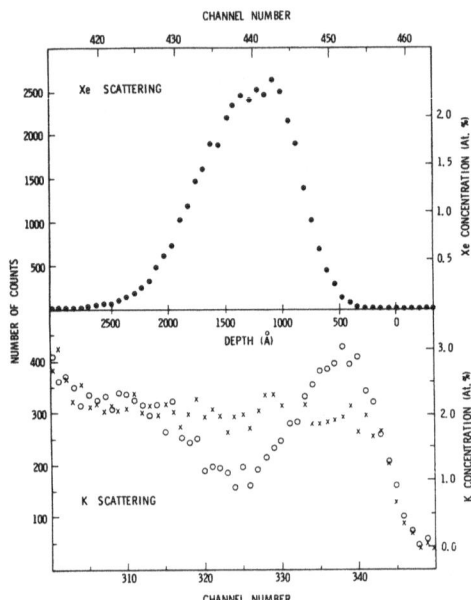

Fig. 7 - Depth profiles of K and implanted Xe plotted on a common depth scale. The K distributions are shown before (x) and after (o) the Xe implantation.

while stopping in a semiconductor or metal it is relatively easy to redistribute the charge within the sample and connect a charge or current-measuring device to measure the implanted ion dose. In insulators, however, the lack of free carriers and/or their low mobility can cause the implanted surface to build up charge leading to 1) high local fields causing beam distortion and local lateral beam non-uniformity and unknown kinetic energies for the ion actually striking the surface, 2) inaccurate measurement of ion dose and/or 3) damage within the sample surface layer due to electrical discharge. These effects are much more noticeable for implantation and analysis beams of up to a few hundred kilovolts than for analysis beams of several megavolts. These problems have been discussed recently[13] and the most effective means of solving the problem appears to be the use of a thin evaporated conductive film over the surface of the sample.

     The preparation of damage-free surfaces can also present problems for insulators. Many of the materials are quite inert and not attacked by gentle chemical etchants. Naguib et al.[14] have studied damage due to Kr bombardment of α-Al_2O_3 by 2 MeV He^+ $\langle 0001 \rangle$ channeling and backscattering. They were able to prepare damage-free surfaces for channeling purposes by high temperature (1200 C) annealing. Channeling along the $\langle 1\bar{1}00 \rangle$ and $\langle 11\bar{2}0 \rangle$ axes also give reasonably low minimum yields after annealing at 1200 C.[15] Studies of impurity locations in insulators using channeling are almost non-existent and they offer rich possibilities for future research.

Finally, we should mention the problem of damage caused by energy lost to ionization. Ion-induced damage in metals and semiconductors is almost entirely related to the energy lost to atomic processes. This is not the case for insulators. In alkali halides the well-known Pooley-Hersh mechanism[16] relates F-center formation to energy lost to ionization processes. For high-energy ion analysis in particular, most of the ion energy is lost to electronic (ionization) processes. However, little is known about ionization-related damage mechanisms[17,18] in materials other than alkali halides, and care should be taken to consider such mechanisms when using high energy ions to analyze or modify insulators.

VI. CONCLUSIONS

The use of ion beams to analyze and/or modify insulators is a research field that has not been extensively investigated. Problems of diffusion in glasses and lattice location in insulators are easily tractable using these methods of combining ion beam techniques with techniques such as optical spectroscopy and EPR. Insulators appear to be a class of materials where stoichiometry changes due to ion irradiation will be an important area of study and insulators are often very nice systems for compositional measurements. There are some problems using ion beams on insulators which are not encountered when studying metals or semiconductors, but these problems appear to be surmountable.

REFERENCES

1. Applications of Ion Beams to Metals, Ed. by S. T. Picraux, E. P. EerNisse and F. L. Vook (Plenum Press, N. Y., 1974).

2. S. Namba et al., J. Vac. Sci. Technol. 10, 936 (1973).

3. G. Linker, O. Meyer and W. Scherber, phys. stat. sol. (a) 16, 377 (1973).

4. Catania Working Data, A Compilation of Tables, Graphs and Formulae for Ion Beam Analysis, Ed. by J. W. Mayer and E. Rimini, Catania, Italy, 1974.

5. R. T. Johnson, R. K. Quinn and J. A. Borders, J. Non-Cryst. Solids 15, 289 (1974).

6. G. W. Arnold, J. Appl. Phys., to be published.

7. G. W. Arnold and J. A. Borders, Proc. of the Intl. Conf. on the Application of Ion Beams to Materials, Warwick, England, Sept. 1975, to be published.

8.   D. K. Brice, unpublished data.

9.   G. W. Arnold and J. A. Borders, to be published.

10. R. S. Nelson in Applications of Ion Beams to Metals, Ed. by
     S. T. Picraux, E. P. EerNisse and F. L. Vook (Plenum Press,
     New York, 1974).

11. S. T. Picraux in Ion Implantation in Semiconductors and Other
     Materials, Ed. by B. L. Crowder (Plenum Press, New York, 1973).

12. H. M. Naguib and R. Kelly, Rad. Effects 25, 1 (1975).

13. W. Beezhold and E. P. EerNisse, Appl. Phys. Lett. 21, 592 (1972).

14. H. M. Naguib, J. F. Singleton, W. A. Grant and G. Carter,
     J. Mat. Sci. 8, 1633 (1973).

15. J. A. Borders, unpublished data.

16. E. Sonder and W. A. Sibley, Point Defects in Solids, Vol. 1,
     Ed. by J. Crawford, Jr., and L. M. Slefkin (Plenum Press,
     New York, 1972), p. 201.

17. G. W. Arnold and F. L. Vook, Rad. Effects 14, 157 (1972).

18. E. P. EerNisse and C. B. Norris, J. Appl. Phys. 45, 5196 (1974).

DISCUSSION

Q: (R.E. Honig) In view of the high mobility of K^+ ions in electric
fields, what was done to take into account the various charge-up
effects occurring with insulators, in order to obtain true potassium
profiles ?

A: (J.A. Borders) Nothing! The data is preliminary and I present it
without explanation to stimulate your interest.

Q: (M.-A. Nicolet) What conducting layer did you use?

A: (J.A. Borders) Aluminum.

Comment: (N. Itoh) Morita, Tsubouchi and Matsunami have found that
the charging-up in insulators produces non-characteristic X-rays
caused by accelerated electrons towards the specimen; the maximum X-ray
energy reaching nearby to 35 keV. These X-rays are easily removed by
placing heated filaments in the proximity of the specimens.

LITHIUM ION BACKSCATTERING AS A NOVEL TOOL FOR THE CHARACTERIZATION OF OXIDIZED PHASES OF ALUMINUM OBTAINED FROM INDUSTRIAL ANODIZATION PROCEDURES

J.P.THOMAS[+]A.CACHARD[++]M.FALLAVIER[+]J.TARDY[++]S.MARSAUD[+]

Institut de physique nucléaire [+]Dpt.de physique des matériaux [++]
Université Cl.Bernard LYON 1 43 Bd du 11 nov.1918
Villeurbanne 69621 FRANCE

ABSTRACT

In the increasing trend to use ions heavier than α - particles in backscattering (MeV energy range) lithium ions lead to a noticeable improvement of the main characteristics of the method (especially depth resolution and selectivity). Together with a comparison of the analytical performances of Li-ions and other particles, the method has been applied to the characterization of oxidized phases of aluminum obtained from industrial processes. Comparisons with other physical methods (^{27}Al (p, γ) resonant reaction, ellipsometry, electrical measurements) are presented.

INTRODUCTION

In the extensive use of elastic backscattering for microanalysis purposes, α-particles up to several MeV have been for long considered as the most convenient projectile. It appears from the last I.B.S.L.A. conference (1) that the performances of the method are very near to the optimum mainly because of the resolution limit for the most commonly used surface barrier detectors (12 to 15 Kev at room temperature). With respect to the basic parameters of backscattering -mass resolution, sensitivity and depth resolution - heavier ions have been recognized as very promising (2). Pioneering this way, ABEL et al (3) in 1972 with nitrogen and carbon beams, PETERSSON et al (4) in 1973 with oxygen beams showed evidence of some very interesting improvements but also severe limitations due to the degradation of the detector resolution v.s. the collected

particles dose. Suggested by the experimental resolution-energy relationships of the already mentioned authors as well as those of
BERGSTRÖM et al (5) for other ions, a insight to "lighter" heavy
ions capabilities has been undertaken in our laboratory using lithium (6 or 7) ions. We thus show, in this paper, that while maintaining fairly good sensitivity, an optimum selectivity is obtained
for medium Z elements with an improved depth resolution and a satisfactory behavior of the detector. These features as well as the
compromise so realised between depth resolution and analyzing depth
are of importance in the study of anodic oxidation layers of aluminum obtained from industrial processes. Oxidised-phase analysis can
then be performed for small thicknesses and with better depth resolution. In the same way more information is obtained on the impurity
distribution, mainly from the oxidation bath (phosphorus) or from
the metal itself (silicon). A discussion of the processes involved
in the basic formation procedures is undertaken, the backscattering
results being compared with those of other physical methods like
^{27}Al (p, γ) resonant reaction, ellipsometry and electrical measurements. Segregation phenomena are pointed out, in the other hand,
confirmed by ESCA experiments.

LITHIUM IONS CHARACTERISTICS IN BACKSCATTERING

From the well known kinematics relation, the selectivity can
be expressed as :

$$d \ (E_{B.S}/E_i) \ / \ dM_2 = f \ (M_2)$$

subscripts B.S. and i refering to backscattered and incident energy,
1 and 2 to the incident and target nucleus.

For a given E_i and $\Delta E_{B.S}= R$ (detection resolution) the corresponding ΔM_2 is the mass resolution. As shown in fig.1 for the selected ^{4}He, ^{6}Li, ^{7}Li and ^{12}C ions (here for θ = 160°), optimum selectivity is obtained for ^{4}He ions ($M_2 < 18$), lithium ions ($18 < M_2 < 34$)and
carbon ions ($M_2 > 34$) respectively. With respect to this parameter the
superiority of ^{4}He ions has to be noticed for light elements. In the
other hand, it is not worthwhile to increase the projectile mass for
the heavy elements, the values of selectivity remaining low and as
mentioned (3) and discussed later on, the resolution increasing.

Considering the backscattering yield, if strong deviations from
the RUTHERFORD law are encountered with α-particles interacting with
light elements in the Mev range, the increase of the Coulomb barrier
when using heavier projectiles allows the classical RUTHERFORD cross
section to be valid with great precision. As demonstrated by ABEL et
al (3) for carbon ions, the sensitivity can thus reach very high values with limited pile-up effects. This feature is not predominant for
our purpose but the sensitivity remains satisfactory for Li ions.
The applicability of the RUTHERFORD formula will appear more powerful
in quantitative analysis and stoichiometry determination.

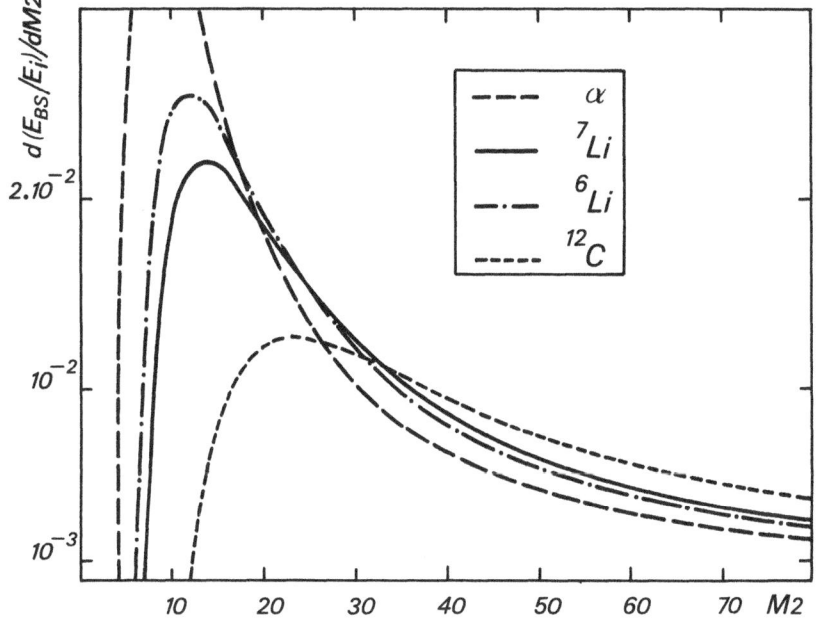

Fig.1. Selectivity versus target mass for different particles

As already mentioned, the energy dependance of the detector resolution must be carefully considered in determining the depth resolution from the backscattering energy loss parameter :

$$[S] = K \left|\frac{dE}{dx}\right|_{Ei} + \frac{1}{|\cos \theta|} \left|\frac{dE}{dx}\right|_{E_{B.S.}}$$

From the results of BERGSTRÖM et al (5) for ^{12}C, ^{20}Ne, ^{40}Ar, and ^{84}Kr ions in agreement with those of ref (3) and (4), a limited resolution increase is expected for ions lighter than ^{12}C. As shown, in fig.2, values less than 25 Kev are obtained with 2 Mev Li ions backscattering at 160° on thin Au films (~ 50 Å) deposited on glass substrates. The constant (16 Kev) α- particle resolution is also shown on fig.2. Another important result, noticeable degradation of the detector performances, is not observed for a long running operation time. Still comparing He, Li and C ions, depth resolutions are then represented on fig.3 for the elements of our study, i.e., 0 and Al in Al_2O_3 and Au in a Au matrix. (resolution values for ^{12}C taken from (5), stopping powers data from NORTHCLIFFE and SCHILLING (6)).The optimum values are, to our knowledge, the best obtained using backscattering technique in this energy range. It should be noticed that up to 3 - 4 Mev they are maintained along a fairly extended energy range.

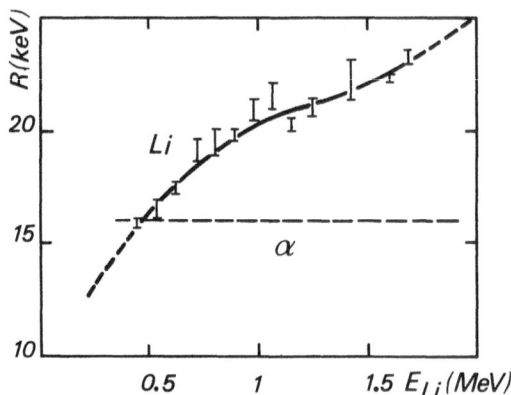

Fig.2. Detector resolution versus detected energy for Li ions and
       α - particles.

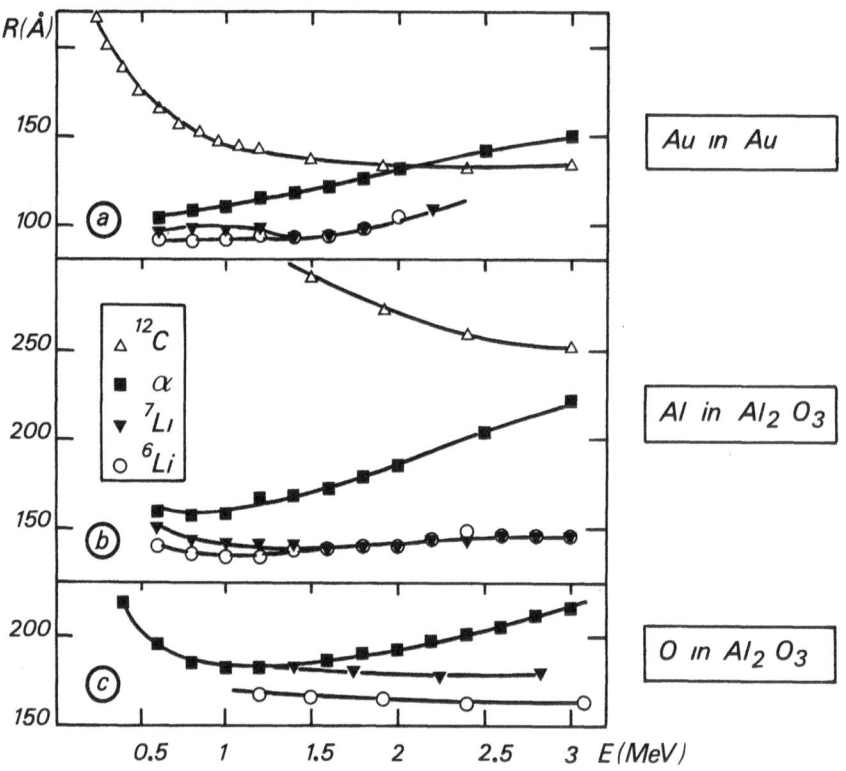

Fig.3. Depth resolution versus incident energy for
       a) Au in Au   b) Al in Al$_2$O$_3$ c) O in Al$_2$O$_3$

This is due to the slow variation of the stopping power near the maximum, which allows, going to the higher energy of conventional small accelerators, an increase in the analyzing depth without reducing significantly the depth resolution.

APPLICATION TO THE STUDY OF ANODIC OXIDATION OF ALUMINUM

Experimental

Sources of ^{7}Li or ^{6}Li (99 % enriched) are prepared in solid form (7) and used with the 2 Mev VAN DE GRAAFF accelerator of the institut de physique nucléaire de Lyon.

The previously described target chamber (8) has ionic pumping, (10^{-6} torr inside the chamber), a turbomolecular pump being used for the beam line inside which a liquid nitrogen trap is interposited. The 25 mm^2 100 μm ORTEC surface barrier detector shows typical values of about 16 Kev when tested with 1 Mev α - particles backscattered from a thin Au target. Conventional TENNELEC electronics (0.5 μS time constant) are associated with a HEWLETT-PACKARD A.D. converter coupled to a 2116 C computer.

The samples to be studied are made from refined aluminum (100μm) which has been mechanically and chemically cleaned and thus annealed. In order to be used as anode in industrial electrolytic capacitors they are subject to different treatments (among them the anodic oxidation called "formation") leading to a barrier layer of required electrical qualities. Three classical steps of elaboration will be considered : A/ Formation at a given voltage B/ Passivation without voltage in a phosphoric solution C/ Reanodization or post-formation. For A and C the electrolytic solutions are here phosphoric (a) and carboxylic (b) for voltages below 300 V. This study will only deal with a 60 V voltage.

Data analysis

At the maximum 1.9 Mev energy and a 160° detection angle, the main information given by the backscattering spectrum is shown in figure 4 (A.b) for a 60 V formation with carboxylic electrolyte. Due to the rapid increase of the aluminum yield, one cannot obtain a precise determination of oxygen from the superimposed peak nor the stoichiometry by comparing the oxygen and aluminum peaks (9) as deconvoluted in this figure. Nevertheless , as already detailed by several authors, among them ABEL et al (10), stoichiometry and thickness determinations are obtained by comparing the steps on the Al spectrum with the ratio of the heights corresponding to unoxidized Al and Al in the oxide. Referring to surface yields and for Al$_2$O$_3$ stoichiometry we have :

$$\frac{h_1}{h_2} = \frac{h\ Al}{h\ Al_2O_3} = \frac{n_{at}/g\ \ in\ \ Al}{n_{at}/g\ \ in\ \ Al_2O_3}\ \frac{\lfloor S\rfloor\ Al_2O_3}{\lfloor S\rfloor\ Al} = 1.973\ for\ 1.9Mev\ ^7Li$$

With a 150 Å depth resolution for Al in Al_2O_3, the stoichiometry can then be determined for layers as thin as 300 Å. The typical
uncertainty, not exceeding 6%, can be lowered if deconvolution programs are used (11). Sharper edges can be then obtained allowing a
better precision in the determination of the plateau heights. The
Al_2O_3 stoichiometry has thus been well established for the A type
formation with a maximum deviation of 3% for samples ranging from 420
to 3700 Å. From the width of the plateau and the energy-depth relationship of ^7Li in Al_2O_3 (assuming the validity of the BRAGG's rule),
the layer thickness is determined within an uncertainty ranging
from 3 to 10 %. This maximum uncertainty will correspond to extreme
values of the thickness, too small (channel width indetermination)
or too large (poor determination of the interface due to straggling
effect). If the deviation from the Al_2O_3 stoichiometry is too large,
the previous relation cannot be accounted since the nature of the
medium is changed. Correspondingly the thickness determination is
affected by this variation.

Still looking at the spectrum of figure 4 (A.b) it appears that
the phosphorus (present as surface contaminant on the picture) can
be determined upon a thickness of about 700 Å (limited by the Al
edge). According to the RUTHERFORD's law no standard is required for
direct quantitative analysis and profiling of this element, provided
its presence does not significantly affect the nature of the medium.

For Z >13 other impurities are simultaneously detected but only
identified by their backscattering energy. The complementary use of
proton- induced X-ray excitation allows unambiguous identification
of S ,Cl, Ca and heavier elements excepted for a peak near the aluminum edge (figure 4 A-b) attributed to silicon. Despite the fact that
this peak could be hardly attributed to pile-up effects, neither (p,x)
reactions nor nuclear reactions (like $^{28}Si(d,p)$ or $^{28}Si(p,p'\ \gamma)$ are
able to assess the presence of this element due to the strong interference of the aluminum matrix. The same behavior is observed on the
unoxidized metal; this suggested segregation phenomenum must be
confirmed by other experiments.

RESULTS AND DISCUSSIONS

Backscattering analysis

The influence of the already described B (a and b) and C (a and
b) treatments can be followed from the other spectra of figure 4.From
a formation using phosphoric electrolyte (a),the B treatment clearly
shows an overstoichiometry in Al, with a reduced thickness of the layer. Correspondingly, at an analyzing depth of 700 Å, a strong concentration graident of phosphorus can be noticed. The mean derivation

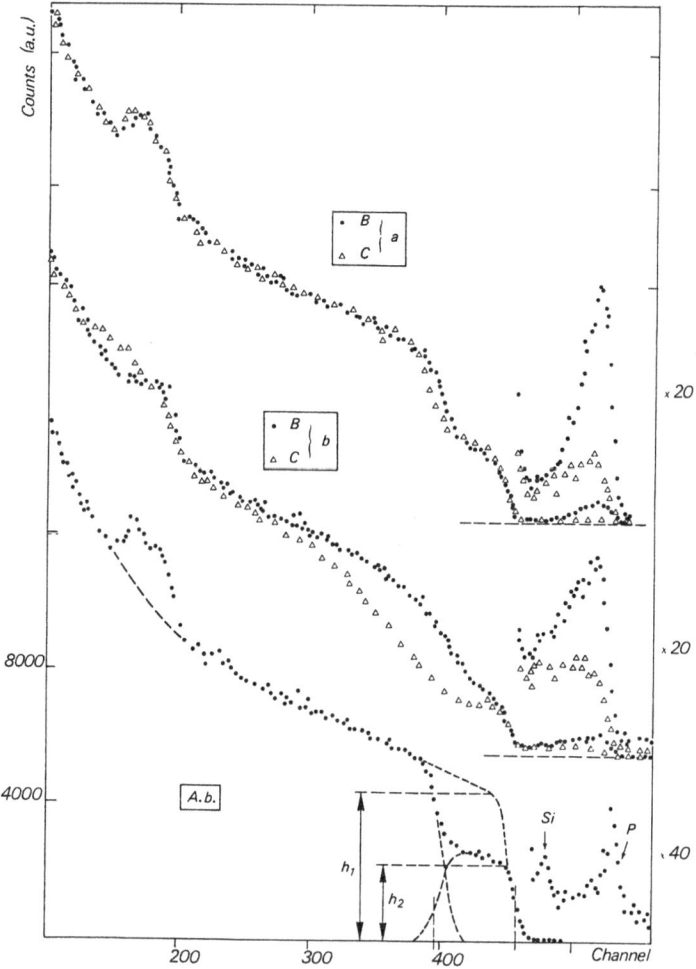

Fig.4. Backscattering spectra (^{7}Li 1, 9 Mev, $\theta = 160°$) for the samples investigated (see text).

to Al_2O_3 stoichiometry is about 10 %.

As expected the thickness is reduced to about 600 Å. The same phenomenum occurs when the formation is made with carboxylic electrolyte (B.b) but the two effects are enhanced. The phosphorus appears as deeply incorporated, but the strong overstoichiometry (~ 24 % deviation) suggests hydrogen incorporation.

With the reformation (C), two distinct final structures are obtained using a or b electrolyte. From the phosphoric electrolyte, complete restoration of the layer is achieved as well for the

stoichiometry as for the thickness while the phosphorus is at a lower
level ($\sim$ 3%) and uniformely distributed,From the carboxylic solution
a strong overstoichiometry ($\sim$ 37% deviation) is pointed out at the
surface together with a complete destruction of the interface (up to
$\sim$ 3000 Å). The phosphorus distribution very similar to the (a)case
(at a 5% level), as well as the oxygen distribution, cannot explain
such a phenomenum. With the widening of the oxygen peak, aluminum
hydroxide formation appears highly probable. Such an hydrogen ana-
lysis is under investigation, the resonant reaction, $_1^1$H (^{11}B,α)$\alpha\alpha$
proposed by LIGEON et al (12) being especially adapted for this stu-
dy. In order to increase the analyzing depth of the phosphorus, backscattering of 3.5 Mev α - particles has also been performed allowing
a 1730 Å analyzing depth. This improvement, for the last case discussed (Bb and Cb), shows a phosphorus distribution deeper than the layer thickness.A rapid decrease is observed after 800 Å. Better quality of information with ^{7}Li, and larger analyzing depth with α -
particles illustrate the complementarity of these particles.A more detailed discussion of this complementarity will be given elsewhere(13).

   The presence of silicon can be an interference in the phosphorus
profiling. Systematic investigation of the metal preparation has been
done with respect to this element. For the 140 Å depth resolution
limit of the method the Si contents are in the % range without marked difference for a 0.5 % Si-doped sample. After formation, a small
decrease is observed but not enough significant. For the samples investigated 1.3% appears as an upper limit of the silicon content. As
discussed later on, a segregation phenomenum has been pointed out
by ESCA experiments.

Other physical methods

   The resonant reaction ^{27}Al(p,γ) (Γ < 100 eV for E_p = 992 Kev)
is particularly useful to study the aluminum distribution near the
surface. AMSEL et al (14) have pointed out the feasibility of the method, the stoichiometry and thickness being determined in the same
way as previously described but from the γ yield of the aluminum.
The depth resolution at the surface is mainly governed by the energy
distribution of the accelerator and can be as low as 50Å(15). However,
with the 1 Kev resolution of our machine,135 Å is obtained.Straggling
effects and higher energy resonances limit the depth range investigation and the precision ; another important restriction is that the method is fairly time consuming (at least one order of magnitude compared with backscattering). Nevertheless the results reported in table
1 are in good agreement with the backscattering ones.
   More specific to aluminum oxide, ellipsometry and capacitance
measurements allow the thickness to be determined. Owing to the
expected uncertainties of these methods, the measurements given in
table 1 can be considered as quite satisfactory excepted for the
sample C(b). The capacitance measurement using liquid electrode leads
to an unexpected result as well as other electrical tests from which

Table 1 : B.S. and other physical methods results (samples identification : see text)

Al$_2$O$_3$		A(b)	B(b)	C(b)	B (a)	C (a)
Stoichiometry deviation	B.S.	~1.8 %	~23.8 %	~37 %	~10 %	~1 %
	^{27}Al(p, γ)	~3 %		~32 %		
Thickness	B.S.	820 ± 25 Å	530 ± 50 Å	<3000 Å	610 ± 50 Å	830 ± 25 Å
	^{27}Al(p, γ)	812 ± 70 Å	540 ± 100 Å	>1000 Å		
	Ellipsometry	1000 Å	650 Å	~1800 Å	~800 Å	~950 Å
	Capacity (electrolyte)	650 Å (11µF/dm^2)			500 Å	700 Å
	Capacity(solid electrode)			650 Å		
				1100 Å	700 Å	750 Å

satisfactory behavior of the barrier layer is observed (polarisation time, tg δ and stability). Identical electrical results from C(a) and C(b) samples are not yet fully explained, but it can be noticed from these comparisons that only Li backscattering is able to give unambiguous informations about the layer structure.

The use of the ESCA technique allows the unambiguous identification of silicon (together with its chemical state) very near the surface ($\sim$ 10 Å). For different metal preparations (5 to 50 ppm in bulk material - 0.5 % for a Si-doped sample) the silicon content is thus determined at a 16 to 34 % level. For the 140 Å corresponding to the backscattering depth resolution, an ion bombardment etching is thus performed. Integrating the Si gradient concentration over this thickness and within the approximations of the method (16), the ESCA results are in fairly good agreement with the backscattering ones (1.1 % to 2.6 % for the silicon content).

CONCLUSION

The use of Li ions up to 2 Mev offers several significant improvements in the backscattering analysis among them depth resolution and selectivity. These features are of interest in the study of the anodic oxidation of aluminum for which 150 Å depth resolution is obtained with a profile up to 3700 Å. Stoichiometry and depth determination, performed at different steps of barrier layer elaboration, point out an unexpected behavior of the samples post-formed in carboxylic solution. These results can be considered more significant than those obtained from other physical methods and especially from electrical measurements. It appears from a tentative interpretation that hydrogen determination is strongly needed. The use of boron ‾ ion-induced nuclear reaction would be of great interest in such a study, as well as backscattering, owing to the performances comparison between Li and heavier ions.

ACKNOWLEDGEMENTS

This work was supported by the D.G.R.S.T. (contrat 73.7.1545) The assistance and results of M. BADIA and P. SPENDER from the P.U.K. company are greatly acknowledged.

REFERENCES

(1) J.W.MAYER J.F.ZIEGLER, Proc. Intern.cont. on ion beam surface layer analysis YORKTOWN HEIGNTS, NEW YORK 1973.

(2) O.MEYER J.GYULAI J.W.MAYER Surf. Sc. 22 (1970) 263

(3) F.ABEL G.AMSEL M.BRUNEAUX C.COHEN A.MAUREL S.RIGO J.ROUSSEL
     J. of Radioanal. Chem 16 (1973) 587

(4) S.PETERSSON P.A.TOVE O.MEYER B.SUNDQVIST A.JOHANSSON
     Thin Solid films, 19 (1973) 157

(5) J.BERGSTRÖM K.BJÖRKQVIST B.DOMEIJ G.FLADDA
     Can J. of Phys 46 (1968) 2679

(6) L.C.NORTHCLIFFE R.F.SCHILLING Nucl. Data tables 7 (1970) 233

(7) J.P.BUCHET   Thesis LYON (1967)

(8) J.P.THOMAS   Thesis LYON (1974)

(9) J.GYULAI O.MEYER J.W.MAYER V.RODRIGUEZ J. of Appl.Phys. 42
     (1971) 451

(10) F.ABEL G.AMSEL E.D'ARTEMARE M.BRUNEAUX C.COHEN B.MAUREL C.ORTEGA
     S.RIGO J.SIEJKA M.CROSET D.DIEUMEGARD J.of Rad. Chem 16(1973)567

(11) J.ENGERRAN Thesis LYON (1975)

(12) E.LIGEON A.GUIVARC'H   Rad.Effects 22 (1974) 101

(13) J.TOUSSET J.P.THOMAS A.CACHARD   International nuclear and ato-
     mic activation analysis conference oct. 14-16 1974 GATLINBURG,
     TENNESSEE U.S.A.

(14) G.AMSEL J.P.NADAI E.D'ARTEMARE D.DAVID E.GIRARD J.MOULIN
     Nucl. Inst. Meth. 92 (1971) 481

(15) K.L.DUNNING G.K. HUBLER J.COMAS W.H.LUCKE H.L. HUGUES
     Thin solid films 19 (1973) 145

(16) G.HOLLINGER Y.JUGNET TRAN MINH DUC
     Annual meeting of soc.chim. de France CAEN mai 1975

DISCUSSION

Q: (A. Turos) How thick layers can you analyze with the reasonable
good depth resolution ?

A: (J.P. Thomas) For our particular case of alumina, we performed analysis of "thin" targets as deposited on light backing and self-suppor-
ting: 2000 to 3000 Å corresponds to a "reasonable" depth resolution
with respect to the straggling phenomenon.

INVESTIGATION OF AN AMINO SUGAR-LIKE COMPOUND FROM THE CELL WALLS OF BACTERIA USING BACKSCATTERING OF MeV PARTICLES

T.G. Finstad[*] and T. Olsen

Institute of Physics, University of Oslo

R. Reistad

Department of Microbiology, Dental Faculty

University of Oslo, Oslo 3, Norway

ABSTRACT

Backscattering of 1.5-2 MeV alpha particles has been used for elemental analysis of an amino sugar-like compound isolated from the cell walls of an extremely halophilic bacterium. The composition of the compound is changing upon exposure to the ion beam. The experiments showed the compound to contain sulphur. This information in conjunction with conventional biological analytical techniques indicates that the compound contains sulphur in an uncommon form. The compound has not been conclusively identified but a tentative structure is suggested.

INTRODUCTION

The interest in the bacteria was partly initiated by observations that fish salted with crude sea-salt turned red. This was caused by bacteria living in the salt and multiplying on the fish. Such salty environments are considered rough even for bacteria and focused the interest upon the bacterial cell walls. One of us[1,2] has previously reported upon amino sugar-like compounds isolated from cell wall hydrolysates of an extremely halophilic bacterium,

[*]Present address: Norwegian Defence Research Establishment, N-2007 Kjeller, Norway

<u>Halococcus</u> sp., strain 24. One of the unidentified compounds
behaved like a strongly acidic compound upon paper electrophoresis
and on the amino acid analyzer. This indicated acidic groups other
than carboxyl.

In strongly acidic carbohydrates the acidity is normally
ascribed to sulphate or phosphate. The presence of P was ruled
out by a negative neutron activation test. Tests for the presence
of sulphate[3] were negative, and a sensitive and more direct test
for S was desired. Neutron activation analysis of sulphur was
not sensitive enough, and incorporation of radioactive S-labelled
precursors for biosynthesis of the acidic amino sugar was rejected
as a possible method, as the bacteria are slow growing and require
a complex growth medium. Other micro-methods for sulphur detection
and determination in biological material[4,5,6] require a cumbersome
conversion to sulphate. Rutherford backscattering requires no such
conversion, and the method might therefore be of advantage in the
study of the chemical composition of the compound.

EXPERIMENTAL

The isolation procedure of the acidic amino sugar from bacterial cell walls is identical to the one previously described[2]. The
procedure,which includes splitting of the polymer by hydrolysis
followed by separation of the products on an amino acid analyzer,
leaves relatively small amounts of the compound for analysis.

Rutherford backscattering as an analytical tool for studying
surfaces has been described by several authors[7]. If one wants to
do elemental analysis of biological samples and possibly study the
relative abundance of the elements, several target preparation
techniques can be used. The sample material can be applied as a
thin film on the surface of a lighter substrate yielding a spectrum
with separated peaks for each element as illustrated in Fig. 1a.
The ratio of the counts in each peak then determine their relative
abundance[8]. Another method would be to apply the material on a thin
foil and thereby detect all elements with masses heavier than the
bombarding ions as shown in Fig. 1b. Thin films of biological
samples can be obtained if they are soluble in a liquid. The liquid
is then applied on the backing and dried. The liquid should have a
low surface tension to prevent it from squeezing into small droplets.
Also the spectrum from a thick target can be used for the analysis
as shown in Fig. 1c. Here each element shows up as a step in the
spectrum, and their relative abundance can be calculated if one
knows the stopping power of the target[7].

In this investigation the amino sugar-like compound was dissolved in water and applied onto substrates of vitreous carbon.
Graphite was also tried but was found too porous, allowing the compound to penetrate. The use of C-backings makes the determination
of the C-content difficult and in most cases impossible. To measure

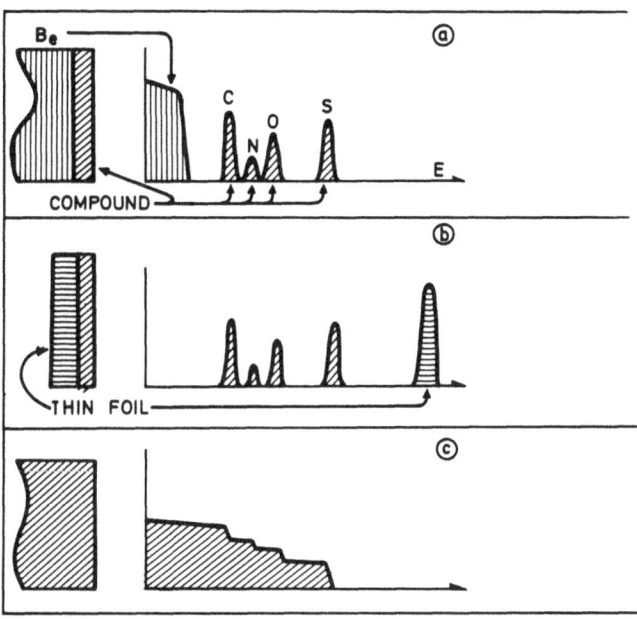

Fig. 1. Schematical backscattering spectra of a biological compound obtained by different target preparation techniques. Thin film on a light substrate (a) and on a thin foil (b). Thick target yield (c).

the carbon content in the compound, even lighter backings such as Be must be used. However, from backscattering measurements done by others, it seems that Be often form an oxide on the surface. This makes the detection of small amounts of oxygen within the compound difficult.

The backscattering spectra in this investigation were obtained with a standard experimental setup for MeV-backscattering measurements using α-particles with energies between 1.5 to 2 MeV. Typical beam currents were a few nA. The beam spot was approximately 1 mm^2, and no special precautions were taken to obtain a homogenous beam.

RESULTS AND DISCUSSION

Fig. 2 shows a spectrum obtained by backscattering from the unidentified compound. This is similar to those obtained from thick targets. The presence of S can clearly be seen. Other elements are N,O and Na. From the step heights in the spectrum the ratio between some of the elements can be estimated. Unfortunately, the

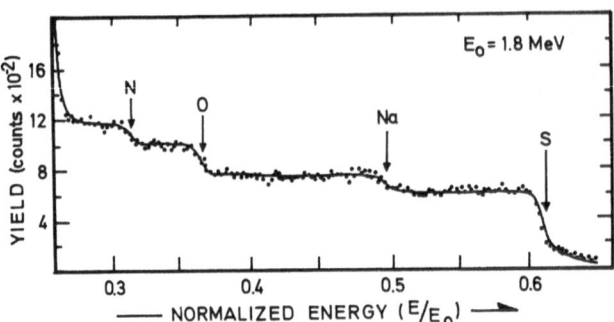

Fig. 2. Energy spectrum of 1.8 MeV $^4He^+$ ions backscattered from
the investigated amino sugar-like compound.

stopping power is unknown but the ratio between N and S atoms which
is of interest in this case, is not very sensitive to this parameter.
The ratio can be estimated within 10% from approximate values of the
stopping power, and the spectrum in Fig. 2 gives 1.6 S atoms to every
N atom.

However, before such ratios are attributed to any molecular
structure one must ask what is happening with the compound during
exposure to the ion beam and to the vacuum. We observed that the
spectra changed upon bombarding the compound as can be seen in
Fig. 3. In these spectra we obtained peaks for the elements corresponding to thin target yields. A Si peak is also observed in this

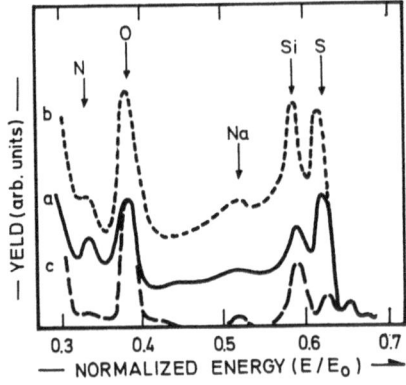

Fig. 3. Energy spectra of 2.0 MeV $^4He^+$ ions backscattered from the
investigated amino sugar-like compound after prebombarding with
different doses. a) 2.5 μC. b) 5 μC. c) 10 μC.

case and the ratio of Si to S increases with bombarding dose. The
Si is probably a contaminant from the diffusion pump oil and the
vacuum grease. The counts under the Si peak corresponds to
approximately 5×10^{14} Si atoms cm^{-2}, and a contamination was also
seen on uncovered carbon substrates though to a lesser extent. The
observed Na atoms are believed to be an impurity from the isolation
procedure. After prolonged bombardment an increase of the yield
from most of the elements was observed. We attribute this to the
difficulty of applying the compound on the substrate in a uniform
way and the inhomogenous analyzing beam. These problems prevented
reproducible measurements of the effect of the bombardment, but
some features are frequently observed like the N peak decreasing
and the ratio between Si and S increasing.

To elucidate the conclusion that changes take place in our
compound during the analysis, we also investigated 2-amino ethane
sulphonic acid, commonly known as taurine, which consists of the
same elements as the amino sugar-like compound. Taurine was dissolved in water, applied upon a vitreous carbon backing and then
dried which gave thin films with good lateral homogeneity. A
backscattering spectrum from taurine is shown in Fig. 4. The Si
contamination is less visible in the spectrum due to the larger
quantity of compound than in Fig. 3. Fig. 4 yields approximately
10^{17} S atoms cm^{-2} within the applied film. The effect of bombardment on this target is shown in Fig. 5, and the effect upon the
constitution of the compound is illustrated in Fig. 6. It is
clearly seen that the ratios between the counts in the N,O and S
peaks are changing upon bombardment. The ratios calculated from
the known constitution of taurine are also shown in Fig. 6. As can
be seen it is not possible to determine the ratio between various

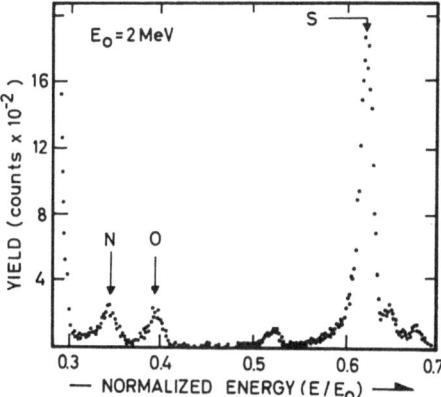

Fig. 4. Energy spectrum of 2.0 MeV ^{4}He$^+$ ions backscattered from a
target prepared by applying taurine ($H_2NCH_2CH_2SO_3H$) on a vitreous
carbon backing.

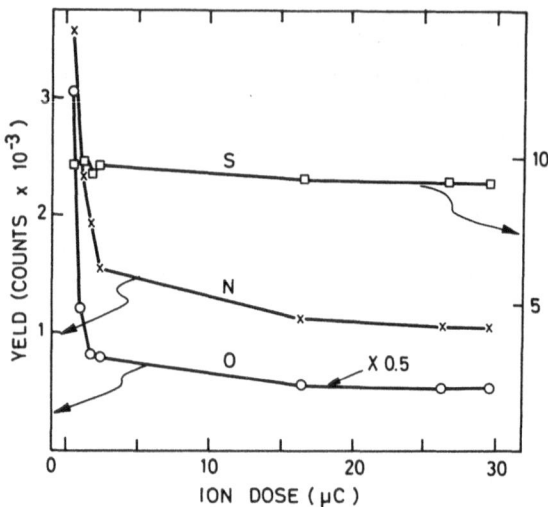

Fig. 5. The backscattering yield from S, N and O versus prebombard-
ment for a target prepared by applying taurine upon vitreous carbon.
The yields have been normalized to the same analyzing dose.

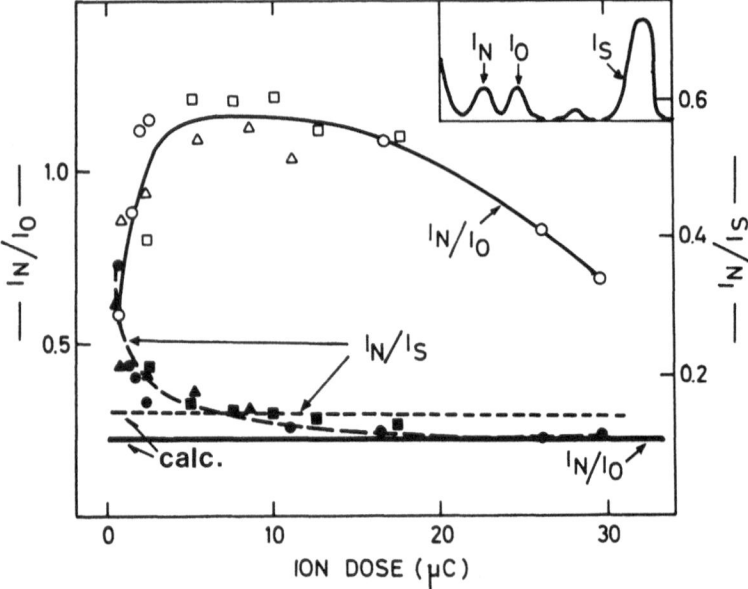

Fig. 6. The variation upon bombardment of the composition of targets
similar to that in Fig. 4. The compositions are deduced from the
ratio of the counts within the peaks in the energy spectra. Different symbols mark different samples. The lines marked "calc." are
the calculated ratios of taurine.

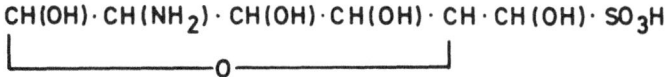

Fig. 7. A tentative molecular model of the investigated amino sugar-like compound.

elements present in taurine from the backscattering spectra.

SUMMARY AND CONCLUSIONS

We have demonstrated that Rutherford backscattering can be used to analyze the constituent elements of a biological compound, and the method may be useful especially when only small amounts of the material are obtainable. We have also investigated the possibility of analyzing molecular structures by backscattering, but structural changes due to the irradiation makes this difficult.

We have positively verified that the amino sugar-like compound from the cell walls of Halococcus sp., strain 24, contains sulphur, but the elemental composition cannot quantitatively be measured by backscattering. The presence of sulphur in the amino sugar, the acidity not beeing caused by sulphate, indicates the presence of sulphonate. A tentative molecular model is shown in Fig. 7 based on this result and several chemical analyses to be described elsewhere[9]. However, the model still has to be conclusively verified.

REFERENCES

1. R. Reistad, Arch. Mikrobiol. 82, 24 (1972)
2. R. Reistad, Carbohydr. Res. 36, 420 (1974)
3. T.T. Terho and K. Hartiala, Anal. Biochem. 41, 471 (1971)
4. P. Stoffyn and W. Keane, Anal. Chem. 36, 397 (1964)
5. M.N. Camien, Anal. Biochem. 15, 127 (1966)
6. D.A. Roe, P.S. Miller, and L. Lutwak, Anal. Biochem. 15, 313 (1966)
7. W.K. Chu, J.W. Mayer, M.-A. Nicolet, T.M. Buck, G. Amsel, and F. Eisen, Thin Solid Films 17, 1 (1973)
8. J.F. Ziegler and R.F. Lever, Thin Solid Films, 19, 291 (1973)
9. R. Reistad, (to be published).

DISCUSSION

Q: (W. Brown) Is it feasible to cover the biological specimen with a thin metal or carbon layer to prevent changes in composition under the beam ?

A: (T. Finstad) Due to only small amounts of the amino sugar-like compound available, we have not been able to do systematical investigations of target preparation techniques and have not tried this yet.

Equipment

VERSATILE APPARATUS FOR REAL-TIME PROFILING OF INTERACTING THIN

FILMS DEPOSITED IN SITU

J. E. E. Baglin and W. N. Hammer

IBM Thomas J. Watson Research Center

Yorktown Heights, New York 10598

ABSTRACT: A system has been built to permit heat treatment and
real-time backscattering analysis of films which have been deposited in the same high vacuum chamber. Several items of new
technology are described. The equipment will enable us to address
a whole range of important thin-film interaction studies requiring
minimized contamination and excellent thermal control.

INTRODUCTION

The kinetics of an interaction between thin films can often be
strongly sensitive to contamination or oxidation of the interfaces
and exposed surface. Interaction properties can also show significant dependence on morphology, grain sizes and annealing history
of the films.

The equipment described here has been built to enable us to
take full advantage of the speed of backscattering analysis by
observing in real time the interaction of films newly deposited in
a cryo-pumped system on several substrates maintained at a series
of well-controlled temperatures during the time of the experiment.
The profiling can be done continuously as heat treatment proceeds,
in the same clean vacuum in which the films were e-beam deposited.
We thereby minimize the possibility of contamination of films
before reaction, and gain substantial freedom to control substrate
temperature, rate of film deposition, and subsequent annealing,
which often need to be recognized as important determinants of the
subsequent interaction rates to be measured. Because of the
facility and speed with which the new system can produce complete

kinetics data for a particular film pair, we expect to be able to
investigate in detail such morphological factors when necessary.
Ultimately, the speed and flexibility of the system should also
permit us to produce meaningful diffusion/interaction data on a
wide variety of materials presently in need of study and
correlation - including in particular, materials which are too
reactive in air to permit normal sample handling, or which require
a well-controlled ambient for long periods of time.

BASIC SYSTEM

Fig. 1a shows a schematic of the system, and Fig. 1b is a
corresponding photograph. Fig. 1c is a view from the left hand end
of the system of Fig. 1a, with the covers removed from the target
chamber. The two principal chambers are interconnected at a 10"
flange. The lower one contains the 4-hearth e-beam evaporator unit
and the accompanying rate monitor, shutter and baffles. The upper
chamber contains the substrate holder and selective mask, together
with equipment related to profiling by backscattering in the ion
beam--(entrance aperture and shutter, Faraday cup, and cooled surface barrier detector). A telescope attached to this chamber is
used for optical alignment of the beam and components.

Vacuum System. The apparatus is pumped initially by sorption
pumps and then by a cryo pump, modified to increase pump speed with
enlarged collecting panels and extended LN_2 trap. Its speed at the
target appears to be in the region of 1000ℓ/sec for N_2. Chamber
walls are either aluminum or stainless steel, and most flanges are
sealed with bakeable Viton rings. These materials were chosen to
facilitate access to the chamber. Manipulator seals are all of the
bellows-sealed type, with the exception of the target-holder shaft,
where a Ferrofluid rotating seal is employed. The system regularly
achieves a pressure of $2x10^{-8}$ Torr in a few hours and better performance is anticipated following more extended bakeout and purge-
pumping procedures. It should be noted that our pump types were
specifically chosen to exclude all oil from the system and provide
high pump speed for oxygen in the vicinity of the sample.

The target-holding carousel (see Fig. 2) holds eight 3" x 1"
fused quartz substrates, clamped around it in squirrel-cage formation. Each substrate is heated by passing current through a resistive
film of 2000A Pt deposited on its back surface, and is maintained at
an independently controlled temperature (up to 1000°C). During evaporation of each film, the carousel is rotated at 200 rpm, thus ensur-
ing that each substrate is exposed to the same incident evaporated
flux. Using a multiple mask, successive depositions may be used to
leave 1/4"-wide stripes of various combinations of film across each
substrate at different positions. Most commonly, this enables us to
examine a stripe of each component alone as the heat treatment proceeds.

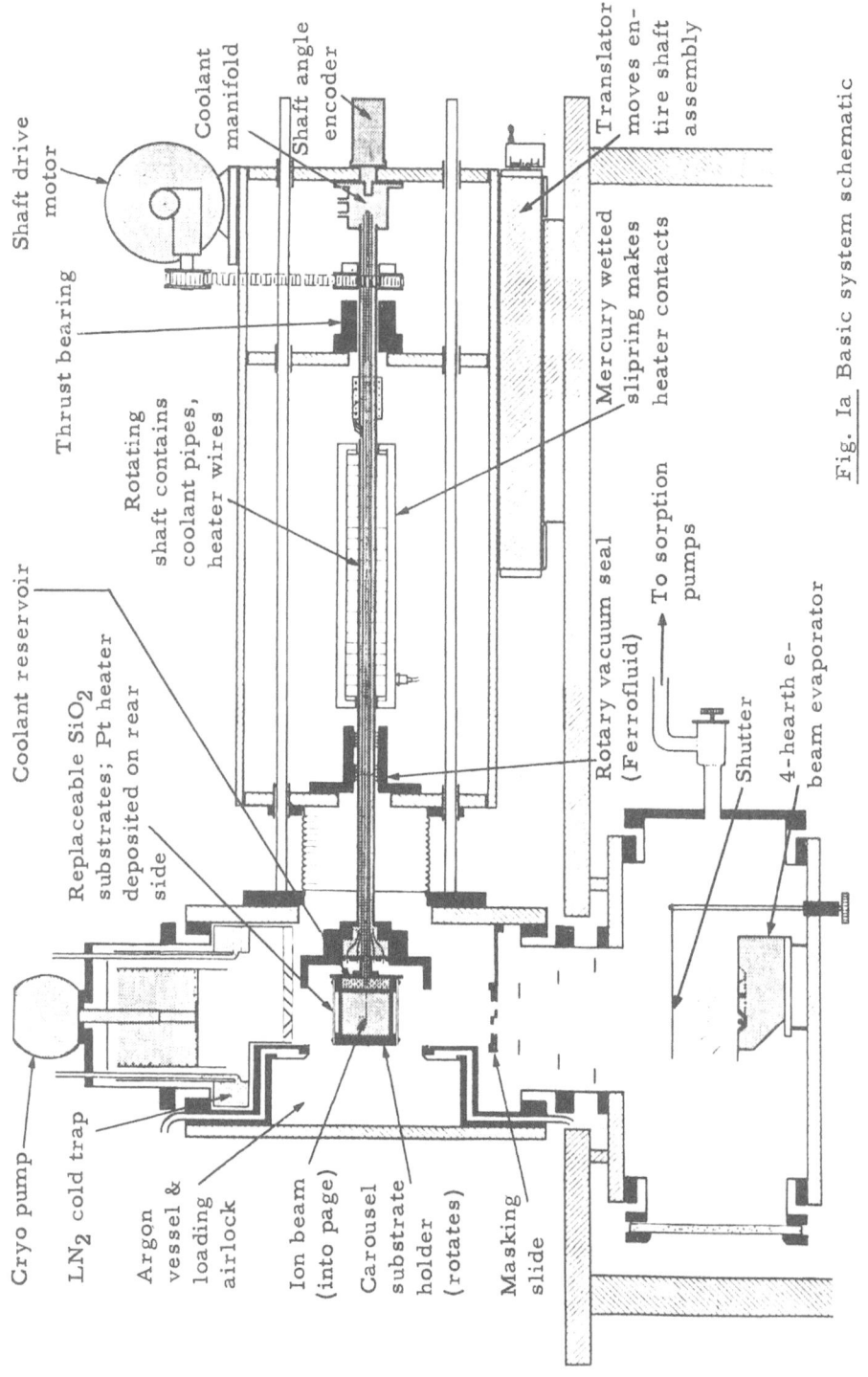

Fig. 1a Basic system schematic

Cryo pump

Target
chamber

Ion beam

Evaporator

Fig. Ib Layout

Preamp port
(re-entrant)

LN$_2$ trap

Ion beam
(stretched wire shows
beam position here)

Detector mount
(cooled)

e-beam
evaporator

Fig. Ic Target chamberr
with cover removed

Fig. 2. Interior of target chamber, showing (L. to R.) Faraday cup,
substrate-holder units in place in carousel above evaporation mask.
and mount for cooled surface barrier detector.

    Profiling can begin at any time (normally, when the films
have been brought to annealing temperature). Rotation of the
carousel is maintained while the ^{4}He or ^{1}H beam is directed so as
to strike the chosen stripe of specimen films. Spectral data from
the surface barrier detector are routed in the multichannel
analyzer according to the identity of the substrate currently in
the beam. A precision shaft angle encoder and programmable limit
switch with microsecond response time is attached to the driving
shaft (see Fig. 1) and it supplies gate signals at preset angles
for this purpose).

    The Detector chosen is a 130mm^2 annular, "ruggedized" (Al-
coated) Ortec surface barrier detector of nominal resolution 17keV
at room temperature. The ion beam passes through the 4mm dia.
hole at its center and backscattered particles are received from
$\theta=170°$ to 175° in a net solid angle of 80 msr. Such efficient
geometry is needed in order to obtain useful statistics in each of
the 8 spectra in a typical run of a few minutes. The aluminum-
coated detector is insensitive to the infrared emission from these
substrates, even when they are kept at 1000°C. In order to improve
both detector resolution and rise time of preamp output pulses, the
detector and preamp are both cooled by attachment to the LN$_2$

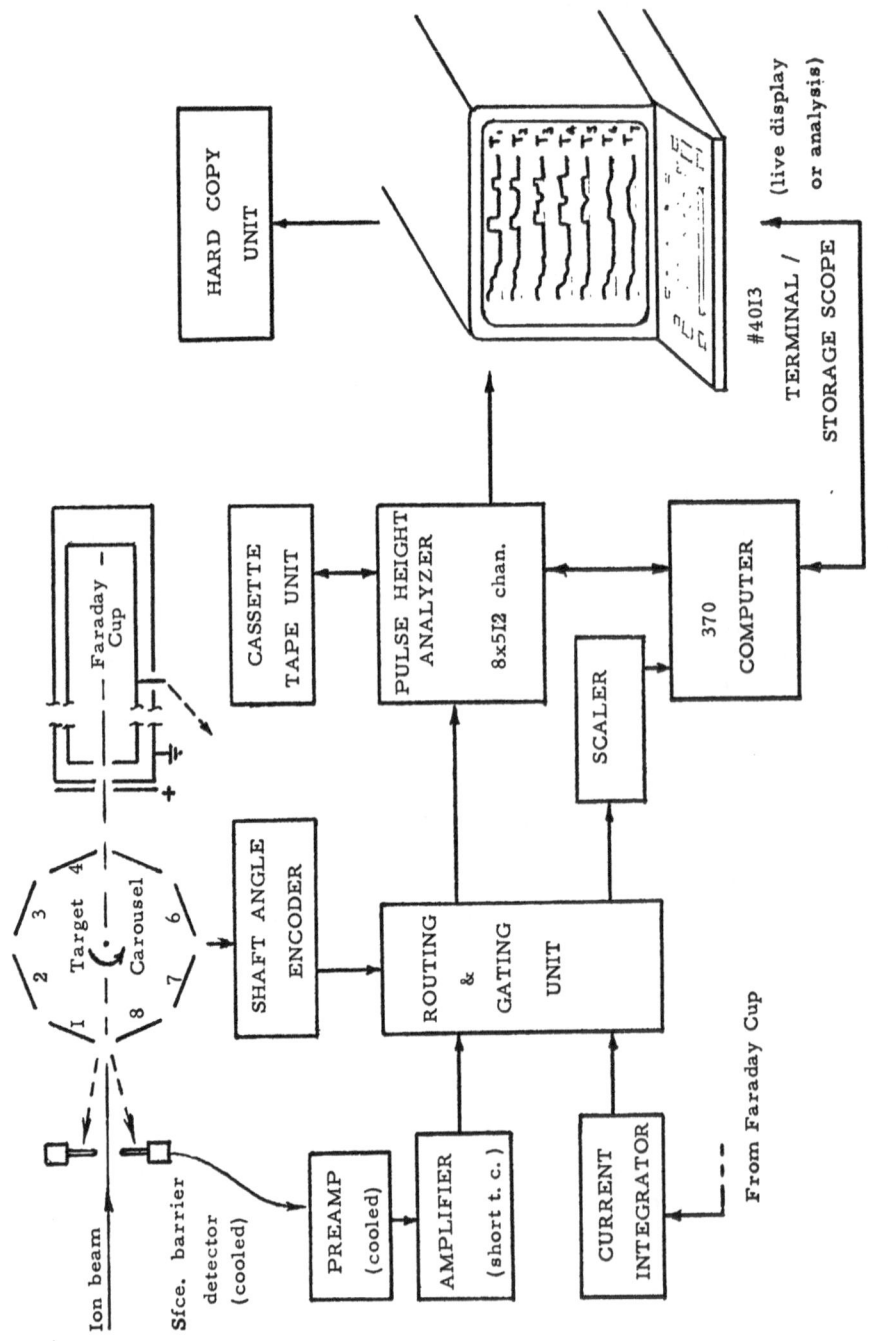

Figure 3. System schematic

reservoir of the cold trap. The preamplifier itself resides in a
re-entrant port in contact with that reservoir (see Fig. 2). It
is hoped that resolution better than 10keV for 2MeV $^4He^+$ can be
obtained in this way.

Multiplexing, Data Display and Storage. Fig. 3 shows the data
collection schematically. While a target is intercepting all the
incident ion beam, as preset in the shaft angle limit switches,
counts from the detector will be routed to that 512-channel location in the 4096-channel MCA which is identified with that
particular target. In addition, when the beam is passing clearly
between frames in the carousel to reach the Faraday cup, input to
the beam current integrator will be gated on. The Faraday cup is
of standard design (see Fig. 1c) and should then provide a
reliable absolute sampling of the profiling beam.

During data accumulation, backscattering spectra are continuously updated on a storage scope, to enable the experimenter
to evaluate the progress of the run constantly. At longer intervals
(of a few minutes), the current spectra will be automatically
stored on cassette tape, ready for computer storage and scanning
later.

Substrates. Initially we have standardized on fused quartz
slides as test substrates having the generally desirable properties
of being amorphous, pure, smooth, and non-reactive. They also
accept well the vapor-deposited Pt heating film which must be prepared and cured before the new slide is loaded into the carousel.
We have found it possible to produce convenient low-power heaters
in this way which can generate surface temperatures uniform to
$\pm$ 10°C to within $\sim$1mm of the edges, at 1000°C, and which have
typical lifetimes of about 40 hrs. The heater is found to serve
reliably as its own resistance thermometer, and temperature
regulating circuits are used to set up the desired temperature of
each test film.

It is nearly always necessary to retain one sample at room-
temperature for reference. Sometimes, as in the case of Pb-In,
the film needs to be cooled in order to slow down a very fast
diffusion to "visible" rates. For cases like this, the carousel
includes a central coolant reservoir. Thermal contact with this
tank is made when the substrate is loaded on top of a copper
bridging block which contacts the reservoir surface. The coolant
used is normally room temperature water. However, the system will
accept a refrigerant instead if needed.

"Argon vessel": A film which is highly susceptible to oxidation such as Cr needs to be heat treated in surroundings freer of
oxygen than our main vessel can be at 10^{-8} Torr. For cases like
this, an adjunct chamber is provided, into which the substrate

carousel can be moved and sealed for immersion in a 'blanket' flow
of inert gas which is constantly purified with hot Ti and a cold
trap. We propose to use argon here, since helium makes a serious
thermal short-circuit from the heaters to the chamber walls, and
since our sorption and cryo pumps can not handle helium, as they
would need to when the target is returned to the main chamber for
profiling.

This subsidiary chamber can also be used as a reaction chamber
to allow controlled amounts of gases to react with test films prior
to or during heat treatment and profiling.

It is planned to use this chamber as a substrate-loading air-
lock also, allowing the main system to remain well evacuated for
long periods.

Target Translation. The carousel needs to be translated
horizontally in order to show different film "stripes" to the
probing beam. It also needs to move further horizontally if it is
to enter the "argon vessel". This is accomplished by moving the
entire framework supporting the rotating shaft, which slides on
precision tracks. To seal the "argon vessel", the shaft rotation
is stopped and the cup behind the carousel is forced into sealing
contact with an O-ring at the vessel's entrance aperture. (See
Fig. 1a.)

Carousel Services; Mounting. Electrical connections needed
for powering the substrate heaters and sensing their resistance are
made from the rotating shaft to the fixed world by means of the
mercury-wetted "slipring" assembly shown in Fig. 1a. Sensing lines
are rated for millivolt noise levels.

Coolant similarly passes through a rotating manifold into
concentric pipes which join directly into the carousel reservoir.

It is important in such a system that solid angle of the backscatter detector should not change from target to target. Hence,
care has been taken to maintain a net 'runout' of <.001" in the
mounted substrate holder as it rotates.

Additional Facilities. The system has been built to facilitate
the addition of other equipment. Obviously desirable attachments
for in situ use would include a sputtering unit, residual gas
analyzer, optical microscope, or a small scanning electron microscope. The new apparatus should provide the basic support for such
developments in the future.

APPLICATION

The system described is about to be put into service in a
variety of applications to thin film interactions. No doubt
experience with its actual use will indicate places where improvements can be made. We believe that this equipment will substantially advance our application of backscattering analysis to thin
film interactions. We will welcome your comments and suggestions.

DISCUSSION

Q: (G. Amsel) Are you not worried by the effects of the probing
beam itself on the very phenomena you want to observe in situ ?

A(J. Baglin) Not worried... in fact, we have considered the potential use of this apparatus to study radiation-enhanced diffusion.
We have the facility by translating the target holder to control
very sensitively the integral dose per unit film area over the
length of a run.

APPLICATION OF A HIGH-RESOLUTION MAGNETIC SPECTROMETER TO NEAR-SURFACE MATERIALS ANALYSIS

James K. Hirvonen and Graham K. Hubler

Naval Research Laboratory

Washington, D. C.   20375, U. S. A.

ABSTRACT

Experimental data obtained using a high-resolution magnetic spectrometer is presented to illustrate the advantages, limitations and/or complications of using such an instrument for various types of surface-analysis experiments.

1.   INTRODUCTION

It has long been realized that the application of a magnetic spectrometer to near-surface analyses offers substantially better (10X-20X) energy resolution than that obtained by use of solid-state detectors. This improved energy resolution provides a correspondingly better depth and mass resolution. It also offers means of avoiding pulse pile-up or interference from other particle groups at the expense of collecting only a small portion of the total energy spectrum at one time. These advantages have not been widely realized partly because of the limited number of such instruments, which is in turn related to their relatively high initial cost.

Although the use of magnetic spectrometers for target analysis is not new (1950)[1] there is still a limited source of data concerning their usage in surface-analysis experiments.[2,3] This paper will present experimental data pointing out the advantages, limitations and/or complications of using a magnetic spectrometer for various types of surface-analysis experiments.

The NRL 180° 20-inch radius double-focusing magnetic spectrometer has recently been utilized for high energy resolution ($\Delta E/E \approx$
10^{-3}) Rutherford Backscattering (RBS) measurements, with the backscattering measurements being done at 135°. Scattering measurements
on this system have been facilitated by the installation of a multi-
purpose scattering chamber onto the spectrometer and by the use of
a position-sensitive detector (PSD) in the image plane of the
spectrometer, depicted in Figure 1.

The scattering chamber consists of a commerical multiported
stainless steel vacuum housing, provided with two additional 1" beam
ports, attached to the magnetic spectrometer. A moveable target
holder accommodates several samples at one time, and samples can be
changed via a vacuum lock on top of the chamber without letting the
target chamber up to atmospheric pressure. A copper cold wall is
connected to an internal liquid-nitrogen reservoir and encloses the
target to ensure no discernable hydrocarbon buildup on the target
during analysis. A moveable collimator within the cold enclosure
allows different target areas to be selected for analysis. A secondary electron grid held at -200 V with respect to the sample repels
low-energy secondary electrons from both the collimator and sample
for accurate charge integration. A conventional surface-barrier
solid-state particle detector ($\Delta\Omega \approx 2.6$ msr) is located at 165° with
respect to the incoming beam. A second detector can be moved in
front of the magnetic spectrometer input slits to allow a direct
comparison of solid-state detector and magnetic spectrometer results.
This second detector also facilitates setting up the spectrometer.
The input slits can be adjusted to provide a solid angle from zero
to a maximum of 25 msr.

The magnetic spectrometer analyzes the momentum (energy) of the
charged particles entering the input slit and focuses the dispersed
charged particles onto a position-sensitive detector (PSD) located
in the image plane of the spectrometer. Particles of slightly
different energies E_0 and $E_0 + \Delta E$ will traverse slightly different
paths through the spectrometer and will be brought to focus at
different positions (x_0 and $x_0 + \Delta x$) along the image plane (or along
the position-sensitive detector). The spectrum of the position
signals from the PSD is directly interpretable as an energy spectrum.
The energy signal from the PSD allows selection of particle type and
charge state by placement of a single channel analyzer window in the
appropriate region of the energy spectrum.

The PSD subtends about 2.3% of the total energy spectrum at a
fixed magnetic field setting. Because of this limited energy interval, only limited depth regions can be examined at one time. It is
for this reason as well as for local differential nonlinearities in
the PSD that some of the spectra presented are actually composite
spectra each taken at a slightly different magnetic field setting.

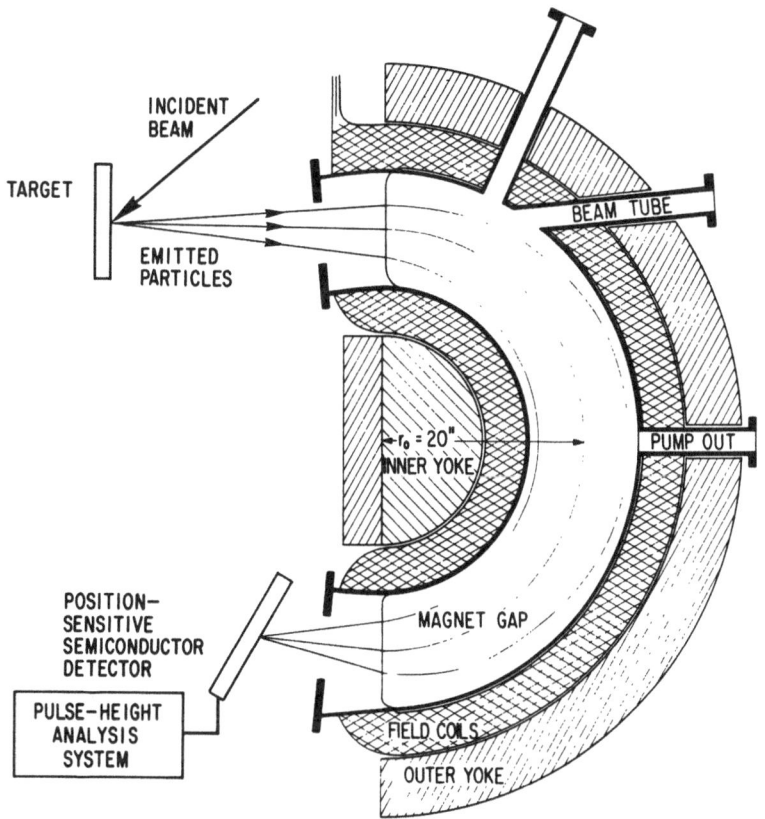

Figure 1.   Schematic drawing of the NRL 180° 20"-radius double-focusing magnetic spectrometer and position-sensitive detector.

The spectrometer system resolution has been measured to be $\Delta E/E \leq 0.05\%$ for very small solid angles and $\leq 0.08\%$ for an input solid angle of $\Omega = 2.5$ msr. The resolution worsens somewhat for larger input solid angles. This resolution (800 eV at 1 MeV or 1.6 keV at 2 MeV) is to be compared with a solid-state detector resolution of approximately 12–16 keV.

3.   RESULTS AND DISCUSSION

This spectrometer has been used in nuclear backscattering, nuclear reaction, and particle channeling experiments. Experimental results are presented which demonstrate the advantages a magnetic spectrometer may have over a conventional solid-state detector for certain types of experiments; viz., (i) superior mass resolution

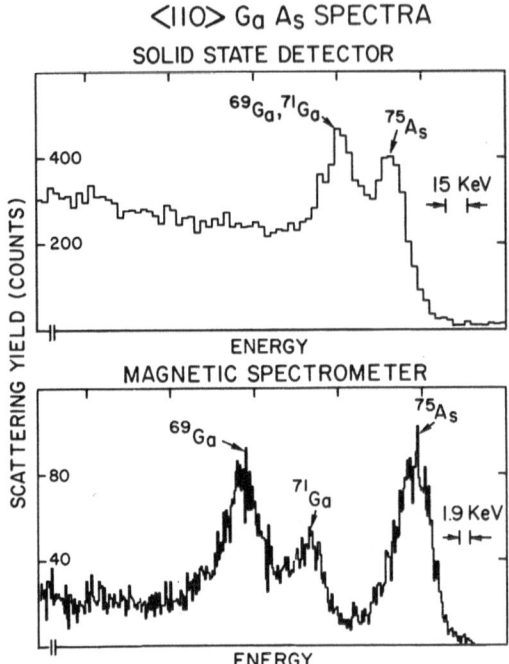

Figure 2. Solid-state detector and magnetic spectrometer back-scattering spectra for He ions incident along the $\langle 100 \rangle$ direction of GaAs.

(ii) superior (X10) depth resolution, and (iii) superior sensitivity (> 10X) for impurity detection or for depth distribution profiling. In practice, the extent to which a spectrometer's superior energy resolution can be exploited for surface analysis may be limited or complicated by several factors, such as energy-loss straggling, kinematic broadening, isotope effects, surface roughness, surface oxides and other contaminants, aberrations of spectrometer optics, and ion-beam-induced target damage. Some of these factors are discussed, and the extent to which they limit or complicate the sensitivity of high-resolution spectrometers for surface analyses are illustrated by experimental data. The particular experiments from which data are taken will not be discussed in any detail with the primary intent of using the data to illustrate different aspects of using high-resolution magnetic spectrometers as a technique in materials analysis.

3.1 Mass and Depth Resolution

Figure 2 illustrates several features of high resolution

backscattering. An obvious difference between the solid-state detector and magnetic spectrometer spectra for channeling along the
$\langle 110 \rangle$ direction of GaAs is the resolving of the ^{69}Ga, ^{71}Ga isotopes
in the latter. This better mass (depth) resolution allows a higher
sensitivity to the near-surface region for channeling investigations
(e.g., minimum yield measurements, surface peak structure, near-
surface damage). In return for this improvement, the increased
energy dispersion requires a corresponding increase in analyzing
beam charge for comparable statistics. In addition, the analyzing
beam spot must be kept relatively small (< 1 mm^2) in order that the
resolution will not suffer as a result of the (magnetic) optics of
the spectrometer. Hence, to ensure that the analyzing beam does
not produce significant target damage, movement of the target or the
beam spot may be necessary.

It is also seen that the width of the various surface peaks are
appreciably wider than the system resolution for the case of the
spectrometer spectrum. This is mostly attributed to the thickness
of the native oxide on the GaAs. However, kinematic broadening can
make a significant contribution to the relative widths of peaks or
edges measured by high resolution backscattering. For example, a
spectrometer input solid angle of 2.5 msr used here contributes a
kinematic energy spread of nearly 3 keV to the Ga or As surface
peaks for incident 2 MeV He ions. The kinematic broadening increases
for lighter target atoms and larger input solid angles and would be,
for example, about 5 keV for 2 MeV He scattering from silicon for
the same solid angle (2.5 msr). This contribution alone is significantly larger than the system resolution and often forces a compromise between resolution achieved and practical counting rates.

The treatment of isotope effects is dealt with in another
paper[4] in this Conference and will not be discussed here except
to note that in some situations the presence of isotopes can complicate the interpretation of scattering spectra and also can considerably reduce the sensitivity of RBS for investigating interactions
between thin films and thick substrates (e.g., thin Ta or Ta$_2$O$_5$ films
on silicon).

Figure 3 shows the well-resolved peaks resulting from incident
2 MeV He$^+$ ions backscattered from i) a ^{197}Au "monolayer" on silicon
and from ii) a ^{209}Bi "monolayer" on silicon. The kinematic energy-
loss difference between ^{197}Au and ^{209}Bi on the surface is approximately 6 keV. Besides illustrating the enhanced mass resolution
available with the spectrometer the peak widths of these spectra
are considerably wider than the system resolution and this result
is attributed to "islanding" of the deposited impurity atoms.
Although this observation has been reported previously [2] it serves
to point out how interpretation of high-resolution scattering data
may depend strongly on the morphology of the target surface being

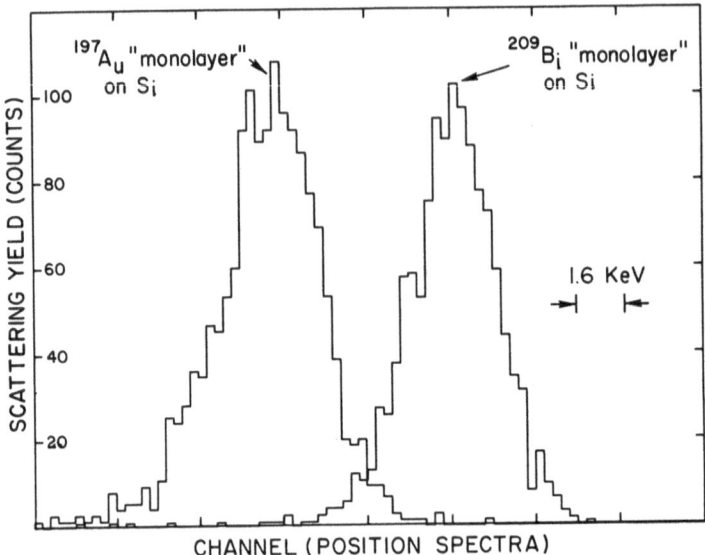

Figure 3. Superimposed spectra from 2 MeV He ions backscattered from ^{197}Au "monolayer" and ^{209}Bi "monolayer" on silicon substrates.

examined. In general, the width (12%-88%) of the leading edges of thick targets is consistently larger than the system resolution and is attributed to native oxides or other surface contaminents. Other workers with in-situ sputter cleaning capabilities in a UHV system have demonstrated this[6]. This shows the desireability of performing high-resolution scattering experiments in a UHV system.

A third example of the depth resolving capabilities of magnetic spectrometers is shown in Figure 4. This Figure shows a series of spectra taken of 2 keV to 60 keV ^{209}Bi implantations into pre-implanted (5×10^{14} Ar/cm^2 - 60 keV) silicon. The depth scale was calculated assuming constant but different stopping powers for the incident 2 MeV He$^+$ ion beam and the scattered He^{++} ion beam, and ignoring the contribution of the implanted ^{209}Bi to the stopping power. This should be a reasonable approximation except for the 40 keV-5×10^{16} Bi/cm^2 implantation. The point of presenting these data here is not to make a detailed comparison with theory or other experiments but to illustrate that projected ranges and straggling of low energy heavy ions in light substrates can be directly measured with an estimated depth resolution less than 30 Å near the surface (in silicon). The energy-loss straggling of the analysis beam will eventually limit the depth resolution attainable in such depth profiling measurements. It is estimated that the straggling width will be equivalent to the spectrometer resolution for scattering

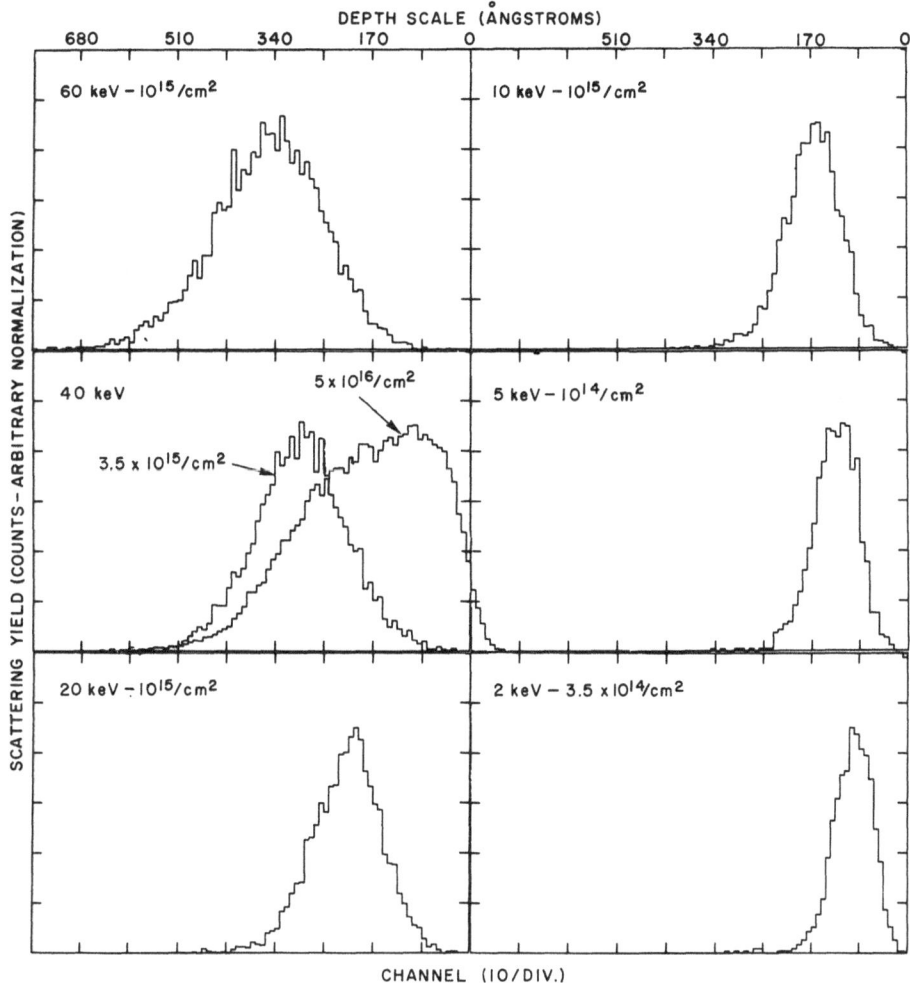

Figure 4.   Spectra for 2 MeV He ions backscattered from 2 keV-60 keV
^{209}Bi atoms implanted into silicon. The depth scale shown at the top
is approximate.

depths of approximately 40 Å.[7]   However, the measured widths of the
Bi profiles shown here are mostly due to the range straggling of the
implanted Bi ions and would only decrease by 5-10% if account were
taken of the energy-loss straggling of the analyzing beam. Other
scattering experiments using solid state detectors and glancing angle incident beams to improve the depth resolution have recently been
reported.[8]   This technique retains the desireable feature of collecting the complete energy spectrum at one time, however, the

attainable depth resolution using this technique is either more
sensitive to or involves additional limiting factors as compared to
these measurements utilizing analyzing beams incident normal to the
target surface (e.g., surface roughness, multiple scattering, and
finite detector acceptance angle).[8]

3.2. Impurity Detection Sensitivity

The fact that only a portion of the scattered ion spectrum is
analyzed at one magnetic field setting can be either disadvantageous
or beneficial. Typically only a 10-15% energy slice of the scattered particles can be well resolved in the image plane of the
spectrometer. If a single conventional position-sensitive detector
is used to detect the particles, the analyzed energy region can be
further reduced (~ 2% in our system). If one is interested in studying only the surface or an interface this limitation may present no
problems, but analysis over an extended energy region necessitates
collecting a number of scattering spectra using a different magnetic
field setting for each. However, the fact that only a small energy
slice is analyzed at one time allows a means of avoiding pulse pile-
up or interference from other particle groups. Although pulse pile-
up can be considerably reduced by electronic means or can sometimes
be circumvented by the use of heavy incident analyzing ions[9] (e.g.,
$^{12}C^+$) the application of a magnetic spectrometer can yield significantly better detection sensitivities for heavy impurity atoms on
light substrates without i) the loss of energy resolution or ii) the
necessity of using heavy analysis beams.

Figure 5 shows the depth distribution of implanted ^{69}Ga atoms
(3.5 x 10^{14} Ga/cm^2 at 60 keV) in silicon both before and after an
annealing treatment in an effort to look for redistribution of the
implanted ^{69}Ga.[10] It was not feasible to do this by means of a
solid-state detector mainly because of pile-up problems and also
because of inadequate depth resolution. However, it proved relatively straight forward using the spectrometer in spite of the relatively large amount of analyzing beam charge accumulated (distributed
over several adjacent target areas).

For a final example, we will briefly describe the use of the
spectrometer to study the depth resolution and detection sensitivity
for determining the concentration of ^9Be in GaAs by means of the
$^9Be(p,\alpha)^6Li$ reaction. This example does not represent an optimum
nuclear reaction for either depth resolution or detection sensitivity
but does illustrate the many factors contributing to these types of
measurements. Many nuclear reactions like the one above have been
studied for use in materials analysis measurements[11,12] and many
of the following comments pertain to these other reactions as well. In

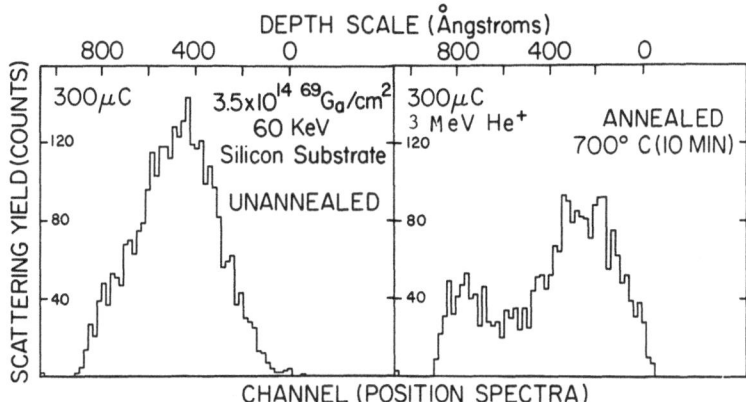

Figure 5. Spectra for 3 MeV He ions backscattered from ^{69}Ga implanted into silicon taken before and after thermal annealing.

these measurements a Si detector is used to detect the emitted reaction particles and an absorber foil is ordinarily placed in front of the detector to stop the predominant elastically scattered incident beam particles. This technique limits the depth resolution of the method to the order of one micron because of energy straggling of the detected ions in the absorber foil. For our example, detecting the α particles with a magnetic spectrometer eliminates the necessity of an absorber foil and increases the depth resolution dramatically (~ 100X).

The experimental geometry is shown in Figure 6. A Be concentration may be profiled in the following way. It is assumed that the excitation cross section σ for the p,α reaction is constant over the range of energies of the protons in the target (σ changes by less than 1%). Therefore, in Figure 6 the protons are creating α particles of the same energy throughout the Be implanted region. The alpha particles, emerging at angle ∅, traverse different path lengths of material for different depths of origin X. Hence, the energy loss of the α particle is a measure of the depth at which the particle was produced.

The parameters of most interest are detection limits (the minimum concentration of atoms that can be measured) and depth resolution. In this study, the depth resolution was calculated from well known parameters, and the detection limit was measured. The depth resolution was evaluated for a 2.25-MeV proton beam and a laboratory scattering angle of 45° by considering the various

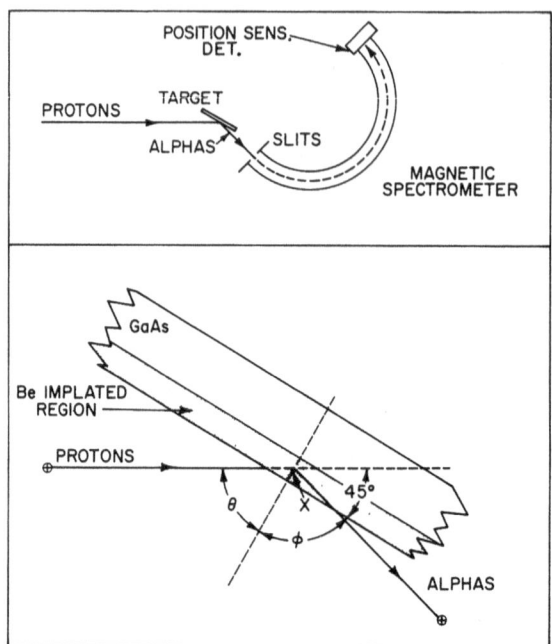

Figure 6. Schematic drawing of the magnetic spectrometer system
(top) and the beam–target geometry (bottom) used to profile ^{9}Be in
GaAs via the ^{9}Be(p,α)^{6}Li reaction.

contributions to the energy spread of the emitted He^{++} particles
shown in Figure 7. The resulting depth resolution for a 1 msr solid
angle varies from approximately 60 Å at the surface to 175 Å at
a 2000 Å depth.

To ascertain the detection limit of the technique, we measured
the cross section by bombarding thin Be targets of known thicknesses
with protons and collecting the yield in the α_0 group from the
^{9}Be(p,α)^{6}Li reaction with a Si detector placed at 45° to the beam
direction. Particle spectra were collected for the same target with
the magnetic spectrometer substituted for the silicon detector at a
scattering angle of 45°. A GaAs target (tellurium impurity of 10^{16}
atoms/cm^3 – Honeywell) was bombarded with 2.25-MeV protons in the
geometry of Figure 6 to determine the background for the magnetic
analyzer detection system, which was set to accept α particles in
the region of 3.16 MeV. The detection limit for Be in GaAs assuming
a constant integrated–charge/solid–angle product of 3600 μC msr is
approximately 1-2 x 10^{14} ^{9}Be atoms/cm^2.

Hence, the ^{9}Be(p,α)^{6}Li reaction profiling of Be implants is
feasible in GaAs to a sensitivity of about 5 x 10^{14} atoms/cm^2

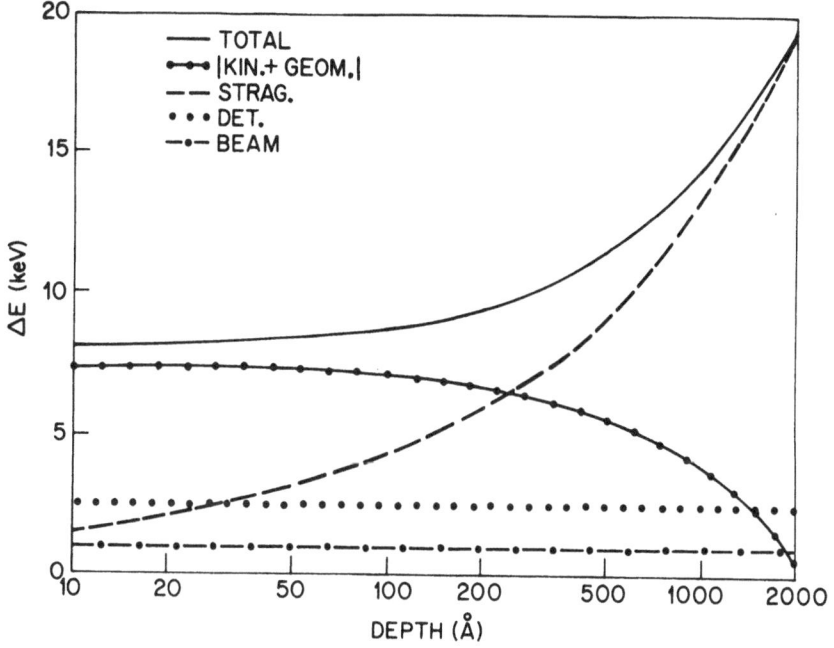

Figure 7. Various contributions to the total energy spread (FWHM in keV) of detected He^{++} ions as a function of depth in GaAs.

implanted in a layer less than 200 Å thick. Depth resolutions of 60 Å-85 Å can be attained near the surface for small solid angles (~ 1 msr). Profiles can be obtained in about two hours for a 5 x 10^{14} Be/cm^2 fluence with quantitative analysis in both measured dose and depth. Other ion species that may be profiled in this manner include Li, B, O, Al, and P, in appropriate substrates where competing reactions do not interfere. The calculations reported here of the depth resolution for profiling Be in GaAs are representative of the technqiue of profiling by means of p,α reactions.

CONCLUSIONS

The application of a magnetic spectrometer to near-surface analyses offers substantially better energy resolution than that obtained by use of solid-state detectors. This improved energy resolution provides a correspondingly better depth and mass resolution. It also offers means of avoiding pulse pile-up or interference from other particle groups. In practice, the extent to which a spectrometer's superior energy resolution can be exploited for surface analysis may be limited or complicated by several factors (e.g., energy-loss straggling, kinematic broadening, etc.) which must be

avoided or accounted for in order to fully exploit the potential
advantages of such instruments.

REFERENCES

[1] C. W. Snyder, S. Rubin, W. A. Fowler, and C. C. Lauritsen, Rev.
Sci. Inst., 21, 852 (1950).

[2] E. Bøgh, Channeling, Ed. D. V. Morgan, John Wiley and Sons, p.
435 (1973).

[3] J. M. Harris, W. K. Chu, and M-A. Nicolet, Thin Solid Films, 19,
259 (1973).

[4] J. F. Ziegler, R. F. Lever, and J. K. Hirvonen, "Computer Analysis of Nuclear Backscattering," this Conference.

[5] J. K. Hirvonen, A. K. Revesz, and T. D. Kirkendall, to be
published.

[6] W. M. Gibson and J. M. Poate, Private communication.

[7] W. K. Chu and J. W. Mayer, Catania Working Data from U. S. -
Italy Seminar, Catania, Italy, June 1974.

[8] J. S. Williams, Nucl. Inst. and Meth., 126, 205 (1975).

[9] R. R. Hart, H. L. Dunlap, A. J. Mohr, and O. J. Marsh, Thin
Solid Films, 19, 139 (1973).

[10] H. B. Dietrich and J. K. Hirvonen, to be published.

[11] G. Amsel, J. P. Nadai, E. D'Artemarte, D. David, E. Girard, and
J. Moulin, Nucl. Inst. and Meth., 92, 481 (1971).

[12] W. W. Lindstrom and A. H. Heuer, Nucl. Inst. and Meth., 116,
145 (1974).

DISCUSSION

Q: (L.C. Feldman) How does the depth resolution of a magnetic
spectrometer compare to the best possible with an electrostatic analyzer using low energy protons?

A: (J. Hirvonen) The "nominal" depth resolution (i.e. ΔE resolution
$\div A(dE/dx)$ is somewhat better for the magnetic spectrometer compared
to the electrostatic analyser at the surface($\sim$5 Å versus 15 Å for Au)
but both yield about the same resolution within the target (i.e. $\approx$25
Å at a 100 Å depth).

Q: (G. Amsel) Nuclear reactions may yield rather high energy particles:
protons, alphas etc. What is the maximum particle energy can you
analyse ?

A: (J. Hirvonen) The NRL magnetic spectrometer can analyze protons
or He^{++} ions at energies up to 20 MeV.

Q: (M.A. Chaudhri) Covering Si detectors with absorber foils for separating protons and alphas in nuclear reactions is not very much used these days. Better techniques like using counter-telescopes or controlling the applied bias on the Si-detector and hence the detector depths are most commonly used in order to separate charged particles of different masses coming out of nuclear reaction.

A: (J. Hirvonen) I was referring only to thick targets where the elastically scattered incident beam would prohibitively limit the count rate for the nuclear reaction of interest if not stopped by an absorber foil.

RUTHERFORD BACKSCATTERING ANALYSIS WITH VERY HIGH DEPTH RESOLUTION USING AN ELECTROSTATIC ANALYSING SYSTEM

A. FEUERSTEIN, H. GRAHMANN, S. KALBITZER AND

H. OETZMANN

MAX-PLANCK-INSTITUT FÜR KERNPHYSIK

HEIDELBERG, GERMANY

ABSTRACT

Backscattered ions with energies in the hundred keV range have been analysed with an electrostatic analyser. Depth resolutions have been measured for ^{1}H-, ^{2}H- and ^{4}He-ions in thin films of SiO_2, Pt and Au. The optimum depth resolution at the surface obtained so far is 5 Å (FWHM).

INTRODUCTION

In the recent years the investigation of thin surface layers by Rutherford backscattering (RBS) has found increasing interest. With silicon detectors, depth resolution in high energy ion scattering (HEIS) is limited to typically 100 - 300 Å (FWHM)[+]. With target tilt up to glancing incidence, 30 - 50 Å depth resolution has been reported (1). These limits may be overcome by using analysing devices of better resolution. Using 500 keV He-scattering and a magnetic spectrometer Bøgh (2) reported a depth resolution of about 20 Å at the surface of tungsten crystals. Thin films of Au and NiO were analysed by Wijngaarden (3) using an electrostatic analyser (ESA) and 80 keV proton scattering.

[+] Throughout the paper, all resolution figures are given in FWHM's.

This work is an approach to optimize RBS techniques
with respect to depth resolution by using an electrostatic
analyser and medium energy ion scattering (MEIS).

BASIC CONSIDERATIONS

Several factors contribute to the depth resolution
of RBS experiments. The relative depth resolution at a
depth t is given by

$$(1) \qquad \frac{\delta t}{t} = \frac{\delta E}{\Delta E}$$

where δE is the total experimental energy spread of the
ions backscattered from a depth t and ΔE is the energy
loss due to ionization along the path. Optimum depth resolution is obtained by making δE minimum and ΔE maximum.
The total energy loss ΔE can be largely increased by tilting the target, i. e. by increasing the pathlength of
the ions in the target (Fig. 1).

ENERGY LOSS ΔE IN A LAYER OF THICKNESS t_0 :

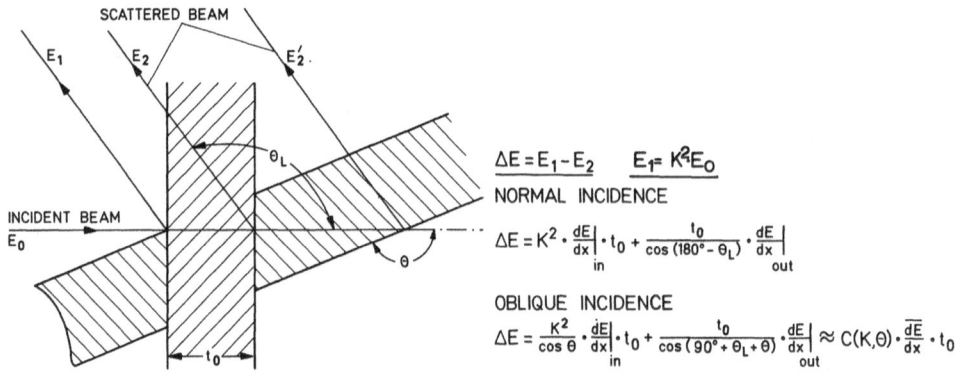

$$\Delta E = E_1 - E_2 \qquad E_1 = K^2 E_0$$

NORMAL INCIDENCE

$$\Delta E = K^2 \cdot \frac{dE}{dx}\Big|_{in} \cdot t_0 + \frac{t_0}{\cos(180° - \theta_L)} \cdot \frac{dE}{dx}\Big|_{out}$$

OBLIQUE INCIDENCE

$$\Delta E = \frac{K^2}{\cos\theta} \cdot \frac{dE}{dx}\Big|_{in} \cdot t_0 + \frac{t_0}{\cos(90° + \theta_L + \theta)} \cdot \frac{dE}{dx}\Big|_{out} \approx C(K,\theta) \cdot \overline{\frac{dE}{dx}} \cdot t_0$$

Fig. 1. Total energy loss ΔE of an ion beam backscattered
from a layer of thickness t_0 with normal incidence of the
beam and target tilt. We call $C(K,\theta)$ the pathlength
factor.

The energy resolution of an ESA is proportional to energy.
Since improvement of spectrometer resolution is considered
to be a solvable technological problem, the choice of the
primary energy of the ion beam depends mainly on the position of the maximum dE/dx with respect to energy. The
optimum energy is about 100 - 150 keV for protons and 400 -
600 keV for helium ions. Helium scattering at E_{opt} is advantageous to proton or deuteron scattering because of the

about twice higher maximum dE/dx. The total experimental
energy spread ĖE is made up by several contributions:

  i) the energy resolution of the analysing system
 ii) the energy straggling of the ion beam with depth
iii) the energy spread due to pathlength straggling (mul-
     tiple scattering).

In the energy range used in this paper, few experimental
data is known about energy straggling and multiple scattering. With normal incidence geometries, the latter process can be neglected for Au-layers < 500 Å thickness with
ion beam energies > 200 keV. This was confirmed by calculating the pathlength straggling due to an angular spread
from multiple scattering, which we calculated according to
the theory of Meyer (4). Pathlength straggling due to multiple scattering may be dominant with extreme target tilts
because of geometrical enhancement.

Using an analysing system with a very good energy
resolution, the influence of energy straggling on depth
resolution can be investigated in experiment. This is best
done by analysing wellknown concentration profiles. Homogeneous thin films are well suited for this kind of investigation, because the backscattering spectra are simple.
Their preparation, however, is far from easy.

EXPERIMENTAL

The RBS arrangement is shown in Fig. 2. The analysing
beam is produced by the Heidelberg 2 MeV van de Graaff
accelerator. After passing a 90° magnet the ion beam is
collimated to 0.064° minimum divergence. The target is
mounted on a three axis goniometer. A vacuum lock system
allows target changes within 2 minutes. The backscattered
ions are analysed by a 90° cylindrical ESA under 127°
laboratory angle. The bending radius of the ESA is 30 cm,
the spacing of the deflection plates is 1 cm. With 1 mm
slitwidth of the ESA, corresponding to 0.7% energy resolution, the solid angle subtended by the analyser is $4 \cdot 10^{-4}$
steradian. The analysed ions are counted with a silicon
detector with an effective length of 5 cm. The voltage
applied to the ESA deflection plates is ± 10 kV at maximum
equivalent to 300 keV ion energy. The high voltage supply
is driven by a linear function generator. A dc-signal proportional to the deflection voltage modulates the am-
plitude of a normalized particle pulse from the detector
electronics. These pulses are processed in a multichannel
analyser. Slow variations of the beam intensity are

averaged by periodically scanning the chosen energy range.
A detailed description of the analysing system will be
published elsewhere. Typically the vacuum in the stainless
steel chamber was about $2 \cdot 10^{-7}$ torr during the measurements.
A cold trap kept at liquid nitrogen temperature prevented
ion beam induced build up of carbon layers on the target,
which is best seen by taking channeling spectra of e.g.
silicon single crystals. Platinum films of 150, 300 and
600 Å thickness were fabricated by BALZERS, Liechtenstein,

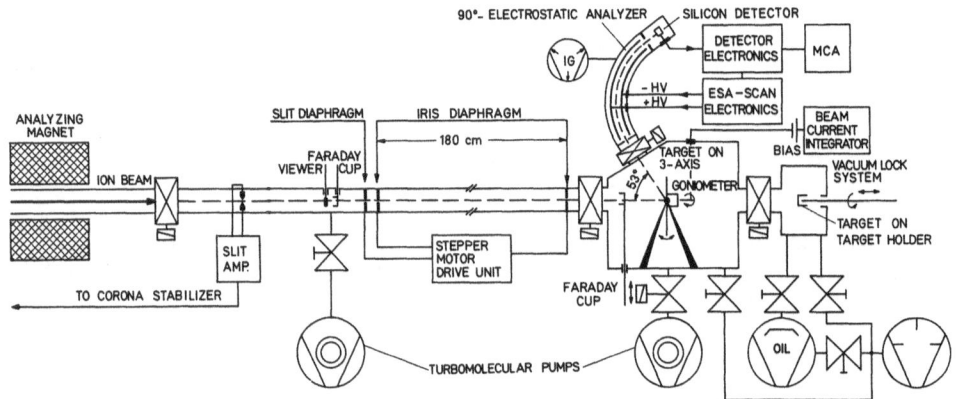

Fig. 2. RBS analysing system with an electrostatic analyser

on electropolished silicon wafers.An uniformity of about
10 Å in thickness was quoted by the manufacturer. Gold
layers 35, 100 and 200 Å thick were vacuum evaporated by
ourselves at a background vacuum of several 10^{-6} torr. The
deposition rates were about 1 to 10 Å per second. SiO_2-
layers of 100, 300 and 1000 Å thickness were grown by AEG-
TELEFUNKEN, Heilbronn, their thickness was determined
by ellipsometric measurement with an error of about 5%.

RESULTS AND DISCUSSION

Backscattering spectra from Pt- and Au-layers have bee
taken with ^{1}H-, ^{2}H- and ^{4}He-ion beams for different target
orientations from normal incidence up to 60° with respect
to the beam direction. With SiO_2-layers, channeling spectra
only with normal beam incidence could be taken because of
the fixed laboratory angle of the ESA. A typical backscattering spectrum of a 200 Å Au layer is shown in Fig. 3.
Fig. 4 shows the backscattering spectra of 100(300) Å SiO_2-
layers. For the extraction of the energy straggling from

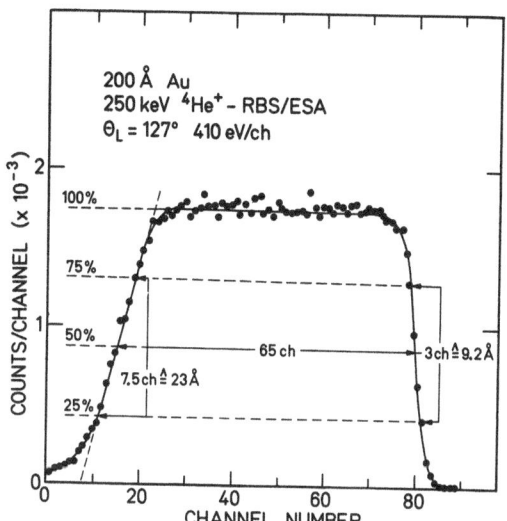

Fig. 3. Backscattering spectrum from a 200 Å gold-layer.
Normal beam incidence.

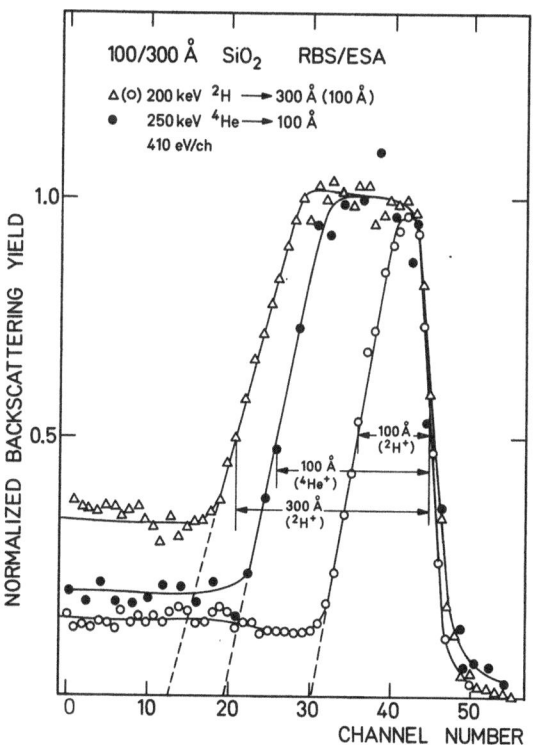

Fig. 4. Channeling spectra from 100(300) Å SiO$_2$-layers.

the spectra it is assumed that the front edges of the nearly rectangular spectra are broadened by the system resolution (about 0.8% energy resolution equivalent to 15 Å depth
resolution for ^{4}He in Au with normal incidence of the analysing beam), whereas the broadening of the rear edges is
attributed to the energy straggling of the ions in the
layer plus the system resolution. By subtracting the FWHMś
of the associated gauss shaped distributions in quadrature,
the energy straggling can be evaluated from the spectra.
In Fig. 5, the thus reduced FWHM´s of the rear edges of the
Au- and Pt-films as well from the SiO_2-layers are plotted
vs. the total energy loss of the analysing ions. Also
shown are the theoretical values for the energy straggling
in Au and Si. Si instead of SiO_2 was chosen, because no
large differences are to be expected and till now no satisfying information about the treatment of the energy straggling in compounds exists. According to Bohr (5) with the
correction factors of Chu et. al. (6)

$$(2) \qquad \Omega^2 = \alpha_c \cdot \Omega_B^2 = \alpha_c \cdot (4\pi z_1^2 z_2 e^4) \cdot N \cdot \Delta R \ ,$$

$$\Omega^2 = \alpha_c \ C_B \cdot N \cdot \Delta R \qquad \text{where } C_B = 4\pi z_1^2 z_2 e^4 \ ;$$

Ω = the energy straggling, i.e. the standard
    deviation of the energy broadening ,

α_c = the correction factor given by Chu (6)

N = the target density ,

ΔR = the target thickness, in our case
     $\Delta R = C(K, \theta) \cdot t .$

The calculation was done for 400 keV ^{4}He and 100 keV ^{1}H,
which represent the range of the measurements. Setting
$\Delta E / \Delta R \approx dE/dx = N \cdot S_e$, S_e = electronic stopping power, one can
substitute $N \Delta R$ by $\Delta E / S_e$, thus (2) reads

$$(3) \qquad \Omega^2 = (\alpha_c \cdot C_B / S_e) \cdot \Delta E .$$

The energy dependence of the straggling vs. energy loss is
weak, especially for ^{4}He-ions, because of the monotonous
increase of α_c and S_e with energy in the energy range considered. For our calculation, $105 (45) \cdot 10^{-15} eVcm^2$ for ^{4}He
(^{1}H) in Au and $70 (22) \cdot 10^{-15} eVcm^2$ for ^{4}He (^{1}H) in Si have
been used as electronic stopping powers. Because the differences between the Si and Au data are less than 10%,
which is small in comparison with the experimental uncertainties , just one straight line has been drawn for the

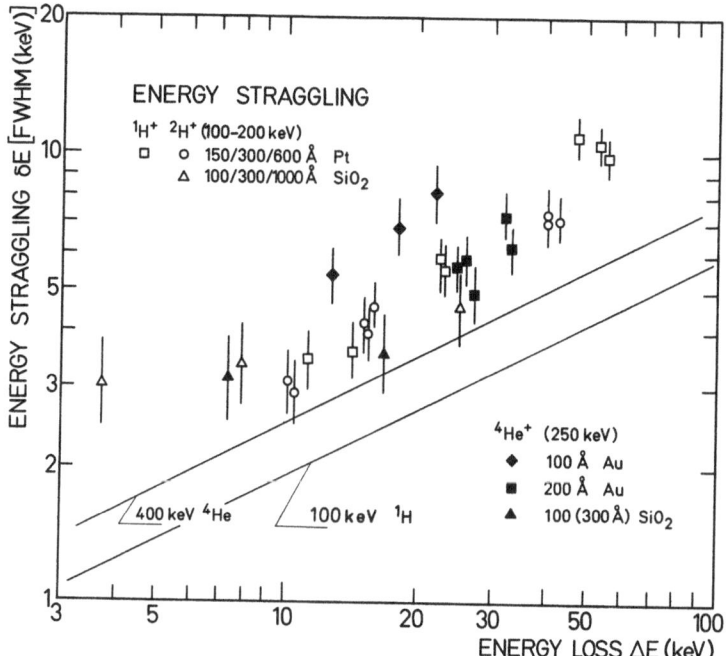

Fig. 5. Energy straggling of $^1H^+$, $^2H^+$ and $^4He^+$ in Pt, Au and SiO_2. Theoretical values according to Chu et al.

representation of their mean value. All experimental straggling data for $^1H^+$ and $^2H^+$ are well above theory, whereas the $^4He^+$-data are closer to theory. Possibly, the extrapolation of the calculation of the $^4He^+$-straggling to the $^1H^+$-straggling performed by Chu (6) is not allowed. The energy dependence could be much weaker. We attribute the finding, that the 100 Å Au straggling data is substantially higher than the 200 Å data, to non-uniformities of the gold layers produced by ourselves. Very likely the formation of islands of varying shape during evaporation is responsible for this effect. We believe the theoretical ^4He-data being confirmed within 30%. For practical applications, we suggest the theoretical ^4He-line to be used as an approximate estimate for the ^1H- and ^2H-data. Using the ^4He-line, a diagram with depth resolution vs. depth was constructed. With the relation

$$(1) \quad \frac{\delta t}{t} = \frac{\delta E}{\Delta E} = \frac{2.355 \cdot \Omega(\Delta E)}{\Delta E}$$

and inserting (3) and $C(K,\theta) \cdot N \cdot S_e \cdot t$ for ΔE, one gets

$$(4) \quad \delta t(t) = \frac{2.355}{S_e} \left(\frac{c_B \cdot d_C}{N \cdot C(K,\Theta)} \right)^{1/2} \cdot t^{1/2}$$

From (4) a net gain should yield for δt by increasing
the pathlength factor $C(K,\Theta)$ in the backscattering geometry. This theoretical net gain is shown in Fig. 6 for
$C(K,\Theta)$ equal 2 and 8. In experiment, $C(K,\Theta)$ was between
2.2 and 3.4 for the data in Fig. 5. So far, these predictions could not yet be proved experimentally, likely
due to not sufficiently good targets. Nevertheless this
diagram may be useful for the estimation of the depth resolution to be expected. For comparison, we have plotted
the data of Williams(7) into our diagram, who has used
2 MeV helium scattering with a glancing incidence geometry.

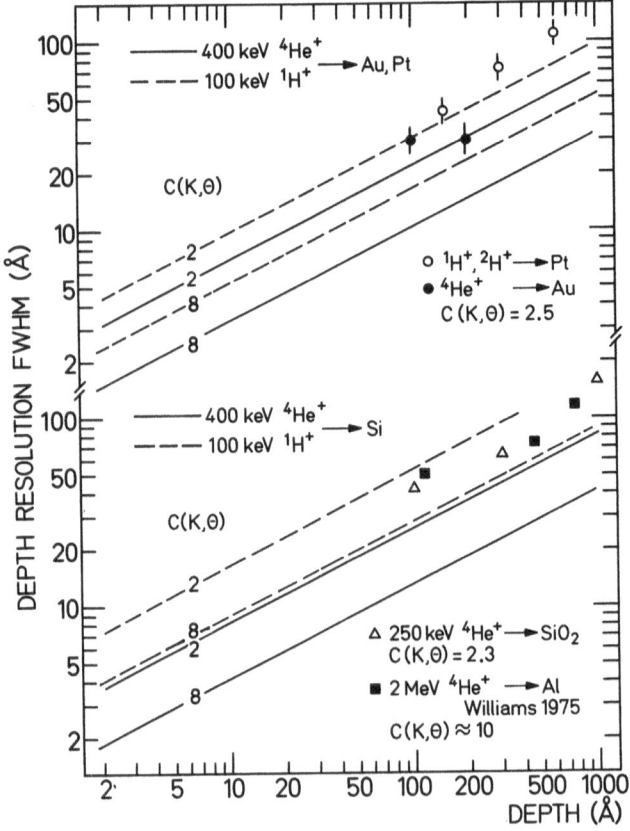

Fig. 6. Theoretical and experimental depth resolution
vs. depth in an RBS experiment.

Target tilt will be advantageous for the analysis of very thin layers, since the energy resolution of our present system is comparable to the energy straggling. In Fig. 7 a "35 Å" Au-layer has been tilted up to 83° with respect to the beam direction. A depth resolution of 5 Å has been reached by RBS in near surface analysis. We refrain from discussing the interesting question whether resolutions equivalent to monolayer thicknesses can be obtained.

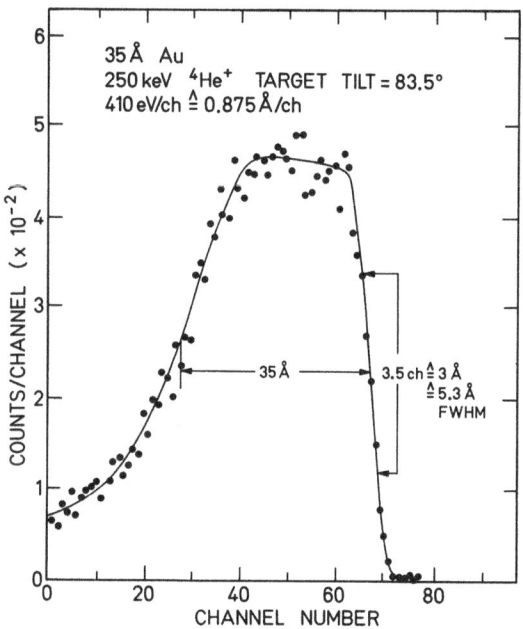

Fig. 7. Backscattering spectrum of a 35 Å Au-layer, $C(K,\theta)=10$

CONCLUSIONS

The combination of ion backscattering and electrostatic energy analysis has been found to yield high depth resolution in thin surface layers. At surface position, the optimum resolution obtained so far has been 5 Å (FWHM) with a thin gold film. The equivalent figure for silicon would be 10 to 15 Å. The maximum range of application is seen at a thickness of about 1000 Å, where resolution is degraded to roughly 100 Å due to energy straggling. The possibilities of glancing incidence techniques have not yet been fully exploited, the reason

being mainly problems of target preparation. Near the
surface, an improvement in depth resolution of at least
a factor of 2·should be possible.

APPLICATIONS

As an application of the combined RBS/ESA-technique, in
Fig. 8, the backscattering spectrum of a gold/aluminum
sandwich is shown. The sandwich has been produced by
successively evaporating about 100 Å Al-layers and Au-
layers corresponding to about 10 Å thickness. The gold
peaks are resolved clearly up to a depth of about 700 Å.

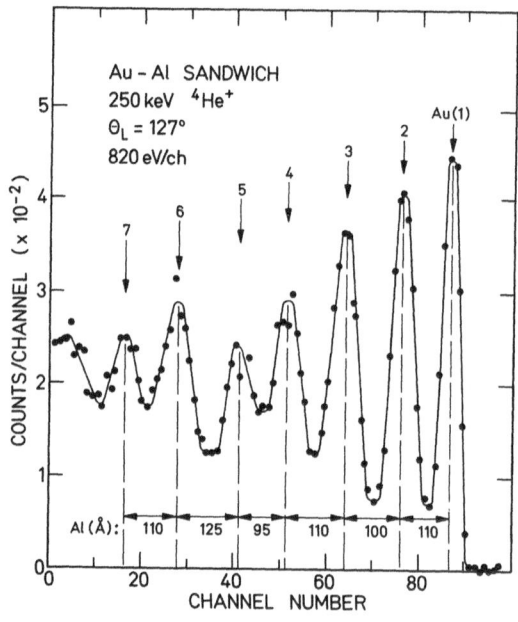

Fig. 8. Backscattering spectrum of a gold/aluminum sandwich.

ACKNOWLEDGEMENTS

The preparation of the SiO_2-layers and the ellipsometric
measurements were performed at AEG-TELEFUNKEN, Heilbronn.
We thank Drs. Goldbach and Kostka for their kind cooperation.

REFERENCES

1. J. S. Williams, Phys. Lett. 51 A, No. 2 (1975) p. 85

2. E. Bøgh, Rad. Eff. 12 (1972) p. 13

3. A. van Wijngaarden, B. Miremadi, W. Baylis, Can. J. Phys. 49 (1971) p. 2440

4. L. Meyer, Phys. Stat. Sol. (b) 44 (1971) p. 253

5. N. Bohr, Mat. Fys. Medd. 18, No. 8 (1948)

6. W. K. Chu, J. W. Mayer, Catania Working data: "Energy Loss and Energy Straggling" (1974)

7. J. S. Williams, Nucl. Instr. Meth. 126 (1975) p. 205

DISCUSSION

Q: (J. Baglin) Your straggling width results are derived by assuming perfectly flat films. Yet we know how difficult it is to obtain perfectly uniform 200 Å films of Au. Have you made independent measurements of film flatness, to ensure that you are not overestimating the straggling widths ?

A: (A. Feuerstein) You are right. It is very difficult to obtain good thin films. So far we have not controlled them with other techniques.

Comment: (W.K. Chu) The low value of straggling value you are obtaining at low energy is very exciting. Even there is 20-30 % difference between experiment and the calculation.

AN APPARATUS FOR THE STUDY OF ION AND PHOTON EMISSION FROM

ION BOMBARDED SURFACES: I. SOME PRELIMINARY RESULTS

R.J. MacDonald, A.R. Bayly and P.J. Martin

Department of Physics, The Australian National University

Canberra, A.C.T., Australia 2600

ABSTRACT

Secondary ion, secondary electron and photon emission represent inelastic effects in the collision processes occurring at ion-bombarded surfaces. Such effects have become of interest to a number of research groups, both for their fundamental importance to our understanding of ion-surface interaction and also for their potential in surface analysis. In this paper we wish to describe a new system for the study of photon and secondary ion emission and to give some preliminary results of experiments we have been conducting.

1. INTRODUCTION

When an ion interacts with a solid surface, there is a very complex collision process occurring. We have all the possibilities of ion reflection or penetration, of atom ejection, of photon, electron and ion formation to consider. The particular aspects of the collision associated with atom ejection (i.e. sputtering) and ion penetration have received detailed consideration in the literature over the last decade (1), though we do not claim to fully understand these aspects of the interaction. Secondary electron emission has been studied for many years and a wealth of information exists in the literature but interest in the properties of secondary electrons, particularly Auger electrons has been revived recently (1). The two processes of photon and secondary ion emission however were largely ignored until the last five years or so, and had only been studied by a few people almost in isolation. Secondary ion emission is the better documented of the two processes, but with the exception of Benninghoven (2) and Blaise and Slodzian (3), most reports have been isolated studies by individuals who have not continued with their experiments.

The potential of these two processes of secondary ion emission and photon excitation as a method of surface analysis was recognised. The ion microprobe was brought to fulfilment as an analytical tool by Anderson (4), Liebl (5), Long (6), Castaing and Slodzian (7), but most of the work was originally qualitative. Some advances have been made in developing quantitative analysis techniques, as in the case of Anderson's model (8) based on a plasma in local thermal equilibrium while individual systems such as the transition metals and their alloys (Blaise and Slodzian)(3) have been considered in detail and models developed which allow the prediction and explanation of experimental results within this system, but the mechanisms of ion formation are still not understood.

Photon excitation is a recently arrived field of interest stemming mainly from the work of White and Tolk (9), and before that of Snoek, van der Weg and Rol (10) but already there is enough evidence to suggest a link between photon excitation and secondary ion emission. There is strong evidence of surface chemical effects in both secondary ion and photon excitation. The photon spectrum is primarily the arc spectrum of the metal, and thus is the physical environment in which the local thermal equilibrium model of secondary ion formation developed by Anderson (8) is likely to prove correct.

Our interest in ion surface interaction and our previous work in secondary ion emission has led us to develop equipment for the study of the inelastic processes occurring at ion bombarded surfaces. We are continuing with various studies in sputtering but in this paper details of the experimental studies of secondary ion and photon emission are given.

2. THE ACCELERATOR

Our basic accelerating device is a 100 kV machine, consisting of an ORTEC duoplasmatron and einzel lens, a uniform gradient field tube and an analysing magnet with a mass energy product of approximately 0.8 amu MeV. The analysing magnet switches the ion beam into tubes at $\pm$ 30° and 0°. The unanalysed beam is used for sputtering studies, while one of the analysed beam lines is used for secondary ion emission experiments, the other for photon excitation experiments. Both analysed beam lines have the same arrangement of electrostatic quadrupole lens and beam deflecting plates, and are identical up to the experimental chambers. The system is wholly oil-diffusion pumped, using Santovac 5 oil, and because of the low back-streaming properties of this fluid, liquid nitrogen trapping has been eliminated except at the target chambers. Water cooled chevron baffles and elbows are sufficient to trap the backstreaming oil. The servicing of the machine is thus reduced markedly. The ion source and beam lines operate in the low 10^{-7} torr region without beam. The pressure rises to about 5×10^{-6} torr near the ion source only when a beam is being extracted. A schematic diagram of the accelerator showing the beam transport sections appears as Figure 1.

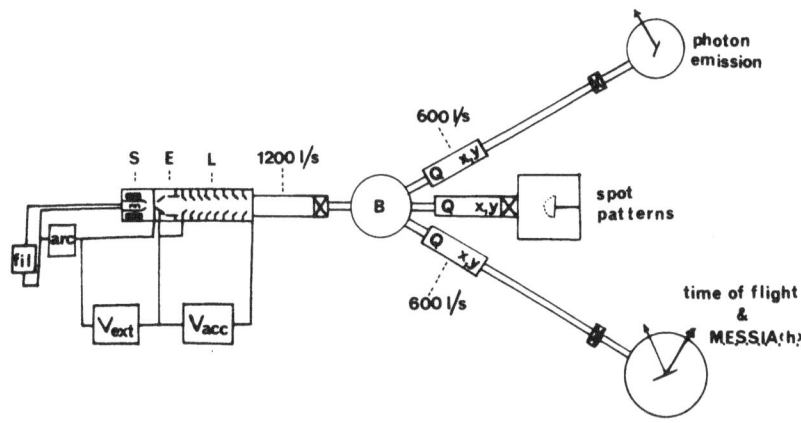

<u>Figure 1.</u>   Schematic diagram of accelerator : S - source;
                 E - extractor; L - acceleration tube; B - analysing
                 magnet; Q - quadrupole; x,y - deflection.

The two main experiments to be described here are the secondary
ion emission experiment and the photon excitation experiment. These
will be described separately below, with some representative results
of these experiments included.

3.   SECONDARY ION EMISSION STUDIES

A complete and detailed report of the secondary ion apparatus
has been submitted for publication elsewhere (Bayly and MacDonald)
(11) so only the essential elements will be indicated here. It is
basically an improved version of the mass and energy spectrometer
developed and reported earlier by Dennis and MacDonald (12). The
major improvements are in energy resolution, and, by more precise
matching between the energy and mass spectrometer, in mass
resolution, plus an off-axis pulse counting detection system feeding
a 4000 channel M.C.A. The inclusion of a 3 axis stepping goniometer and a variable primary beam-to-spectrometer angle allows a
wider range of spatial emission experiments. The system is such
that automatic step-scanning and data acquisition controlled on a
constant ion-dose per step basis is possible for mass, energy or
angular scanning experiments. The system surveys a 1mm diameter
area of the irradiated target and is therefore a "macroprobe"
although in principle it could function with micron sized beams and
become a "microprobe". Referring to Figure (2) the energy analyser,
a 90° spherical condensor, and mass analyser, an E.A.I. quadrupole
mass spectrometer operate at the same pass energy which is set to
a fixed value for any particular experiment. A retarding/
accelerating field at the entrance to the energy spectrometer alters

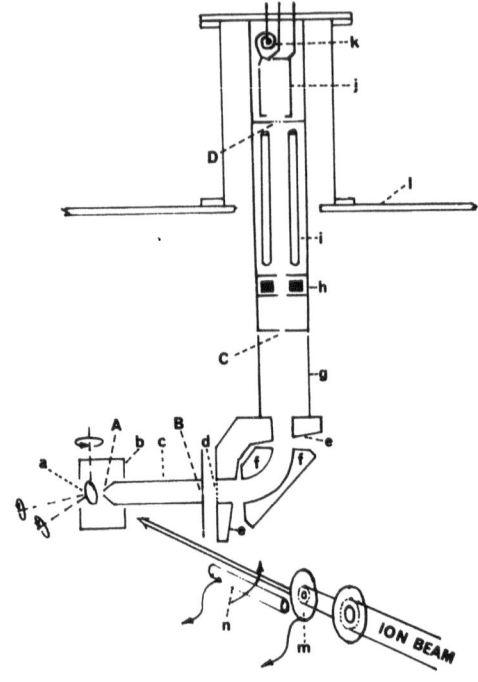

Figure 2. Mass-energy
analyser : A,B,C,D,
apertures; a - target;
b - screen; c - drift tube;
d - accel/decel grid;
e - Hertzog plates; f -
electrostatic energy
analyser; g - spacer;
h - lens; i - quadrupole
mass analyser; j - deflection plates; k - channeltron;
l - mounting plate; n -
faraday cup; m - beam
monitor.

the energy of the ions that drift from the target through a field
free collimator and thus by applying a linear scan to this field
strength, energy spectrum scanning is effected. Thus the mass and
energy spectrometer has two basic operating modes:-

   (a)   in mass or angular scanning a selected energy band (about
        1 eV wide) of the total energy spectrum is passed and the
        variation of the intensity of this narrow band is monitored.

   (b)   in energy scanning the required ion mass is fixed and the
        constant width energy band is swept through the energy
        spectrum.

Extensive testing of this spectrometer, including the use of
very low energy spread thermal ion sources, has shown that within
known limits no appreciable distortion is introduced to energy
spectra by the spectrometer.

3.1. Typical Experiments

The performance of the instrument in mass scanning mode is
demonstrated in Figure (3) by the mass spectrum emitted by
Aluminium irradiated by 40 kVA$^+$. Note that the background counts per
channel is typically zero while the highest peak shown is 100,000.
In fact background count rates of less than 1 count/min. are
typical while the major peak from oxidised aluminium, for example,
can approach 10^5 counts/sec. Thus a signal to noise ratio
approaching 10^7 is obtainable. This, plus the mass resolution

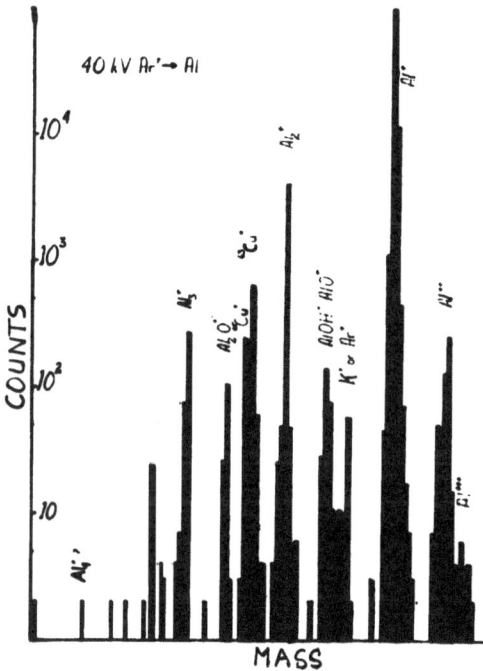

Figure 3. Mass spectrum of Al foil bombarded with 40kV Ar$^+$ ions.

($\simeq$ 60 at 10% peak width) allow interesting changes in the complex spectra shown, to be followed. The mass resolution is well illustrated in Figure (4) for $^{63}Cu^+$ and $^{65}Cu^+$ isotopes.

Figure (5) shows the energy spectrum of Al$^+$ sputtered by a 45 kV Ar$^+$ beam. The energy resolution of the system, $\frac{E}{\Delta E}$ at 10% peak height, is approximately 50, so by varying the pass energy E_p we can vary the absolute energy resolution. At 90 eV, $\Delta E \simeq 1.8$ eV while at 45 eV, $\Delta E \simeq 0.9$ eV. The two spectra in Figure (5) were taken with E_p = 45 eV and 90 eV, and we note the similarity of the spectra first, then the difference in sharpness of the low energy peak. This suggests the spectrum obtained with E_p = 90 eV is being smoothed by the lower absolute resolution, and that the low energy peak is perhaps even sharper than that shown by the 45 eV measurement.

In the energy scan mode Figure 6 shows a sequence of consecutive energy scans of $^{63}Cu^+$ emitted from polycrystalline copper with a natural oxide layer. Such scanning is achieved by an automatic recycling of the scanner, thus allowing changes in the spectra as the oxide layer is removed. These changes are shown in more detail in Figure 7 which shows several sequential spectra from Figure 6 and the ratio of two spectra point for point. Important changes are obvious, especially in the 0-10 eV region.

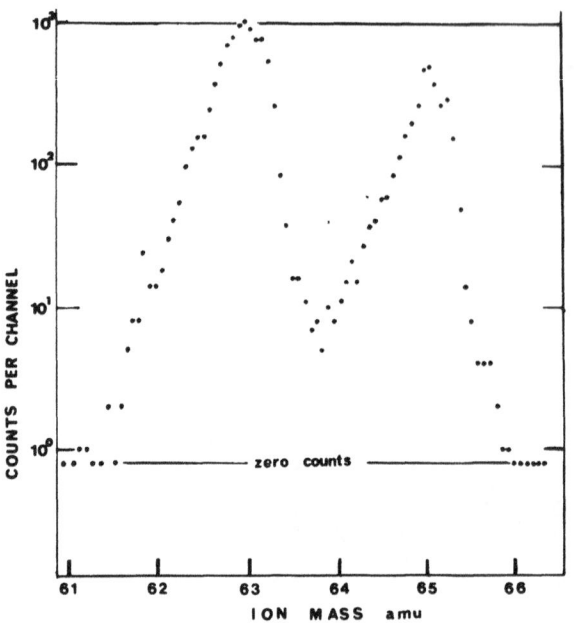

Figure 4. Typical mass resolution, showing separation of ^{63}Cu and ^{65}Cu from Cu bombarded with 45kV Ar$^+$.

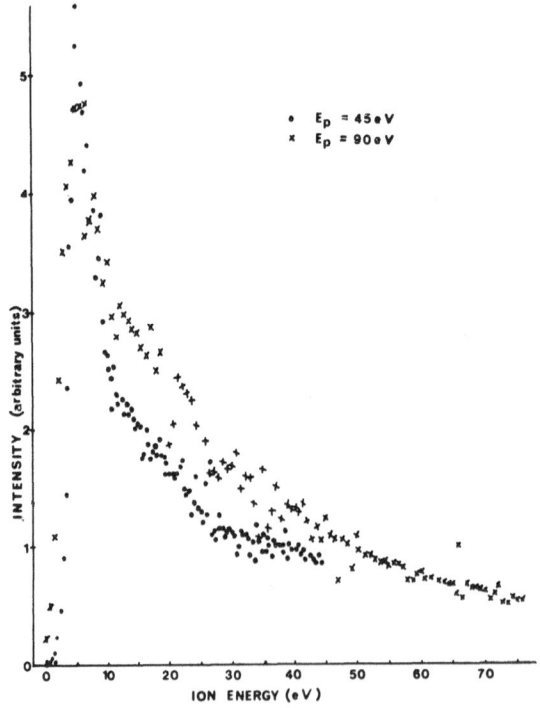

Figure 5. Energy spectrum of Al : (●) pass energy of analyser 45eV energy resolution approximately ± 0·5eV; (X) pass energy of analyser 90eV energy resolution approximately ± 1eV.

Figure 6. A sequence of energy scans of ^{63}Cu from Cu bombarded with 45kV Ar$^+$. Each spectrum covers the range -1 to 69eV and takes approximately 37secs to collect. Each spectrum is automatically normalised to incident ion dose. Horizontal axis is channel number.

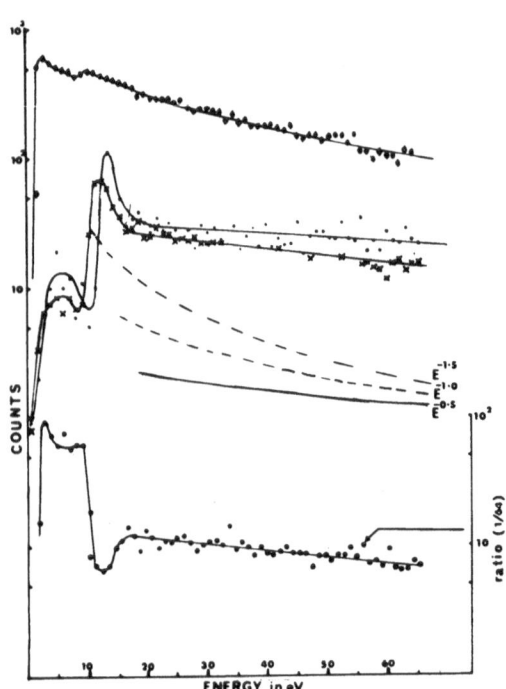

Figure 7. Expanded plots of
the 1st, 15th and 64th
spectrum of Figure 6.
The ratio of the 1st to
64th spectrum is also
shown at the bottom.

These 64 step scans cover the energy spectrum in steps roughly
equal to the energy bandwidth and thus represent approximately the
useful minimum step width limit. It is worth noting that for a
high yield system such as an aluminium oxide layer on aluminium,
the system sensitivity is great enough to allow acquisition of
such spectra at the rate of one energy spectrum per sputtered
monolayer giving approximately 100 counts in the energy peak. This
would be sufficient to allow observation of the changes in the major
features of the spectrum as the oxide film was stripped progressively
layer by layer.

4. PHOTON EXCITATION STUDIES

The apparatus used for photon excitation studies is inherently
simpler than that described above. The experimental chamber contains a simple target holder and a faraday cup for current
measurement. The chamber is normally operated with background
pressures of 5×10^{-8} to 2×10^{-7} torr with the beam on. The target is irradiated either at $45°$ to the surface normal, or normal
to the surface. The photons are analysed at $90°$ to the incident
beam. Photon decay occurs within a centimetre or so from the
surface.

Two monochromators are used. One is a Bausch and Lomb 0.5m
monochromator, with a wavelength resolution of $\sim 3\text{Å}$ while the
other is a MacPherson monochromator with a wavelength resolution

TABLE 1. A summary of target studies, showing the correlation between cohesive energy and observation of a continuum.

Element	Ground State	Structure	Ionisation 1)	Potential 2)	Cohesion Energy Monatomic metals (298°K) Kcal/g-atom	ev/per atom	Ion Spectra	Strong (30KeV) G	Continuum (10ev-100KeV) T	(40-50KeV) V
Mg	^1S	$(Ne)\,3s$	7.64	15.03	35.6	-	Strong	X		X
Al	$^2P_{1/2}$	$-\ 3s^23p$	5.98	18.82	77.5	-	Strong	X		X
Si	$^3P_0$	$-\ 3s^23p^2$	8.15	16.34	108	-	Strong		?	X
Sc	$^2D_{3/2}$	$(A)\,3d4s^2$	6.54	12.80	80.45			X		✓
Ti	$^3F_2$	$-\ 3d^24s^2$	6.82	13.57	112.7	4.9			V	✓
V	$^4F_{3/2}$	$-\ 3d^34s^2$	6.74	14.65	112.8				V	X
Cr	^7S	$-\ 3d^54S$	6.76	16.49	95.0				V	✓
Fe	$^5D_4$	$-\ 3d^64s^2$	7.87	14.18	100	4.34		G	V	X
Ni	$^3F_4$	$-\ 3d^84s^2$	7.63	18.15	102.8	4.50			T	X
Cu	^2S	$-\ 3d^{10}4s$	7.72	20.29	81.1			X	X	X
Ge	$^3P_0$	$-\ 3d^{10}4s^24p^2$	7.88	15.93	89.5	6.4	Strong		V	X
Zr	$^3F_3$	$(Kr)\,4d^25s^2$	6.84	13.13	146.0					
Nb	$^6D_{1/2}$	$-\ 4d^45s$	6.88	14.32	175					
Mo	^7S	$-\ 4d^55s$	7.10	16.15	157.5	6.80		G	V	X
Pd	^1S	$-\ 4d^{10}$	8.33	19.42	90.1			G	X	✓
Ag	^2S	$-\ 4d^{10}5s$	7.57	21.48	68.4			X	X	X
Hf	$^3F_2$	$(Xe)\,4f^{14}5d^26s^2$	7.0	14.9	145	6.3		G		✓
Ta	$^4F_{3/2}$	$-\ 4f^{14}5d^36s^2$	7.88	16.2	186.8	8.1		G	V	✓
W	$^5D_0$	$-\ 4f^{14}5d^46s^2$	7.98	17.7	200	8.7		G	V	X
Pt	$^3D_3$	$-\ 4f^{14}5d^96s$	9.0	18.56	135			G	X	X
Au	^2S	$-\ 4f^{14}5d^{10}6s$	9.22	20.5	87.6				X	X
Pb	$^3P_0$	$-\ 5d^{10}6s^26p^2$	7.42	15.03	46.8				X	X

X = no continuum reported; G = Gritsyna, T = Tolk, V = van der Weg; / = group.

of 0.4Å and the possibility of vacuum UV operation if suitably
pumped. Both spectrometers have been set up for photon counting
using an E.M.I. 9784QB phototube, either a custom built preamp
or an ORTEC 9301 and standard NIM electronics. The wavelength
drives or both monochromators have been converted to accept
stepping motors and data logging is by means of multichannel
analyser or chart recorder. Current densities are typically up to
100 $\mu A/cm^2$.

4.1.   Photon Excitation ·Experiments

The experimental work has concentrated mainly on surveying the
spectra of light emitted by a variety of surfaces irradiated with
a variety of ions. The aim has been to develop a general class-
ification and familiarise ourselves with the general behaviour of
photon excitation events. We find that it is possible to broadly
classify the pure elements into three groups.

(a)   the light elements such as Si, Al, Mg which typically
      show strong lines associated with charged states as well
      as excitation of the neutral atoms. These elements have
      very pronounced emission from ionised atoms (lines of the
      SiIII spectrum have been identified) and react strongly
      to the type of ion used to irradiate the experiment.
      Figure 8(a) shows a typical spectrum of Si irradiated
      with 45 kV Ar^+.

(b)   the medium mass group such as Cu, Ge, Ni whose spectra
      are dominated by lines from the neutral excited atom.
      Lines associated with excited states of at least the
      second spectrum (i.e. M^+) can be detected but at very low
      intensities compared with those of lines of the neutral
      atom spectrum. Figure 8(b) shows a typical Cu spectrum.

(c)   the heavy elements such as W, Hf etc., which show the
      very marked continuum distribution extending from as high
      as 2000Å to as low as 6000Å (i.e. the whole spectral range
      studied). A spectrum of lines emitted predominantly from
      the excited neutral atoms is superimposed on the continuum.
      Some elements of group (b) showed slight evidence of a
      continuum in certain circumstances but the heavy elements
      of group (c) show the effect so markedly. In some cir-
      cumstances the integrated photon yield from the continuum
      is of the same order as the total photon yield from the
      superimposed line spectrum. A typical spectrum from W
      is shown in Figure 8(c).

The origin of the continuum radiation is not understood. It
has been suggested that the origin lies in the ejection of molecular
components from the irradiated surface (Tolk)(13). It is interesting
to note that those targets giving rise to the high continuum are
the targets of high cohesive energy. This is shown in Table 1, which
tabulates the cohesive energies for a number of common metals and
indicates those for which a continuum spectrum has been observed.

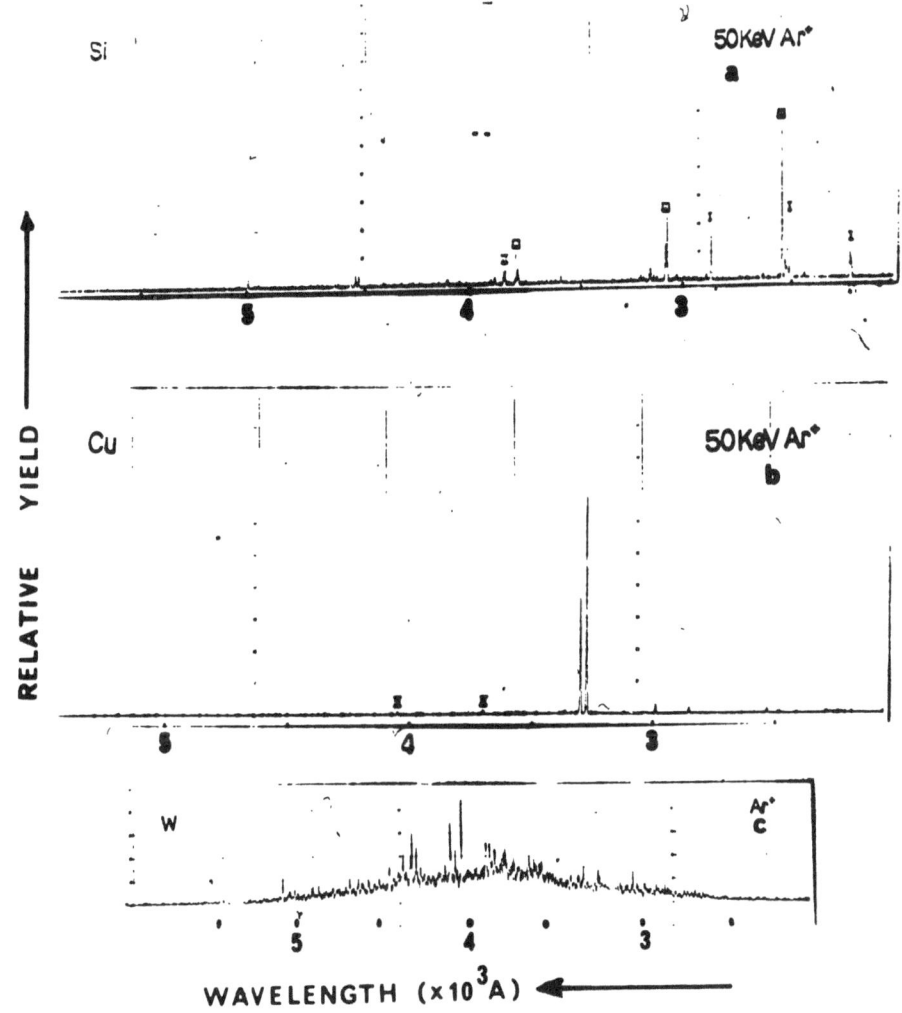

Figure 8. Photon spectra of Si, Cu and W excited by bombardment with 50kV Ar⁺.

A variety of other experiments involving studies of photon excitation have been carried out or are in progress. The effect of various bombarding ions on the photon yield will be described at the Amsterdam Conference (14) and details and results of excitation function measurements and line shape measurements are in preparation.

IN CONCLUSION

We have presented here a brief description of apparatus for the study of secondary ion and photon emission from ion bombarded surfaces. More detailed descriptions of the apparatus and results will be published shortly.

ACKNOWLEDGEMENT

This research was made possible largely as a result of a grant from the A.R.G.C. for a research fellowship to one of us (A.R. Bayly) and a Commonwealth Post-Graduate Fellowship to another (P.J. Martin). The Australian Research Grants Commission (A.R.G.C.) provided grants for some of the equipment. The efforts of our technical staff, particularly G. Horwood, C. Woodland , F. Buckley is gratefully noted.

REFERENCES

(1)   See for example: "Ion Surface Interaction, Sputtering and Related Phenomena", Ed. R. Behrisch et al, Gordon and Breach (1973).

(2)   A Benninghoven, Z. Phys. $\underline{220}$, 159 (1969).

(3)   G. Blaise and G. Slodzian, J. Phys. (Paris) $\underline{31}$, 93 (1970).

(4)   C.A. Anderson, Int. J. Mass Spectrom and Ion Physics $\underline{2}$, 61 (1969) ibid $\underline{3}$, 413 (1970).

(5)   H. Liebl, J. Appl. Phys. $\underline{38}$, 5277 (1967) - see also H. Liebl, Anal. Chem. $\underline{46}$, 22A (1974).

(6)   J.V.P. Long, Brit J. Appl. Physics $\underline{16}$, 1277 (1965).

(7)   R. Castaing and G. Slodzian, Journ. de Microscopic (Paris) $\underline{1}$, 395 (1962).

(8)   C.A. Anderson and J.R. Hinthorne, Anal. Chem. $\underline{45}$, 1421 (1973).

(9)   C.W. White and N.H. Tolk, Phys. Rev. Letts. $\underline{26}$, 456 (1971).

(10) C. Snoek, W.F. van der Weg and P.K. Rol, Physica $\underline{30}$, 341 (1964).

(11) A.R. Bayly and R.J. MacDonald, to be published.

(12) E. Dennis and R.J. MacDonald, Rad. Effects $\underline{13}$, 243 (1972).

(13) N.H. Tolk, private communication.

(14) A.R. Bayly, P.J. Martin and R.J. MacDonald, Proceedings of the International Conference on Atomic Collisions in Solids, Amsterdam, 1975; to be published.

AUTHOR INDEX

SUBJECT INDEX